Targeting Epigenetics in Inflammatory Lung Diseases

Gaurav Gupta • Brian G. Oliver •
Kamal Dua • Md Khadem Ali • Piyush Dave
Editors

Targeting Epigenetics in Inflammatory Lung Diseases

Springer

Editors
Gaurav Gupta
School of Pharmacy
Suresh Gyan University
Jaipur, Rajasthan, India

Brian G. Oliver
Department of Science
University of Technology Sydney
Ultimo, NSW, Australia

Kamal Dua
Discipline of Pharmacy
University of Technology Sydney
Schofields, NSW, Australia

Md Khadem Ali
Division of Pulmonary and Critical Care Medicine
Stanford Medicine
Stanford, CA, USA

Piyush Dave
School of Pharmacy
Gyan Vihar University
Jaipur, Rajasthan, India

ISBN 978-981-99-4779-9 ISBN 978-981-99-4780-5 (eBook)
https://doi.org/10.1007/978-981-99-4780-5

This Springer imprint is published by the registered company Springer Nature Singapore Pte Ltd.
The registered company address is: 152 Beach Road, #21-01/04 Gateway East, Singapore 189721, Singapore

Preface

In recent years, our understanding of human health and disease has undergone a paradigm shift. Traditional approaches focused on genetic factors as the primary determinants of disease susceptibility and progression. However, it has become increasingly clear that environmental factors and epigenetic mechanisms play a crucial role in shaping our health and disease outcomes. This realization has opened up new avenues of research and therapeutic interventions for a wide range of disorders, including inflammatory lung diseases.

Epigenetics refers to heritable changes in gene expression that do not involve alterations in the DNA sequence itself. Instead, epigenetic modifications, such as DNA methylation, histone modifications, and noncoding RNA molecules, regulate gene expression patterns and can be influenced by various environmental factors. These modifications provide a molecular memory of environmental exposures and can have long-lasting effects on cellular function and health.

Inflammatory lung diseases encompass a diverse group of disorders characterized by chronic inflammation and tissue damage in the respiratory system. Conditions such as asthma, chronic obstructive pulmonary disease (COPD), and pulmonary fibrosis pose significant health burdens worldwide, affecting millions of individuals and often leading to substantial morbidity and mortality. While traditional therapies targeting the immune system have been moderately successful, there is a pressing need for novel therapeutic strategies that can provide better outcomes for patients.

This book delves into the emerging field of targeting epigenetics in inflammatory lung diseases. It aims to provide a comprehensive overview of the current knowledge, cutting-edge research, and potential therapeutic avenues that harness the power of epigenetic modifications to mitigate the burden of these devastating disorders. By focusing on the epigenetic mechanisms underlying disease development and progression, we hope to shed light on new possibilities for targeted interventions and personalized medicine.

The chapters in this book bring together the expertise of leading scientists and clinicians who have made significant contributions to the field of epigenetics and respiratory medicine. They explore a range of topics, starting with an introduction to the basic principles of epigenetics and its relevance to inflammatory lung diseases. From there, the book delves into specific epigenetic modifications and their roles in asthma, COPD, pulmonary fibrosis, and other related conditions. The chapters also discuss the impact of environmental factors, such as air pollution and cigarette smoke, on epigenetic regulation in the lungs and explore how these insights can inform preventive measures and public health policies. Furthermore, the book addresses the emerging field of epigenetic therapeutics. It examines the potential of small molecule inhibitors, gene editing technologies, and targeted drug delivery systems to modulate epigenetic marks and restore normal gene expression patterns in lung diseases. The chapters also discuss the challenges associated with developing epigenetic-based therapies and propose strategies to overcome these obstacles.

We believe that this book will serve as a valuable resource for researchers, clinicians, and students interested in the intersection of epigenetics and respiratory medicine. It offers a comprehensive overview of the current state of knowledge, highlighting the exciting advances in understanding the epigenetic mechanisms driving inflammatory lung diseases. By bringing together diverse perspectives and expertise, we hope to inspire collaborations and accelerate the translation of epigenetic research into clinical practice. Ultimately, our goal is to improve patient outcomes and enhance our ability to prevent, diagnose, and treat inflammatory lung diseases effectively. Through the exploration of epigenetic targets and the development of innovative therapeutic approaches, we envision a future where personalized medicine based on an individual's unique epigenetic profile becomes a reality. We invite readers to embark on this journey with us and explore the immense potential of targeting epigenetics in the fight against inflammatory lung diseases.

Jaipur, India Gaurav Gupta
Ultimo, NSW, Australia Brian G. Oliver
Schofields, NSW, Australia Kamal Dua
Stanford, CA, USA Md Khadem Ali
Jaipur, India Piyush Dave

Contents

Editors and Contributors

About the Editors

Gaurav Gupta holds a doctoral degree from Pacific University, Udaipur, Rajasthan, India. He has more than nine years of experience in molecular and biochemical pharmacology, including phytochemistry, respiratory diseases, psycho-pharmacology, and cancer biology, by employing experimental animal models to understand the cellular and molecular mechanism. Dr. Gupta is dedicated to improving outcomes in healthcare through many initiatives in pharmacology and phytochemistry research and effective teaching in the field of Pharmaceutical Sciences. Dr. Gupta is currently associated as an Associate Professor with Suresh Gyan Vihar University, Jaipur, Rajasthan, India. Dr. Gupta has more than 250 research and review articles in the national and international journals of repute.

Brian G. Oliver is a translational researcher who aims to identify and develop new ways of treating respiratory diseases. His scientific training began at the National Heart and Lung Institute, UK, where he mastered the isolation and in vitro culture of several types of human lung cells. He then had further training in molecular biology (University of Leeds) and then respiratory virology at Prof Sebastian Johnston's laboratory at Imperial College, UK, before commencing his PhD at the University of Sydney (supervised by Prof Judith Black). He now leads the Respiratory Cellular and Molecular Biology Group with laboratory facilities at UTS and the Woolcock Institute. The work from his group is recognized to be among the best in the world, evidenced by selection for presentation at symposia at both national and large international conferences and resulting in prestigious publications. He is currently the co-director of the Respiratory, Sleep, Environmental and Occupational Health clinical academic group of Maridulu Budyari Gumal, the Sydney Partnership for Health, Education, Research and Enterprise (SPHERE), A NHMRC AHRTC. He is also the President of the Thoracic Society of Australia and New Zealand's NSW branch. Brian's team was the first to demonstrate that primary human lung cells from people with asthma have an increased inflammatory response to rhinovirus infection (Resp Res 2006). This increased response is virus-specific and related to differential transcription factor recruitment to inflammatory gene promoters. Since rhinovirus-induced inflammation correlates with the occurrence and severity of asthma exacerbations, this finding helps explain why exacerbations occur.

Kamal Dua Senior Lecturer, Discipline of Pharmacy, Graduate School of Health, University of Technology Sydney (UTS), has a research experience of over 12 years working in drug delivery targeting inflammatory diseases. Dr. Dua is also a Node Leader of Drug Delivery Research in the Centre for Inflammation at Centenary Institute/UTS, where the targets identified from the research projects are pursued to develop novel formulations as the first step toward translation into clinics. In addition, Dr. Dua researches two complementary areas: drug delivery and immunology, specifically addressing how these disciplines can advance one another, helping the community live longer and healthier. His extensive publication record evidences this in reputed journals. Dr. Dua's research interests focus on harnessing the pharmaceutical potential of modulating critical regulators such as Interleukins and microRNAs and developing new and effective drug delivery formulations for the management of inflammation in chronic airway diseases and cancer.

Md Khadem Ali received a B.Sc. in Biotechnology and Genetic Engineering in 2010 from Khulna University, Bangladesh, an M.Sc. in Systems Biotechnology in 2013 from Chung-Ang University, South Korea, and a PhD in Immunology and Microbiology in 2018 from the University of Newcastle, Australia. During his PhD with Associate Professor Jay Horvat, he investigated the role of iron in lung disease. In Nov 2018, he joined Spiekerkoetter laboratory at Stanford to identify clinically significant novel bone morphogenic protein receptor 2 (BMPR2) signaling modifier genes that could be targeted with repurposed drugs to increase BMPR2 expression and signaling, one of the key pathways and potential master switch in PAH. In addition to lessons learned from a related genetic disease (PAH), he worked on another disease PAVM/HHT associated with dysfunctional TGF-β/BMPR2 signaling. Dr. Ali's key expertise includes airway/tissue remodeling, pulmonary vascular remodeling, iron metabolism, noncoding RNA biology, and in vitro, in vivo, and ex vivo lung disease modeling.

Piyush Dave did his master's degree and Ph.D. from the Department of Pharmacology and Toxicology at the National Institute of Pharmaceutical Education and Research (NIPER), Mohali, Punjab, a premier pharmaceutical institute of national importance in India and currently working as an Assistant professor in the School of Pharmacy, Suresh Gyan Vihar University, Jaipur, India. His research work of more than eight years primarily oriented in the investigation of the role of epigenetic modification in insulin-dependent (Type-I) and insulin-non-dependent (type II) diabetes mellitus and in the neglected infectious diseases (malaria and leishmaniasis). Dr. Piyush Dave has investigated the role of sodium butyrate and valproic acid (nonspecific HDAC inhibitors) in the progression of diabetes mellitus (type I and type II) associated complications and the immunomodulatory protection against the infectious agents for the discovery and development of novel pharmaceuticals for the betterment of the life of humankind.

Contributors

Md Abubakar Department of Pharmacology and Toxicology, NIPER, Hajipur, India

Neetu Agrawal Institute of Pharmaceutical Research, GLA University, Mathura, India

Sattam Khulaif Alenezi Department of Pharmacology and Toxicology, Unaizah College of Pharmacy, Qassim University, Qassim, Saudi Arabia

Khalid Saad Alharbi Department of Pharmacology and Toxicology, Unaizah College of Pharmacy, Qassim University, Qassim, Saudi Arabia

Md Khadem Ali Pre-Health Academic Program, California State University, East Bay, Hayward, California, USA

Waleed Hassan Almalki Department of Pharmacology, College of Pharmacy, Umm Al-Qura University, Makkah, Saudi Arabia

Samiyah Mohammed Alshehri Department of Pharmacology and Toxicology, College of Pharmacy, King Saud University, Riyadh, Saudi Arabia

Abeer Asif Department of Biochemistry, Faculty of Science, King Abdulaziz University, Jeddah, Saudi Arabia

Asif Ahmad Bhat School of Pharmacy, Suresh Gyan Vihar University, Jaipur, Rajasthan, India

Gurjit Kaur Bhatti Department of Medical Lab Technology, University Institute of Applied Health Sciences, Chandigarh University, Mohali, India

Jasvinder Singh Bhatti Laboratory of Translational Medicine and Nanotherapeutics, Department of Human Genetics and Molecular Medicine, School of Health Sciences, Central University of Punjab, Bathinda, India
Department of Human Genetics and Molecular Medicine, Central University of Punjab, Bathinda, India

Nicholas C. Butzin Department of Biology and Microbiology, South Dakota State University, Brookings, SD, USA
Department of Chemistry and Biochemistry, South Dakota State University, Brookings, SD, USA

Roberto G. Carbone Department of Internal Medicine, University of Genoa, Genoa, Italy

Rajiv Dahiya School of Pharmacy, Faculty of Medical Sciences, The University of the West Indies, St. Augustine, Trinidad and Tobago

Deepika Deopa School of Pharmacy, Suresh Gyan Vihar University, Jaipur, India
Six sigma Institute of Technology and Science, Rudrapur, Uttarakhand, India

Karuna Dhaundhiyal Amrapali Group of Institute, Haldwani, Nanital, Uttarakhand, India

Ishwar Singh Dhramshaktu Dr. Sushila Tiwari Medical college and Hospital, Haldwani, Nainital, Uttarakhand, India

Kamal Dua Discipline of Pharmacy, Graduate School of Health, University of Technology Sydney, Ultimo, NSW, Australia

Bhaskar Jyoti Dutta Department of Pharmacology and Toxicology, NIPER, Guwahati, India

Neeraj Kumar Fuloria Faculty of Pharmacy, AIMST University, Bedong, Kedah, Malaysia

Shivkanya Fuloria Faculty of Pharmacy, AIMST University, Bedong, Kedah, Malaysia

Ritu Gilhotra School of Pharmacy, Suresh Gyan Vihar University, Jaipur, Rajasthan, India

Ahsas Goyal Institute of Pharmaceutical Research, GLA University, Mathura, India

Gaurav Gupta School of Pharmacy, Suresh Gyan Vihar University, Jaipur, Rajasthan, India

Madan Mohan Gupta School of Pharmacy, Faculty of Medical Sciences, School of Pharmacy, The University of the West Indies, Saint Augustine, Trinidad and Tobago

Saurabh Gupta Chameli Devi Institute of Pharmacy, Indore, Madhya Pradesh, India

Md Fahmid Islam Department of Pediatrics, College of Medicine, University of Saskatchewan, Saskatoon, SK, Canada

Tanim Islam Department of Veterinary Pathobiology, Oklahoma State University, Stillwater, Oklahoma, USA

Shikha Jakhotiya Department of Pharmaceutics, Maharishi Arvind College of Pharmacy, Ambabari, Jaipur, India

Neha Kanojia Chitkara University School of Pharmacy, Chitkara University, Baddi, India

Gagandeep Kaur Chitkara University School of Pharmacy, Chitkara University, Baddi, India

Satinder Kaur Laboratory of Translational Medicine and Nanotherapeutics, Department of Human Genetics and Molecular Medicine, School of Health Sciences, Central University of Punjab, Bathinda, India

Imran Kazmi Department of Biochemistry, Faculty of Science, King Abdulaziz University, Jeddah, Saudi Arabia

Abdullah Khan Faculty of Pharmacy, Quest International University, Ipoh, Perak, Malaysia

Amit Khurana Institute of Molecular Pathobiochemistry, Experimental Gene Therapy and Clinical Chemistry (IFMPEGKC), RWTH Aachen University Hospital, Aachen, Germany

Rashi Kulshrestha School of Pharmacy, Suresh Gyan Vihar University, Jaipur, India

Usha Kumari Faculty of Medicine, AIMST University, Kedah, Malaysia

Narender Kumar Kumawat Department of Pharmacology, Maharishi Arvind College of Pharmacy, Ambabari, Jaipur, India

Krushna Ch. Maharana Department of Pharmacology and Toxicology, NIPER, Guwahati, India

Mohammad Nazmul Hasan Maziz Graduate School of Medicine, Perdana University, Kuala Lumpur, Malaysia

Dhanalekshmi Unnikrishnan Meenakshi College of Pharmacy, National University of Science and Technology, Muscat, Oman

Jayapriya Mishra Laboratory of Translational Medicine and Nanotherapeutics, Department of Human Genetics and Molecular Medicine, School of Health Sciences, Central University of Punjab, Bathinda, India

Olorunfemi R. Molehin Department of Science Laboratory Technology, Faculty of Science, Ekiti State University, Ado-Ekiti, Nigeria

Muhammad Mustafa KAM_School of Life Sciences, Forman Christian College (A Chartered University), Lahore, Pakistan

Muhammad Shahid Nadeem Department of Biochemistry, Faculty of Science, King Abdulaziz University, Jeddah, Saudi Arabia

Kamal Narain Faculty of Medicine, AIMST University, Bedong, Kedah, Malaysia

Umashanker Navik Department of Pharmacology, School of Health Sciences, Central University of Punjab, Bathinda, India

Frank A. Ogundolie Department of Biotechnology, Baze University, Abuja, Nigeria

Adeniyi S. Ohunayo Department of Biochemistry, Faculty of Science, Ekiti State University, Ado-Ekiti, Nigeria

Abhijeet Ojha Amrapali Group of Institute, Haldwani, Nanital, Uttarakhand, India

Dhaneshvaree Patel Department of Pharmacology, Central University of Punjab, Bathinda, India

Eswara Rao Puppala Department of Pharmacology and Toxicology, NIPER, Guwahati, India

Manish Purohit School of Pharmacy, Suresh Gyan Vihar University, Jagatpura, Jaipur, India

K. M. Taufiqur Rahman Department of Biology and Microbiology, South Dakota State University, Brookings, SD, USA

Sonu Rajput Department of Pharmacology, Central University of Punjab, Bathinda, India

Lata Rani Chitkara University School of Pharmacy, Chitkara University, Baddi, India

Sarita Rawat Amrapali Group of Institute, Haldwani, Nanital, Uttarakhand, India
School of Pharmacy, Suresh Gyan vihar University, Jaipur, Rajasthan, India

P. Hemachandra Reddy Department of Internal Medicine, Texas Tech University Health Sciences Center, Lubbock, TX, USA
Department of Pharmacology and Neuroscience and Garrison Institute on Aging, Texas Tech University Health Sciences Center, Lubbock, TX, USA
Department of Public Health, Graduate School of Biomedical Sciences, Texas Tech University Health Sciences Center, Lubbock, TX, USA
Department of Neurology, Texas Tech University Health Sciences Center, Lubbock, TX, USA
Department of Speech, Language, and Hearing Sciences, Texas Tech University Health Sciences Center, Lubbock, TX, USA
Nutritional Sciences Department, College of Human Sciences, Texas Tech University, Lubbock, TX, USA

S. Roshan Deccan College of Pharmacy, Hyderabad, India

Abhishek Sehrawat Laboratory of Translational Medicine and Nanotherapeutics, Department of Human Genetics and Molecular Medicine, School of Health Sciences, Central University of Punjab, Bathinda, India

Mahendran Sekar School of Pharmacy, MONASH University Malaysia, Subang Jaya, Selangor, Malaysia

Shalini Shanmugavelu Faculty of Pharmacy, AIMST University, Kedah, Malaysia

Anjali Sharma Department of Pharmacolog, Central University of Punjab, Bathinda, India

Manisha Singh Department of Pharmaceutical Chemistry, Maharishi Arvind College of Pharmacy, Ambabari, Jaipur, India

Santosh Kumar Singh School of Pharmacy, Suresh Gyan Vihar University, Jaipur, Rajasthan, India

Sumeet Kumar Singh Department of Pharmacology, Central University of Punjab, Bathinda, India

Yogendra Singh Department of Pharmacology, Maharishi Arvind College of Pharmacy, Ambabari, Jaipur, India

Neelam Singla School of Pharmacy, Suresh Gyan Vihar University, Jaipur, India

Parul Sood Chitkara University School of Pharmacy, Chitkara University, Baddi, India

Vetriselvan Subramaniyan Department Unit, Jeffrey Cheah School of Medicine & Health Sciences, MONASH University, Subang Jaya, Malaysia

Sampat Singh Tanwar Department of Pharmacology, Central University of Punjab, Bathinda, India

B. Tazneem Deccan College of Pharmacy, Hyderabad, India

Komal Thapa Chitkara University School of Pharmacy, Chitkara University, Baddi, India

Riya Thapa School of Pharmacy, Suresh Gyan Vihar University, Jaipur, India

Vibhav Varshney Institute of Pharmaceutical Research, GLA University, Mathura, India

Nitin Verma Chitkara University School of Pharmacy, Chitkara University, Baddi, India

Kamini Vijeepallam Faculty of Pharmacy, AIMST University, Bedong, Kedah, Malaysia

Poonam Yadav Department of Pharmacology, Central University of Punjab, Bathinda, India

Introduction to Lung Disease

1

Vetriselvan Subramaniyan, Shivkanya Fuloria, Mahendran Sekar, Shalini Shanmugavelu, Kamini Vijeepallam, Usha Kumari, Kamal Narain, Dhanalekshmi Unnikrishnan Meenakshi, Mohammad Nazmul Hasan Maziz, and Neeraj Kumar Fuloria

Abstract

There has been a rise in the number of documented instances of lung illness in both children and adults around the globe. Both the morbidity and mortality rates associated with lung illness are from the unknown source. However, estimates from the GHO and other officialdoms imply that over 400 million people throughout the world are affected by asthma and COPD, most of them with very moderate symptoms. Non-infectious lung disease (LD) includes COPD, cystic fibrosis, asthma, lung cancer, and idiopathic pulmonary fibrosis (IPF), whereas infectious lung diseases include influenza, TB, and COPD. A

V. Subramaniyan
Jeffrey Cheah School of Medicine and Health Sciences, MONASH University, Jalan Lagoon Selatan, Bandar Sunway, Selangor Darul Ehsan, Malaysia
e-mail: subramaniyan.vetriselvan@monash.edu

S. Fuloria · S. Shanmugavelu · K. Vijeepallam · N. K. Fuloria (✉)
Faculty of Pharmacy, AIMST University, Bedong, Kedah, Malaysia
e-mail: shivkanya_fuloria@aimst.edu.my; shalini.shanmugavelu@aimst.edu.my; kamini@aimst.edu.my; neerajkumar@aimst.edu.my

M. Sekar
School of Pharmacy, MONASH University Malaysia, Subang Jaya, Selangor, Malaysia
e-mail: Mahendran.Sekar@monash.edu

U. Kumari · K. Narain
Faculty of Medicine, AIMST University, Bedong, Kedah, Malaysia
e-mail: usha_harischandran@aimst.edu.my

D. U. Meenakshi
College of Pharmacy, National University of Science and Technology, Muscat, Oman
e-mail: dhanalekshmi@nu.edu.om

M. N. H. Maziz
Graduate School of Medicine, Perdana University, Kuala Lumpur, Malaysia

G. Gupta et al. (eds.), *Targeting Epigenetics in Inflammatory Lung Diseases*,
https://doi.org/10.1007/978-981-99-4780-5_1

combination of environmental circumstances, including an aging population and the absence of effective medications to the mild factor of LD, increased the chances of severe lung illnesses. With the widespread prevalence and devastating effects of lung diseases such as COPD, asthma, COVID-19, fibrosis, and flu-like symptoms, studies aimed at developing effective treatments and prevention measures are now being funded on a high-priority basis.

1.1 Introduction

There has been a worldwide increase in the incidence of LD, which may affect persons of any age [1]. Estimates from the GHO and other organizations imply that more than 400 million individuals around the world suffer from mild to severe asthma and COPD, yet the causes of lung disease mortality and morbidity remain unknown [2]. In addition, lower respiratory tract infections due to *Hemophilus influenzae* lies between 250,000 and 500,000 fatalities yearly. In 2015, 10.4 million people throughout the globe were affected with the lower respiratory tract *Mycobacterium* TB infection, and approximately 14% of those infected ultimately died because of the disease [3, 4]. Increasing rates of illnesses, i.e., non-communicable, like lung cancer brought on by cigarette smoking or exposure to environmental pollutants, kill an additional 1.6 million people per year [5, 6]. Noncommunicable pulmonary illnesses include asthma, COPD, cystic fibrosis, lung cancer, and interstitial lung disease (ILD), whereas communicable pulmonary disorders include tuberculosis (TB), influenza (flu), and COPD [7, 8] (Fig. 1.1).

1.2 Overview of Lung Diseases

We have included a brief synopsis of both communicable and noncontagious lung diseases.

1.2.1 Asthma

Airflow restriction, respiratory hyperactivity, and critically increased inflammation in the airways are some of the signs of asthma, basically a complicated lung illness with many other symptoms [9]. Asthma affects almost 10% population of adults in each country, or nearly 300 million people worldwide [10]. And more than 80% of those fatalities will be in low- and middle-income nations, making asthma the leading cause of death for an estimated 383,000 people globally [11]. The yearly expense of managing asthma might reach $3100 per sufferer [12, 13]. Asthma is a common respiratory condition that has a negative public perception since it is not well understood [14]. Dust mites, mold, cigarette smoke, pollen, hazardous compounds in the environment, and even stale air may all act as asthma triggers,

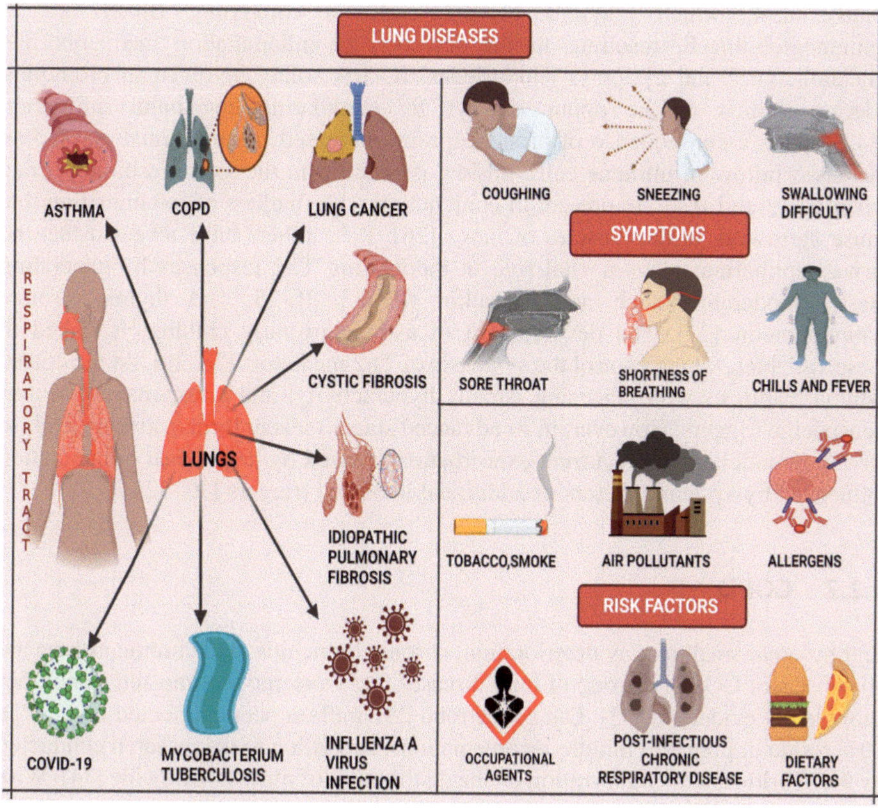

Fig. 1.1 Lung diseases

both inside and outside [15]. In certain cases, asthma symptoms may worsen before improving with medication [16]. In addition, the persistence of characteristic asthmatic clinical symptoms implies that the sickness may be long-lasting in some people. Rapid breathing, wheezing, shortness of breath, and coughing are all signs of asthma [17, 18]. Asthma may be triggered or made worse by a wide variety of factors, including exposure to allergens or irritants, lung infections (bacterial or viral), sinusitis, physical activity, loud noises (thunder and lightning), and extreme cold [19]. Because of recent discoveries into its root causes, asthma is now classified according to endotypes and/or phenotypes [20]. A recent panel in *The Lancet* argues that this is essential for effective asthma medicine because it allows for the identification of "treatable features" in individuals and the precise targeting of these qualities for disease management [21, 22].

Studies have demonstrated that Th2 activation in response to endogenous or extrinsic stimulation leads to the recruitment of Type 2 T-helper cells to the lungs, resulting in the production of significant amounts of cytokines, including IL-9, IL-13, IL-5, and IL-4, which has been linked to the main triggers of asthma [23].

Unlike IL-4, which plays a complex role in converting B-cell IgE to immunoglobulin E, resulting in the secretion of inflammation mediators like catecholamines and cysteinyl leukotrienes, IL-5 is solely involved in promoting allergic rhinitis in the upper airways by stimulating eosinophil infiltration [24, 25]. As a consequence of bronchospasm, increased mucus formation, and an increased inflow of immune cells, airflow is reduced in the lower respiratory tract when IL-13 and IL-4 are present in conjunction with markers of inflammation that cause narrowing in the muscles of lungs [26]. Researchers have revealed that the airway epithelium plays a vital role in modulating Th2 responses by generating master moderators such as interleukin (IL)-33, IL-25, and thymic stromal lymphopoietin [27]. The development of asthma in early children is linked to these variables, which control the synthesis of Th2 mediators [28, 29]. At the outset, asthma manifests with wheezing, airway hyperactivity, and a response to various nonspecific triggers. However, in its advanced stages (severe forms), asthma leads to airway remodeling and recurrent exacerbations caused by heightened inflammation influenced by systemic factors or additional localized triggers [30–32].

1.2.2 COPD

Emphysema, small airway deterioration, chronic bronchitis, and chronic asthma are all subsets of COPD, a group of lung diseases that worsen over time and sometimes show lethal evidences [33]. Out of the total 251 million people affected by COPD, 90% reside in poor and middle-income nations, according to the research conducted by the World Health Organization on the distribution of illness worldwide [34]. With an estimated 3.17 million fatalities in 2015, the global mortality rate rose to 5.0% from 4.8% in 1990. Instead of taking lives, chronic obstructive pulmonary disease is expected to cost $32.1 billion in medical costs and lost productivity in the United States in 2010 [35, 36]. Speculations put this figure at $49.0 billion in 2020 [37–39]. Asthma in the elderly is often misdiagnosed as COPD, and data sets are lacking from undeveloped or emerging Middle Eastern and Asian nations, which might raise the mortality rate by several millions [40]. Dyspnea, or shortness of breath, is the most prevalent symptom of COPD and increases with time and causes slight exertion [41]. More issues arise for those with severe COPD, and they often need to make emergency hospital trips throughout the year. Smokers and ex-smokers with COPD often experience hypoxemia and subsequent organ failure due to the lungs' reduced gaseous exchange abilities and airflow obstruction caused by temporarily destroyed alveoli or an accumulation of cells due to inflammation and a huge portion of phlegm in the bronchioles [42, 43]. Chronic cough (wet or dry), fatigue, wheezing, and chest tightness are among symptoms that could be experienced by someone with COPD and are commonly misdiagnosed as the result of simple aging [44]. In some patients, the symptoms have not shown until the illness become severe. In spite of extensive study, there is currently no effective treatment for repairing damaged lung tissue and regaining normal breathing [45, 46]. Moreover, COPD worsens with time as it is a chronic condition. Although present treatments may alleviate symptoms and slow the development of the disease [47].

Cigarette smoking has been identified as the foremost hazardous factor for COPD in the current form of research [48]. Non-smoking irritants, such as anthropogenic particulates, airborne grit particles, and metal pollutants, have been connected to the exacerbation (smokers) or the beginning (non-smokers) of COPD, a still relatively largely unexplored disorder [49].Genetic predispositions, like alpha-1-antitrypsin (AAT) deficiency, have been demonstrated to have a significant role in the onset of COPD, while the processes behind this association remain unclear [50]. Although genetics have a role in 1% of instances of COPD, it is vital to remember that cigarette smoking and environmental pollution are also key contributors [51, 52]. People who are AAT deficient are at a greater risk for acquiring lung infections. Smokers and ex-smokers are not equally susceptible to the disease; only around 20–30% of those who use tobacco will get the condition [53, 54]. Although the exact processes are unknown, researchers found that passive smoking was a risk factor for COPD (51.2%, $n = 87$) [55, 56]. COPD is caused by the obstruction of airways by tobacco smoke or other airborne particles/gases. CD8+ T lymphocytes, neutrophils, and macrophages, which are inflammatory cells, are released into the airways and surrounding tissues, inducing apoptosis. Thus far research has indicated that various COPD subtypes elicit varying immune responses [57–59].

Increased mucus production, inflammatory cell numbers, MUC5AC gene expression in retort to concealed serine proteases, reactive oxygen species (ROS) from smoking, and macrophages activation all contribute to airway obstruction and cellular destruction, which in turn causes fibrosis and a loss of pulmonary elasticity in patients with chronic bronchitis [60, 61]. Alveolar sac and bronchial inflammation brought on by smoking leads to airway wall constriction and progressive alveolar sac degradation, ultimately resulting in a loss of alveolar structure and a deterioration in lung function [62]. There is still insufficient understanding of the triggers that initiate adaptive and innate immune responses in patients with COPD [63, 64]. Disease progression may be attributed to the overexpression of immune cell IL8, MMPs, and CXCs, along with proinflammatory mediators like leukotriene B4, TGF−, IL1 (Th1 responses), and TNF, which promote fibrosis and disrupt the balance in oxidant-antioxidant ratio (ROS/RNS) [65–67].

1.2.3 Lung Cancer

Seventy percent of lung carcinoma patients die as a result of the disease progression due to a lack of appropriate treatment options and a delay in detection of lung cancer (stage III or IV) [68]. This cancerous illness strikes both genders with the same frequency [69, 70]. When compared to the next four most prevalent types of carcinomas in the United States, the death rate from lung cancer is much greater (colon, pancreas, breast, and prostate) [71]. Twenty years or more of smoking antiquity appears to be related to an elevated risk of progression and death. Based on current theories, exposure to tobacco induces DNA damage, procarcinogens initiate lung cancer, and this process results in incompatible gene-enzyme interactions [72, 73]. So, it is thought that quitting smoking and making other lifestyle changes may greatly increase the odds of survival from lung cancer. An increase in lung cancer cases is a certain assumption in countries where the smoking rate has increased; thus, concerted public measures to curb tobacco use are crucial [74, 75]. Lung cancer is the important cause of cancer-related mortality across the board for both genders [76, 77]. According to research, there were around 1.8 million new cases of lung cancer detected in 2012, accounting for 12.9% of novel cancer incidences [78, 79]. Lung cancer is a leading cause of healthcare expenditures and loads worldwide, according to the 2020 Global Burden of Disease Study. Male cancer mortality was shown to be independent of economic level [80, 81]. One of the most striking findings of the research is the correlation between a country's level of economic development and the rate of female deaths from lung cancer. Lung cancer has a wide range of subtypes, and its complexity comes from the fact that it may start anywhere in the bronchial tree and that various subtype of lung cancer manifest in different ways both in terms of symptoms and physical location [82]. Lung cancer can be classified into two main types, with NSCLC accounting for 85% of cases and SCLC constituting the remaining 15% (15% of all lung malignancies). There are approximately three potential subtypes of NSCLC, namely squamous cell carcinoma, large cell carcinoma, and adenocarcinoma [54, 63]. To further increase this classification of lung cancer, we included specific histological traits and reliable immunohistochemical biomarkers, enabling us to distinguish between preinvasive tumor and aggressive adenocarcinomas in a convincing way [83, 84]. Improvements in molecular characterization of lung malignancies and an expanding toolbox of successful medicines have contributed significantly to the current system for classifying lung carcinoma. Even within the same histopathological category, lung cancer seems to be an assembly of illnesses with varying molecular and histological features [85, 86].

1.2.4 Cystic Fibrosis

Cystic fibrosis (CF) is the most predominant autosomal retreating illness, affecting about one in every three thousand newborns [87]. The majority of affected infants display signs of the condition shortly after birth, with lung infections and poor weight gain being the most prevalent early indicators. Chronic lung infections and pancreatic insufficiency are diagnostic criteria for CF [88]. However, before CF newborn testing became available, a clinical crusade with a perspiration test was often undertaken before a definitive diagnosis could be made [89]. A perspiration chloride level of more than 60 mmol/L is diagnostic of CF. Male infertility is an additional symptom, along with excessive perspiration leading to a significant salt loss [90]. Chronic pulmonary infections brought on by certain bacteria, together with substantial inflammation, may lead to the development of bronchiectasis, impaired lung performance, and, ultimately, pulmonary failure [91]. Despite *Staphylococcus aureus* and *Pseudomonas aeruginosa* being the most common CF pathogens, when

CF worsens, some people may develop infections with *Burkholderia cepacia*, *Achromobacter xylosoxidans*, *Stenotrophomonas maltophilia*, and *Mycobacterium*, all of which are rare and difficult to treat [92, 93]. There are a wide variety of age-related disorders that may affect almost every organ, such as allergic hemoptysis, bronchopulmonary aspergillosis, CF-related hyperglycemia, gastrointestinal obstructions, nasal polyps, and liver sickness. The underlying cause of this autosomal receding illness is alterations in the CF transmembrane conductance regulator (CFTR) gene, that is located on the long arm of chromosome 7 [94, 95]. Due to the fact that more than 1400 distinct CF alterations have been recognized and documented in the Cystic Fibrosis Alteration Database, widespread genetic screening of the population is now unachievable. While lung symptoms account for the majority of CF-related hospitalizations and deaths, the typical CF phenotype is highly intricate and encompasses multiple organs covered by epithelium [96]. In cystic fibrosis (CF), which is caused by mutations in the CFTR gene, there has been a significant progress in recent decades toward a complete understanding of the mechanism(s) underlying the manifest failure of lung defense [97, 98].

1.2.5 Idiopathic Pulmonary Fibrosis

Idiopathic pulmonary fibrosis refers to scarring in the lungs for which there is no established etiology. As a severe type of idiopathic intermittent pneumonia, idiopathic pulmonary fibrosis (IPF) is characterized by increasing fibrosis, a gradual reduction in pulmonary function, a worsening of respiratory distress, and an increased risk of death [99]. Accurate prognosis and treatment options need a precise diagnosis [64, 69]. When looking at the worldwide prevalence of IPF, researchers found that the rate was highest in North America and Europe (at 2.8–9.3 per 100,000 persons per year), with much lower rates in Asia and South America. The wide disparity in outcomes across nations may be due to contact with ecological or working risk factors [100]. The typical survival time for those with IPF is just 2–3 years, according to previous research. Despite hopes, there has been little improvement in survival rates in recent years [101]. Increases in both screening and diagnosis may be contributing to this apparent increase in fatalities. UIP is a histological marker for IPF [70, 74]. Honeycomb cyst development, clustering of fibroblasts and myofibroblasts, may or may not accompany fibrosis, as it is a significant deposition of disorganized collagen and extracellular matrix [102]. While the exact reasons for these events are uncertain, one leading theory is that IPF is brought on by an abnormal immune reaction to the host's environment [103]. Multiple stressors, both genetic and environmental, are implicated in the development of IPF, according to the "multiple strike hypothesis." Asbestos, immunological complexes, medicines, and radiation exposure are only few of the recognized causes of fibrotic disorders, the long-term consequences of which may lead to fibrosis [104, 105].

1.2.6 Tuberculosis

Compared to the human-caused acid-fast *M. tuberculosis* strain that caused TB, *M. bovis* strain is responsible for the reduced proportion of zoonotic cases (143,000 in 2018) [106]. Although M. tuberculosis poses a significant risk when transmitted through aerosols expelled from the lungs of patients with active TB via spitting, sneezing, and coughing, it lacks several typical elements found in other pathogenic bacteria, such as Exotoxins [107, 108]. Inhaling bacterial droplets containing the infection, which settle in the air sacs of the lower airways, can lead to the disease in a susceptible host [75, 78]. Local macrophages are the primary cells of the immune system that contact with *M. tuberculosis* in the airways, and these cells are responsible for ingesting the pathogen [109, 110]. Inhibiting a group of damaging genes in the bacterium reduces virulence in an investigational model of tuberculosis; then it does not stop mycobacterial proliferation under perfect in vitro circumstances, where there is no stress or scarcity [111, 112]. Metal transporters, enzymes involved in lipid metabolism, protein kinases, gene regulators, macrophage activity inhibitors, proteases, cellular membrane proteins, and proteins of unknown function, such as PE PGRS and PE proteins, are all infective causes of *Mycobacterium tuberculosis* [113]. Alveolar macrophages can swallow *M. tuberculosis*, but the bacteria are resistant to further destruction by RNS and reactive oxygen species (ROS), along with lysosomal fusion and phagosome acidification [114, 115]. Each of these steps is essential for the pathogen to endure in the host throughout latent TB, to multiply, disperse, and die in the tissues of active TB patients, and to spread from one person to another. Parenchymal degeneration, fibrosis, bronchostenosis, cavitation, traction bronchiectasis, and emphysema are all potential outcomes of pulmonary tuberculosis's impact on the lung's design [116, 117].

1.2.7 Influenza A Virus Infection

Lung illness triggered by influenza A and B viruses is very contagious. Centers for Disease Control and Prevention statistics show that between 2010 and 2017, influenza-related hospitalizations in the United States accounted for between 140,000 and 710,000 admissions (CDC) [118]. Annual deaths from pandemic seasonal infections caused by influenza A viruses are now estimated to range between 291,243 and 645,832. When the IV or a secondary bacterial contagion of the lower airways develops from a slight upper lung illness marked by tiredness, muscle aches, sore throat, runny nose, headache, coughing, and fever, it may produce severe and, in rare circumstances, deadly pneumonia [119]. Flu infections may cause symptoms outside of the respiratory system, such as problems with the heart, the brain, and other organs. Unlike the common seasonal epidemics that occur every year, a pandemic of avian flu is very rare [120]. The possibility of and evidence for A virus subtypes is compelling. Approximately, once every 10–50 years, a pandemic begins with the introduction of a novel influenza strain. A new virus strain emerges that is genetically different from previously socializing viruses, increasing the severity of sickness and the likelihood of mortality in people [121, 122].

In contrast to the pulmonary route employed by humans, avian influenza viruses may be transmitted by the fecal-oral, fecal-respiratory, and fecal-fecal pathways, and can also be spread between free-living birds. Depending on where it enters the body, the virus contaminates and multiplies in the epithelial cells that line the respiratory or intestinal systems. Besides, the H7 subtype of avian influenza A has been linked to contaminations of the human eye and conjunctivitis (inflammation of the conjunctiva). Virus replication in the lower airways is associated with the severity of an infection in humans, as is the subsequent severe inflammation caused by the infiltration of immune cells [120, 123].

1.2.8 COVID-19

In late 2019, a novel virus called SARS-CoV-2 emerged in Wuhan, China, and has since spread throughout the country, resulting in a widespread outbreak of unusual epidemiologic pneumonia. The newly discovered coronavirus infection, known as COVID-19, has swiftly spread worldwide due to its highly transmissible nature. It has surpassed both MERS and SARS in terms of the number of affected individuals and the number of countries impacted [124]. It is of great public health concern that COVID-19 continues to spread over the world. The first known SARS-CoV-2 infected individual was hospitalized for pneumonia of unknown origin; this individual had symptoms identical to those of SARS-CoV and MERS-CoV infections; this individual subsequently died. Additionally, patients hospitalized to the intensive care unit had raised levels of cytokines such as TNF-α, MIP-1, IP-10, and G-CSF [125].

Anyone may become ill with SARS-CoV-2, although people over the age of 50 are more at risk. On the other hand, the clinical symptoms shift with age. Maximum young persons and teenagers have slight infections (moderate pneumonia or non-pneumonia) or are symptomless, whereas older men (>60 years old) with co-morbidities are more likely to agonize from serious respiratory infections that need hospitalization or even death. According to the results of this research, pregnant women are not at increased risk for any adverse health outcomes when compared to women who are not pregnant [126, 127]. Although this was an exceptional instance, it was confirmed that SARS-CoV-2 might be transmitted from an infected woman to her baby during pregnancy. An elevated body temperature, intense exhaustion, and a dry cough are the most noticeable signs of contamination [128]. Chest discomfort, vomiting, bloody stools, sore throat, sputum discharge, dry cough, loss of appetite, fever, and fatigue were all reported by patients in China who were examined. Patients in Italy often mentioned changes in smell and taste [129]. After an incubation period of 1–14 days, most affected people had symptoms of viral infection, pneumonia, and distressed breathing within a median of 8 days (most typically around 5 days). COVID-19 causes a "cytokine storm" that causes damage and inflammation, particularly in the lungs, and is associated with severe symptoms

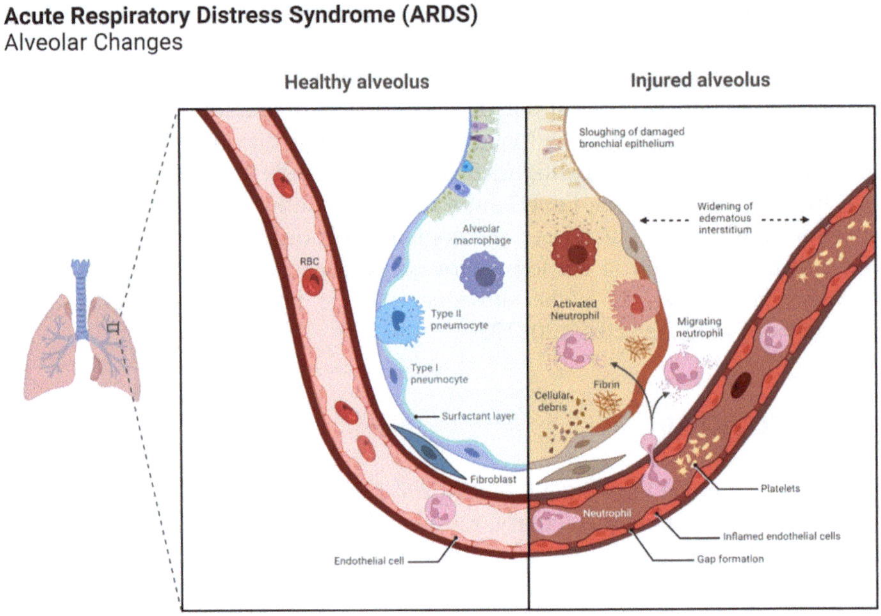

Fig. 1.2 Pathophysiology of acute respiratory distress syndrome

such as acute respiratory distress syndrome (ARDS) and severe pneumonia. COVID-19 produces LTs, IL-2, IL-6, IL-12, IL-1, TNF-α, GM-CSF, and other chemokines since NF-B is expressed by many cells, including those of the lungs, kidney, central nervous system, digestive tract, liver, and cardiovascular system, which might cause a rise in mortality or other problems [130–132] (Fig. 1.2).

1.3 Conclusion

Due to the aging of people and the absence of adequate therapies to reduce hazard factors that contribute to the course of these illnesses, lung diseases are having an increasingly large impact across the globe. Research on efficient treatment methods and sufficient prevention of lung disorders such as COVID-19, fibrosis, COPD, asthma, and influenza is becoming more urgent as these conditions become increasingly life-threatening.

References

1. Alharbi KS, Fuloria NK, Fuloria S, Rahman SB, Al-Malki WH, Javed Shaikh MA, et al. Nuclear factor-kappa B and its role in inflammatory lung disease. Chem Biol Interact. 2021;345:109568.

2. Chivima B. Lung cancer. Nurs Stand (Royal College of Nursing (Great Britain): 1987). 2015;29(22):61.
3. Abolfathi H, Sheikhpour M, Shahraeini SS, Khatami S, Nojoumi SA. Studies in lung cancer cytokine proteomics: a review. Expert Rev Proteomics. 2021;18(1):49–64.
4. Abu Rous F, Singhi EK, Sridhar A, Faisal MS, Desai A. Lung cancer treatment advances in 2022. Cancer Investig. 2023;41(1):12–24.
5. Amann A, Corradi M, Mazzone P, Mutti A. Lung cancer biomarkers in exhaled breath. Expert Rev Mol Diagn. 2011;11(2):207–17.
6. Avasarala SK, Rickman OB. Endobronchial therapies for diagnosis, staging, and treatment of lung cancer. Surg Clin North Am. 2022;102(3):393–412.
7. Bade BC, Dela Cruz CS. Lung cancer 2020: epidemiology, etiology, and prevention. Clin Chest Med. 2020;41(1):1–24.
8. Beattie EJ Jr. Lung cancer. CA Cancer J Clin. 1974;24(2):96–9.
9. Bitsko MJ, Everhart RS, Rubin BK. The adolescent with asthma. Paediatr Respir Rev. 2014;15 (2):146–53.
10. Barnthouse M, Jones BL. The impact of environmental chronic and toxic stress on asthma. Clin Rev Allergy Immunol. 2019;57(3):427–38.
11. Anwar ET, Gupta N, Porwal O, Sharma A, Malviya R, Singh A, et al. Skin diseases and their treatment strategies in Sub-Saharan African Regions. Infect Disord Drug Targets. 2022;22(2): e270921196808.
12. Agache I, Palmer E, Sanver D, Kirtland M, Shamji MH. Molecular allergology approach to allergic asthma. Mol Asp Med. 2022;85:101027.
13. Boulet LP, Boulay M. Asthma-related comorbidities. Expert Rev Respir Med. 2011;5(3): 377–93.
14. Dahiya R, Dahiya S, Fuloria NK, Kumar S, Mourya R, Chennupati SV, et al. Natural bioactive thiazole-based peptides from marine resources: structural and pharmacological aspects. Mar Drugs. 2020;18(6):329.
15. Busse WW, Melén E, Menzies-Gow AN. Holy Grail: the journey towards disease modification in asthma. Eur Respir Rev. 2022;31(163):210183.
16. Dahiya S, Dahiya R, Fuloria NK, Mourya R, Dahiya S, Fuloria S, et al. Natural bridged bicyclic peptide macrobiomolecules from celosia argentea and amanita phalloides. Mini Rev Med Chem. 2022;22(13):1772–88.
17. Chung KF. Clinical management of severe therapy-resistant asthma. Expert Rev Respir Med. 2017;11(5):395–402.
18. Jones TL, Neville DM, Chauhan AJ. Diagnosis and treatment of severe asthma: a phenotype-based approach. Clin Med. 2018;18(Suppl 2):s36–40.
19. Gans MD, Gavrilova T. Understanding the immunology of asthma: Pathophysiology, biomarkers, and treatments for asthma endotypes. Paediatr Respir Rev. 2020;36:118–27.
20. Fuloria NK, Raheja RK, Shah KH, Oza MJ, Kulkarni YA, Subramaniyan V, et al. Biological activities of meroterpenoids isolated from different sources. Front Pharmacol. 2022;13: 830103.
21. Kaur R, Chupp G. Phenotypes and endotypes of adult asthma: moving toward precision medicine. J Allergy Clin Immunol. 2019;144(1):1–12.
22. Kwah JH, Peters AT. Asthma in adults: principles of treatment. Allergy Asthma Proc. 2019;40 (6):396–402.
23. Larenas-Linnemann D. [Asthma treatment]. Revista alergia Mexico (Tecamachalco, Puebla, Mexico: 1993). 2009;56 Suppl 1:S64–78.
24. Lemanske RF Jr, Busse WW. Asthma. JAMA. 1997;278(22):1855–73.
25. Miller RL, Grayson MH, Strothman K. Advances in asthma: new understandings of asthma's natural history, risk factors, underlying mechanisms, and clinical management. J Allergy Clin Immunol. 2021;148(6):1430–41.
26. Mims JW. Asthma: definitions and pathophysiology. Int Forum Allergy Rhinol. 2015;5(Suppl 1):S2–6.

27. Ntontsi P, Photiades A, Zervas E, Xanthou G, Samitas K. Genetics and epigenetics in asthma. Int J Mol Sci. 2021;22(5):2412.
28. Fuloria S, Mehta J, Talukdar MP, Sekar M, Gan SH, Subramaniyan V, et al. Synbiotic effects of fermented rice on human health and wellness: a natural beverage that boosts immunity. Front Microbiol. 2022;13:950913.
29. Fuloria S, Sekar M, Khattulanuar FS, Gan SH, Rani N, Ravi S, et al. Chemistry, biosynthesis and pharmacology of viniferin: potential resveratrol-derived molecules for new drug discovery, development and therapy. Molecules. 2022;27(16):5072.
30. Papadopoulos NG, Miligkos M, Xepapadaki P. A current perspective of allergic asthma: from mechanisms to management. Handb Exp Pharmacol. 2022;268:69–93.
31. Sockrider M, Fussner L. What is asthma? Am J Respir Crit Care Med. 2020;202(9):P25–6.
32. Wu TD, Brigham EP, McCormack MC. Asthma in the primary care setting. Med Clin North Am. 2019;103(3):435–52.
33. COPD. Nurs Stand (Royal College of Nursing (Great Britain): 1987). 2015;29(42):16.
34. López-Campos JL, Tan W, Soriano JB. Global burden of COPD. Respirology. 2016;21(1): 14–23.
35. Fuloria S, Subramaniyan V, Dahiya R, Dahiya S, Sudhakar K, Kumari U, et al. Mesenchymal stem cell-derived extracellular vesicles: regenerative potential and challenges. Biology. 2021;10(3):172.
36. Fuloria S, Yusri MAA, Sekar M, Gan SH, Rani N, Lum PT, et al. Genistein: a potential natural lead molecule for new drug design and development for treating memory impairment. Molecules. 2022;27(1):265.
37. Agustí A, Vogelmeier C, Faner R. COPD 2020: changes and challenges. Am J Physiol Lung Cell Mol Physiol. 2020;319(5):L879–l83.
38. Bagdonas E, Raudoniute J, Bruzauskaite I, Aldonyte R. Novel aspects of pathogenesis and regeneration mechanisms in COPD. Int J Chron Obstruct Pulmon Dis. 2015;10:995–1013.
39. Raherison C, Girodet PO. Epidemiology of COPD. Eur Respir Rev. 2009;18(114):213–21.
40. Barnes PJ. COPD 2020: new directions needed. Am J Physiol Lung Cell Mol Physiol. 2020;319(5):L884–l6.
41. Erhabor GE, Adeniyi B, Arawomo AO, Akinwalere O, Adetona G, Fagbohun FT, et al. Acute exacerbation of COPD: clinical perspectives and literature review. West Afr J Med. 2021;38 (11):1129–42.
42. Fischer BM, Voynow JA, Ghio AJ. COPD: balancing oxidants and antioxidants. Int J Chron Obstruct Pulmon Dis. 2015;10:261–76.
43. Guerreiro I, Soccal PM. [COPD and phenotypes]. Rev Med Suisse 2019;15(671):2082–6.
44. Tantucci C, Modina D. Lung function decline in COPD. Int J Chron Obstruct Pulmon Dis. 2012;7:95–9.
45. Gupta G, Almalki WH, Kazmi I, Fuloria NK, Fuloria S, Subramaniyan V, et al. Current update on the protective effect of naringin in inflammatory lung diseases. EXCLI J. 2022;21:573–9.
46. Hamid UZ, Sim MS, Guad RM, Subramaniyan V, Sekar M, Fuloria NK, et al. Molecular regulatory roles of long non-coding RNA HOTTIP: an overview in gastrointestinal cancers. Curr Mol Med. 2022;22(6):478–90.
47. Uwagboe I, Adcock IM, Lo Bello F, Caramori G, Mumby S. New drugs under development for COPD. Minerva Med. 2022;113(3):471–96.
48. Agusti A, Soriano JB. COPD as a systemic disease. COPD. 2008;5(2):133–8.
49. Aryal S, Diaz-Guzman E, Mannino DM. COPD and gender differences: an update. Transl Res. 2013;162(4):208–18.
50. Barnes PJ. Endo-phenotyping of COPD patients. Expert Rev Respir Med. 2021;15(1):27–37.
51. Huqh MZU, Abdullah JY, Wong LS, Jamayet NB, Alam MK, Rashid QF, et al. Clinical applications of artificial intelligence and machine learning in children with cleft lip and palate-a systematic review. Int J Environ Res Public Health. 2022;19(17):10860.

52. Jha S, Malviya R, Fuloria S, Sundram S, Subramaniyan V, Sekar M, et al. Characterization of microwave-controlled polyacrylamide graft copolymer of tamarind seed polysaccharide. Polymers. 2022;14(5):1037.
53. Kaur I, Behl T, Sehgal A, Singh S, Sharma N, Subramanian V, et al. A motley of possible therapies of the COVID-19: reminiscing the origin of the pandemic. Environ Sci Pollut Res Int. 2022;29(45):67685–703.
54. Kumar S, Behl T, Sehgal A, Chigurupati S, Singh S, Mani V, et al. Exploring the focal role of LRRK2 kinase in Parkinson's disease. Environ Sci Pollut Res Int. 2022;29(22):32368–82.
55. Chakinala RC, Khatri A, Gupta K, Koike K, Epelbaum O. Sphingolipids in COPD. Eur Respir Rev. 2019;28(154):190047.
56. Duffy SP, Criner GJ. Chronic Obstructive Pulmonary Disease: Evaluation and Management. Med Clin North Am. 2019;103(3):453–61.
57. Gupta N, Agrawal S, Chakrabarti S, Ish P. COPD 2020 guidelines - what is new and why? Adv Respir Med. 2020;88(1):38–40.
58. Hanania NA, Sharma G, Sharafkhaneh A. COPD in the elderly patient. Semin Respir Crit Care Med. 2010;31(5):596–606.
59. Hattab Y, Alhassan S, Balaan M, Lega M, Singh AC. Chronic obstructive pulmonary disease. Crit Care Nurs Q. 2016;39(2):124–30.
60. Hering T. [COPD: Capture an exacerbation with just five questions]. MMW Fortschr Med 2022;164(10):61–3.
61. Labaki WW, Rosenberg SR. Chronic obstructive pulmonary disease. Ann Intern Med. 2020;173(3):Itc17–32.
62. Mortensen J, Berg RMG. Lung scintigraphy in COPD. Semin Nucl Med. 2019;49(1):16–21.
63. Lum PT, Sekar M, Gan SH, Jeyabalan S, Bonam SR, Rani N, et al. Therapeutic potential of mangiferin against kidney disorders and its mechanism of action: a review. Saudi J Biol Sci. 2022;29(3):1530–42.
64. Malviya R, Fuloria S, Verma S, Subramaniyan V, Sathasivam KV, Kumarasamy V, et al. Commercial utilities and future perspective of nanomedicines. PeerJ. 2021;9:e12392.
65. Petty TL. COPD in perspective. Chest. 2002;121(5 Suppl):116s–20s.
66. Rabe KF, Watz H. Chronic obstructive pulmonary disease. Lancet. 2017;389(10082): 1931–40.
67. Tan WSD, Shen HM, Wong WSF. Dysregulated autophagy in COPD: a pathogenic process to be deciphered. Pharmacol Res. 2019;144:1–7.
68. Sethi T. Lung cancer.Introduction. Thorax. 2002;57(11):992–3.
69. Mustafa NH, Sekar M, Fuloria S, Begum MY, Gan SH, Rani N, et al. Chemistry, biosynthesis and pharmacology of sarsasapogenin: a potential natural steroid molecule for new drug design, development and therapy. Molecules. 2022;27(6):2032.
70. Nasir NN, Sekar M, Fuloria S, Gan SH, Rani N, Ravi S, et al. Kirenol: a potential natural lead molecule for a new drug design, development, and therapy for inflammation. Molecules. 2022;27(3):734.
71. Seale DD, Beaver BM. Pathophysiology of lung cancer. Nurs Clin North Am. 1992;27(3): 603–13.
72. Rodescu D. Lung cancer. Med Clin North Am. 1977;61(6):1205–18.
73. Ruppert AM, Amrioui F, Fallet V. [Risk factors and prevention of lung cancer]. Rev Prat 2020;70(8):852–6.
74. Selvaraj LK, Jeyabalan S, Wong LS, Sekar M, Logeshwari B, Umamaheswari S, et al. Baicalein prevents stress-induced anxiety behaviors in zebrafish model. Front Pharmacol. 2022;13:990799.
75. Selvaraj S, Naing NN, Wan-Arfah N, Djearamane S, Wong LS, Subramaniyan V, et al. Epidemiological factors of periodontal disease among south indian adults. J Multidiscip Healthc. 2022;15:1547–57.
76. McComb BL, Ko JP. Lung cancer imaging. J Thorac Imaging. 2011;26(2):83–4.
77. Novello S, Vavalà T. Lung cancer and women. Future Oncol. 2008;4(5):705–16.

78. Singh S, Hema, Sharma N, Sachdeva M, Behl T, Zahoor I, et al. Focusing the pivotal role of nanotechnology in Huntington's disease: an insight into the recent advancements. Environ Sci Pollut Res Int. 2022;29(49):73809–27.
79. Subramaniyan V, Chakravarthi S, Jegasothy R, Seng WY, Fuloria NK, Fuloria S, et al. Alcohol-associated liver disease: a review on its pathophysiology, diagnosis and drug therapy. Toxicol Rep. 2021;8:376–85.
80. Chellappan DK, Chellian J, Leong JQ, Liaw YY, Gupta G, Dua K, et al. Biological and therapeutic potential of the edible brown marine seaweed Padina australis and their pharmacological mechanisms. J Trop Biol Conserv. 2020;17:251–71–71.
81. Krasna MJ. Lung cancer. Surg Oncol Clin N Am. 2011;20(4):xv–xvi.
82. Gudbjartsson T, Smáradottir A, Skúladóttir H, Grímsson HN, Hardardóttir H, Björnsson J, et al. [Lung cancer - review]. Laeknabladid 2008;94(4):297–311.
83. Giaccone G, Smit E. Lung cancer. Cancer Chemother Biol Response Modif. 2005;22:413–42.
84. Gilliland FD, Samet JM. Lung cancer. Cancer Surv. 1994;19–20:175–95.
85. Ernster VL. Female lung cancer. Annu Rev Public Health. 1996;17:97–114.
86. Vavalà T, Catino A, Pizzutilo P, Longo V, Galetta D. Gender Differences and Immunotherapy Outcome in Advanced Lung Cancer. Int J Mol Sci. 2021;22(21):11942.
87. Stanton BF. Cystic fibrosis. Pediatr Clin N Am. 2016;63(4):xv.
88. Savant AP, McColley SA. Cystic fibrosis year in review 2018, part 2. Pediatr Pulmonol. 2019;54(8):1129–40.
89. Savant AP, McColley SA. Cystic fibrosis year in review 2018, part 1. Pediatr Pulmonol. 2019;54(8):1117–28.
90. Savant AP, McColley SA. Cystic fibrosis year in review 2016. Pediatr Pulmonol. 2017;52(8): 1092–102.
91. Rafeeq MM, Murad HAS. Cystic fibrosis: current therapeutic targets and future approaches. J Transl Med. 2017;15(1):84.
92. Paranjape SM, Mogayzel PJ Jr. Cystic fibrosis. Pediatr Rev. 2014;35(5):194–205.
93. Radlović N. Cystic fibrosis. Srp Arh Celok Lek. 2012;140(3–4):244–9.
94. Leitch AE, Rodgers HC. Cystic fibrosis. J R Coll Physicians Edinb. 2013;43(2):144–50.
95. Ooi CY, Durie PR. Cystic fibrosis from the gastroenterologist's perspective. Nat Rev Gastroenterol Hepatol. 2016;13(3):175–85.
96. Klimova B, Kuca K, Novotny M, Maresova P. Cystic fibrosis revisited - a review study. Med Chem. 2017;13(2):102–9.
97. De Boeck K. Cystic fibrosis in the year 2020: a disease with a new face. Acta Paediatr. 2020;109(5):893–9.
98. Endres TM, Konstan MW. What is cystic fibrosis? JAMA. 2022;327(2):191.
99. Hewlett JC, Kropski JA, Blackwell TS. Idiopathic pulmonary fibrosis: epithelial-mesenchymal interactions and emerging therapeutic targets. Matrix Biol. 2018;71–72:112–27.
100. King TE Jr, Pardo A, Selman M. Idiopathic pulmonary fibrosis. Lancet. 2011;378(9807): 1949–61.
101. Kishaba T. Acute exacerbation of idiopathic pulmonary fibrosis. Medicina (Kaunas). 2019;55 (3):70.
102. Lederer DJ, Martinez FJ. Idiopathic pulmonary fibrosis. N Engl J Med. 2018;378(19): 1811–23.
103. León-Román F, Valenzuela C, Molina-Molina M. Idiopathic pulmonary fibrosis. Med Clin. 2022;159(4):189–94.
104. Martinez FJ, Collard HR, Pardo A, Raghu G, Richeldi L, Selman M, et al. Idiopathic pulmonary fibrosis. Nat Rev Dis Primers. 2017;3:17074.
105. Moss BJ, Ryter SW, Rosas IO. Pathogenic mechanisms underlying idiopathic pulmonary fibrosis. Annu Rev Pathol. 2022;17:515–46.
106. Cardona PJ. Pathogenesis of tuberculosis and other mycobacteriosis. Enferm Infecc Microbiol Clin (Engl Ed). 2018;36(1):38–46.

107. Casas I, Dominguez J, Rodríguez S, Matllo J, Altet N. [Guidelines for the prevention and control of tuberculosis in health care workers]. Med Clin. 2015;145(12):534.e1–13.
108. Jové N, Masdeu E, Brugueras S, Millet JP, Ospina JE, Orcau À, et al. Threats and interventions during the treatment of tuberculosis in an inner-city district. Arch Bronconeumol. 2021;57(5): 330–7.
109. Ting BYS, Fuloria NK, Subrimanyan V, Bajaj S, Chinni SV, Reddy LV, et al. Biosynthesis and response of zinc oxide nanoparticles against periimplantitis triggering pathogens. Materials (Basel). 2022;15(9):3170.
110. Tune BXJ, Sim MS, Poh CL, Guad RM, Woon CK, Hazarika I, et al. Matrix metalloproteinases in chemoresistance: regulatory roles, molecular interactions, and potential inhibitors. J Oncol. 2022;2022:3249766.
111. Ketata W, Rekik WK, Ayadi H, Kammoun S. [Extrapulmonary tuberculosis]. Rev Pneumol Clin 2015;71(2–3):83–92.
112. Margarit A, Simó S, Rozas L, Deyà-Martínez À, Barrabeig I, Gené A, et al. [Adolescent tuberculosis; a challenge and opportunity to prevent community transmission]. An Pediatr (Barc). 2017;86(3):110–4.
113. Orcau À, Caylà JA, Martínez JA. Present epidemiology of tuberculosis. Prevention and control programs. Enferm Infecc Microbiol Clin. 2011;29(Suppl 1):2–7.
114. Yap KM, Sekar M, Wu YS, Gan SH, Rani N, Seow LJ, et al. Hesperidin and its aglycone hesperetin in breast cancer therapy: a review of recent developments and future prospects. Saudi J Biol Sci. 2021;28(12):6730–47.
115. Zahoor I, Singh S, Behl T, Sharma N, Naved T, Subramaniyan V, et al. Emergence of microneedles as a potential therapeutics in diabetes mellitus. Environ Sci Pollut Res Int. 2022;29(3):3302–22.
116. Schito M, Migliori GB, Fletcher HA, McNerney R, Centis R, D'Ambrosio L, et al. Perspectives on advances in tuberculosis diagnostics, drugs, and vaccines. Clin Infect Dis. 2015;61Suppl 3(Suppl 3):S102–18.
117. Solsona Peiró J, de Souza Galvão ML, Altet Gómez MN. Inactive fibrotic lesions versus pulmonary tuberculosis with negative bacteriology. Arch Bronconeumol. 2014;50(11):484–9.
118. Hoa LNM, Sullivan SG, Mai LQ, Khvorov A, Phuong HVM, Hang NLK, et al. Influenza A (H1N1)pdm09 But Not A(H3N2) virus infection induces durable seroprotection: results from the ha nam cohort. J Infect Dis. 2022;226(1):59–69.
119. Zuraini NZA, Sekar M, Wu YS, Gan SH, Bonam SR, Mat Rani NNI, et al. Promising nutritional fruits against cardiovascular diseases: an overview of experimental evidence and understanding their mechanisms of action. Vasc Health Risk Manag. 2021;17:739–69.
120. Itoh Y. Translational research on influenza virus infection using a nonhuman primate model. Pathol Int. 2016;66(3):132–41.
121. Mifsud EJ, Kuba M, Barr IG. Innate immune responses to influenza virus infections in the upper respiratory tract. Viruses. 2021;13(10):2090.
122. Sura T, Gering V, Cammann C, Hammerschmidt S, Maaß S, Seifert U, et al. Streptococcus pneumoniae and influenza A virus co-infection induces altered polyubiquitination in A549 cells. Front Cell Infect Microbiol. 2022;12:817532.
123. Pantin-Jackwood MJ, Swayne DE. Pathogenesis and pathobiology of avian influenza virus infection in birds. Rev Sci Tech. 2009;28(1):113–36.
124. Alsharif W, Qurashi A. Effectiveness of COVID-19 diagnosis and management tools: a review. Radiography (Lond). 2021;27(2):682–7.
125. Amit S, Beni SA, Biber A, Grinberg A, Leshem E, Regev-Yochay G. Postvaccination COVID-19 among Healthcare Workers, Israel. Emerg Infect Dis. 2021;27(4):1220–2.
126. Fernandes Q, Inchakalody VP, Merhi M, Mestiri S, Taib N, Moustafa Abo El-Ella D, et al. Emerging COVID-19 variants and their impact on SARS-CoV-2 diagnosis, therapeutics and vaccines. Ann Med. 2022;54(1):524–40.
127. Habas K, Nganwuchu C, Shahzad F, Gopalan R, Haque M, Rahman S, et al. Resolution of coronavirus disease 2019 (COVID-19). Expert Rev Anti-Infect Ther. 2020;18(12):1201–11.

128. Majumder J, Minko T. Recent developments on therapeutic and diagnostic approaches for COVID-19. AAPS J. 2021;23(1):14.
129. Safiabadi Tali SH, LeBlanc JJ, Sadiq Z, Oyewunmi OD, Camargo C, Nikpour B, et al. Tools and techniques for severe acute respiratory syndrome coronavirus 2 (SARS-CoV-2)/COVID-19 detection. Clin Microbiol Rev. 2021;34(3):e00228–0.
130. Sidiq Z, Hanif M, Dwivedi KK, Chopra KK. Benefits and limitations of serological assays in COVID-19 infection. Indian J Tuberc. 2020;67(4s):S163–s6.
131. Umakanthan S, Sahu P, Ranade AV, Bukelo MM, Rao JS, Abrahao-Machado LF, et al. Origin, transmission, diagnosis and management of coronavirus disease 2019 (COVID-19). Postgrad Med J. 2020;96(1142):753–8.
132. Yüce M, Filiztekin E, Özkaya KG. COVID-19 diagnosis - a review of current methods. Biosens Bioelectron. 2021;172:112752.

Introduction to Epigenetics 2

Neelam Singla, Riya Thapa, Rashi Kulshrestha, Asif Ahmad Bhat, Saurabh Gupta, Manish Purohit, Santosh Kumar Singh, and Gaurav Gupta

Abstract

Epigenetics is the study of inherited variations in gene expression that do not involve changes to DNA sequences. Over 200 unique cell types in a human adult have a nearly similar genomic sequence. DNA and histones undergo reversible chemical alterations that contribute greatly to cellular variety via dynamic control of global gene expression, in addition to the genomic sequence. Epigenetics is based on chemical modifications of macromolecules rather than alterations to the chromosomal sequence. Epigenetic modifications, such as DNA methylation, chromatin modifications, nucleosome placement, and changes in noncoding RNA profiles, are reversible and fundamental processes in the regulation of gene expression. Alterations in gene activity and neoplastic transformation of cells may result from epigenetic disruptions. Epigenetic alterations occur at an early stage in neoplastic growth, before the corresponding genetic alterations have actually occurred. Researchers in this area will discover information on a wide variety of topics, all grouped together under the explanatory framework of chemical changes that affect gene expression regulation. In this chapter, we provide information on epigenetics, DNA methylation, and histone modifications.

Keywords

Epigenetics · Histone acetylation · HDAC inhibitors · DNA methylation

N. Singla · R. Thapa · R. Kulshrestha · A. A. Bhat · M. Purohit · S. K. Singh · G. Gupta (✉)
School of Pharmacy, Suresh Gyan Vihar University, Jagatpura, Jaipur, India

S. Gupta
Chameli Devi Institute of Pharmacy, Indore, Madhya Pradesh, India

G. Gupta et al. (eds.), *Targeting Epigenetics in Inflammatory Lung Diseases*,
https://doi.org/10.1007/978-981-99-4780-5_2

2.1 Introduction

Reproduction is crucial for the continuation of life and also aids in the transmission of characteristics from one generation to the next. Non-genetic variables, like epigenetics, also have a role in determining heritability and the genetic information stored in gametes. The reproductive processes most important for epigenetic landscape construction or maintenance are germ cell development and early embryo development. There is evidence that prenatal exposure to a certain lifestyle and environmental factors may alter the epigenetic blueprint of gametes, hence altering the phenotype of offspring. It's common knowledge that people's genes contribute to their susceptibility to a wide range of ailments. As time passes, more and more proof emerges that an organism's health state is determined by more than just hereditary elements; environmental influences may also play a role in shaping health via epigenetic modulations. Like genetic characteristics, epigenetic variables affect reproductive health and fertility [1, 2]. So, it's crucial to learn about epigenetic alterations, cell physiology, and the root of disease. In the last three decades, new developments in sequencing the human epigenome have helped scientists understand the origins of several diseases. This opens opportunities for understanding the epigenetic code and creating more effective therapies [3–5].

This section summarizes the most important epigenetic processes and introduces the many epigenetic analysis methods and cutting-edge epigenetic technologies that are now accessible.

2.2 Epigenetics

2.2.1 Definition

Although approximately 200 cell types in humans have the same DNA sequences, their gene expression patterns and phenotypes are very different. Developmental scientist Conrad Waddington first used the term "epigenetic landscape" to characterize the dynamic cellular phenotypic changes that occur throughout the evolution of a multicellular organism. Holliday later defined "epigenetics" as a branch of nuclear inheritance that does not rely on variations in DNA sequence. According to current molecular and mechanistic definitions, epigenetics is the combination of modifications to the chromatin template that generate and maintain separate gene expression patterns and silencing within a single genome. If genes are words, then epigenetics specifies how those words should be interpreted [6–9].

On the other hand, if we see sets of genes as computer hardware, then epigenetic regulation would be analogous to the operating system. That's why epigenetics is so important; it's a window into how cellular processes are orchestrated, functioning as an extra layer of regulation. Three primary processes cause epigenetic alterations to chromatin: DNA methylation, noncoding RNAs, and post-translational histone modifications. This chapter will try to give you a taste of these ideas (Fig. 2.1) [10, 11].

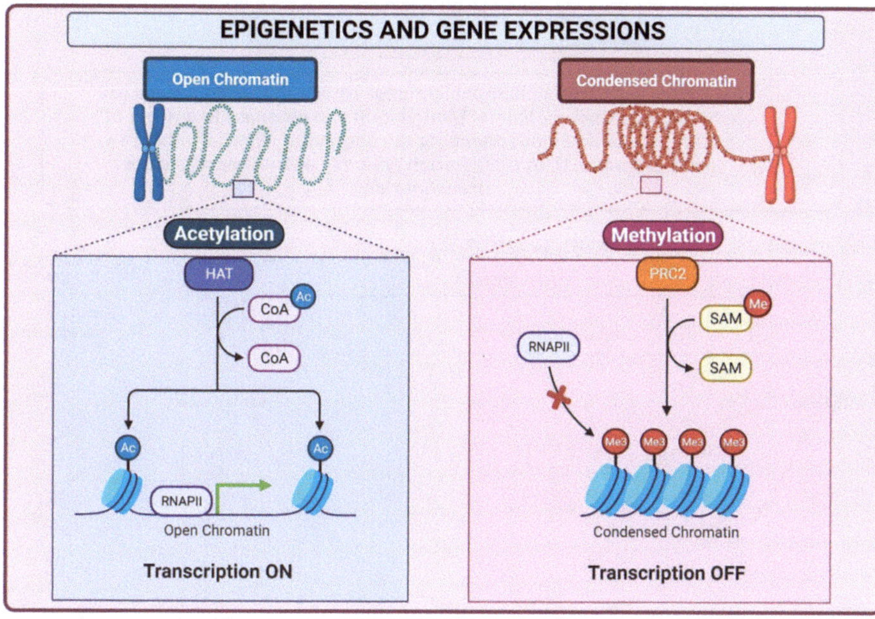

Fig. 2.1 Epigenetics and gene expressions

2.2.2 DNA Methylation

Over 35 years after its discovery, DNA methylation at cytosine residues is still widely considered a fundamental epigenetic mechanism controlling gene expression [6]. 5mC, or 5-methylcytosine, is converted to a cytosine residue in the DNA template by adding a methyl group. The most common location is the dinucleotide sequence 50CpG30 (CpG stands for cytosine and guanine, which a phosphate group separates in DNA). Both strands of DNA may carry it. Therefore, it can be passed down through generations of a family. As a kind of epigenetic marking for the genome that is conserved across cell division, DNA methylation patterns contribute to the formation of cellular memory. Clusters of CpGs, or CpG islands, are common. They tend to be abundant in interspersed repetitive elements and noncoding areas (like centromeric heterochromatin) (e.g., retrotransposons). In addition to their prevalence in gene enhancers, CpG islands are often located in the upstream area of gene promoters [12–15].

The methyl group of cytosine, found in the DNA helix's major groove, is the primary site of interaction for many DNA-binding proteins. Then, different DNA-binding proteins are attracted to or repelled by methylation DNA. Methyl-CpG-binding domain (MBD) proteins attach to methylated CpGs, which in turn recruit repressor complexes (such as histone deacetylases, which remove activating histone acetylation marks) to methylated promoter areas, resulting in transcriptional suppression. CpG methylation, in contrast, prevents transcription by inhibiting the

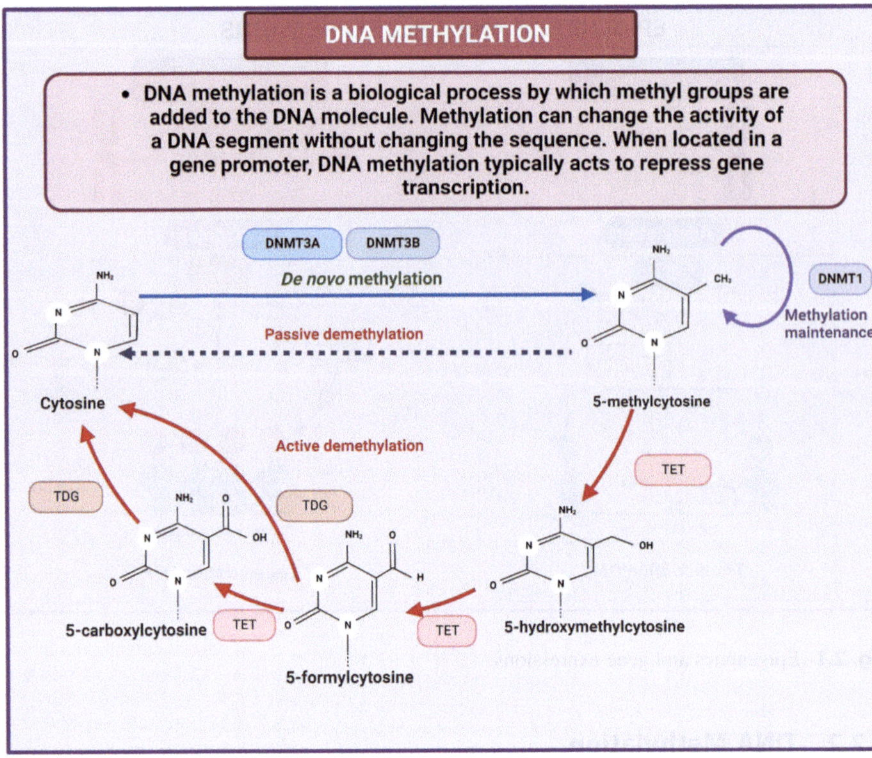

Fig. 2.2 DNA methylation

binding of certain transcriptional regulators like CTCF. Cellular differentiation, genomic imprinting, and inactivation of the X chromosome all rely on DNA methylation as a critical process [16, 17]. DNA methylation is essential for maintaining genomic stability and protecting against the spread of harmful transposable elements. Cell division results in the transmission of DNA methylation patterns from one cell to the next, but these patterns are not permanent. Researchers have discovered that DNA methylation patterns change throughout an individual's life. These alterations may be a healthy reaction to external stimuli or related to pathogenic processes such as neoplastic transformation or the natural aging process [18, 19] (Fig. 2.2).

2.2.3 DNA Methylation: Establishment and Erasure

DNA methylation in mammals is mediated by DNA methyltransferases (Dnmts), which consist of three proteins from two families with very different structures and functions. The first family of maintenance methyltransferases includes Dnmt1, an enzyme that preferentially methylates hemimethylated CpG dinucleotides (i.e., DNA methylated at CpG in one of the two strands). This way, Dnmt1 is responsible for semiconservative DNA methylation patterns during replication. Dnmt3a and 3b belong to a different family of de novo methyltransferases that establish de novo DNA methylation patterns in the developing embryo. Even though it is inactive as a methyltransferase, Dnmt3l plays a crucial role in de novo methylation activities at certain DNA sequences and directs the activities of Dnmt3a and 3b. The ten-eleven translocation (Tet) protein family of DNA hydroxylases is involved in the active demethylation process [20, 21].

In contrast, the suppression of Dnmt1 during cell division is responsible for the passive demethylation process. Tet enzymes may convert the methyl group on CpG to 5-hydroxymethylcytosine (5hmC). This DNA alteration is abundant in the active chromatin areas of the genome and has a role in regulating gene expression. 5hmC may be oxidized further into 5-carboxyl cytosine (5caC) and 5-formyl cytosine (5fC) by the Tet family of proteins, which uses ATP in the process. This triggers the base excision repair process, which removes the mutated base and converts the 5mC back to an unmethylated cytosine, reactivating transcription [22–26].

2.2.4 Histone Modifications

Chromatin, found in multicellular organisms, is a nucleoprotein structure that houses DNA. DNA is wrapped around a ring of histones, which are uncomplicated proteins. Nucleosomes, the basic building blocks of chromatin, comprise 146 bp of DNA wrapped around an octamer of core histones. Two copies of the main histones H2A, H2B, H3, and H4 make up each nucleosome. The H1 linker protein connects each nucleosome's DNA and histone octamer core. The histone core is attached to the DNA through weak ionic contacts between positively charged residues on histone proteins and phosphate groups on the DNA [27–31].

2.2.5 Higher-Order Chromatin Organization

Nucleosomes are the fundamental units of chromatin structure. When the chromatin is fully unfolded, the structures may be seen as "beads on a string" made up of polymers with a size of 11 nm. According to the solenoid model of the chromatin fiber, nucleosomes are arranged in a helical array of around six to eight nucleosomes per turn, with the histone H1 on the inner of the fiber. Since the nucleosomes run in a spiral pattern, the linker DNA must be twisted to join them. This leads to a tighter chromatin shape that is transcriptionally inept at 30 nm. This allows the chromatin to be structured into bigger looping domains (300–700 nm). In the metaphase of mitosis or meiosis, chromosomes develop with the most compacted chromatin structure to ensure accurate genetic material distribution [32–35].

The chromatin structure is very flexible and may take on a wide variety of shapes. Traditionally, chromatin has been split into euchromatin and heterochromatin. Coding and regulatory (for example, promoters and enhancers) portions of the genome are found in euchromatin, often known as "active" chromatin. It is in an accessible, decondensed confirmation that is "poised" for gene expression because it encourages active transcription. Heterochromatin describes the tight, highly compacted form of the "inactive" sections of the genome. It consists mostly of noncoding sequences and repetitive regions, yet it still makes up the vast bulk of the genome (e.g., retrotransposons, satellite repeats, and LINEs). Permanently suppressed "constitutive" heterochromatin is often located in pericentric and subtelomeric areas [36–39], whereas "facultative" heterochromatin allows for temporary derepression of genes within a certain cell cycle or developmental stage.

Histones in the nucleus are among the most conserved proteins in eukaryotes. The N-terminal portion is simple, whereas the C-terminal resembles a histone fold. The DNA is wrapped around the histone's globular domain after heterodimerized with another histone (H3 with H4, H2A with H2B). The N-terminal "tail" domain is unstructured and exists outside the nucleosome. Although many post-translational modifications (PTMs) occur at specific residues in the tails of histones, notably histones H3 and H4, other modifications occur at specific residues in the more organized globular domains as well. These modifications foster nucleosomal and, by extension, chromatin variety. The first covalent modifications studied were histone H3 and H4 acetylation and methylation. Additional modifications to histones, including ubiquitination, phosphorylation, ADP-ribosylation, sumoylation, crotonylation, and biotinylation, have also been discovered [40–42].

2.2.6 Histone Acetylation

Lysine (K) residues at the N terminus of histone may be acetylated. Although acetylation of all core histones is possible in vivo, H3 and H4 acetylation have received the most attention. Lysines 9, 14, 18, and 23 in H3 and lysines 5, 8, 12, and 16 in H4 are available for acetylation. The affinity of histone tails for DNA is reduced when acetyl groups are added, which neutralizes the basic charge of the tails. Furthermore, it affects the contacts between histones inside nucleosomes and the interactions of histones with other regulatory proteins. The acetylation of histones is a hallmark of euchromatin and generates an "open" chromatin environment conducive to gene transcription [43, 44]. Histone acetyltransferases and histone deacetylases work in opposition to one another to regulate the dynamic process of histone acetylation (HDACs) [9, 31].

2.2.7 Histone Methylation

Lysine (K) and arginine (R) residues on histone may be methylated on their side chains. In contrast to acetylation, methylation of histones does not affect the protein's charge. Lysines may be mono-, di-, or tri-methylated; arginines can be mono-, symmetrical, or asymmetric di-methylated. This improves the already high complexity of this update. Several lysine residues are at positions 4, 9, 27, and 36 on

H3 and at lysine 20 on H4 [45]. H3R2, H3R8, H3R17, H3R26, and H4R3 are all arginine methylation sites. However, histone methylation may signify either transcriptional activation or repression, depending on the methylation sites, in contrast to acetylation, which invariably leads to transcriptional activation. Methylation at H3K4 activates genes, whereas H3K27 methylation is known to silence them. The result is that most of these alterations occur in the promoter and enhancer elements of genes. Closed chromatin, caused by H3K9 methylation, is characteristic of heterochromatin areas in the genome. Heterochromatin-mediated transcriptional silence, X chromosome inactivation, and the DNA damage response are just a few of the many biological processes in which histone lysine methylation plays a role. Kinase methyltransferases (KMTs) and lysine demethylases (KDMs) coordinately control histone lysine methylation to preserve cell destiny and genomic integrity [46–49].

2.2.8 Other Histone Modifications

Phosphorylation of histones happens most often on N-terminal serine, threonine, and tyrosine residues. Phosphorylation of histones gives them a negative charge, which leads to a less closed chromatin structure. Thus, it is linked to gene expression and plays a role in chromatin remodeling and DNA damage repair. Histones may acquire a negative charge by the reversible mono- and poly-ADP ribosylation of glutamate and arginine residues, which aids in maintaining a state of relaxed chromatin. When DNA is damaged, these alterations become more common and play a role in the body's repair process [50–53]. A large ubiquitin moiety (a polypeptide of 76 amino acids) is attached to lysine residues in histone ubiquitylation. Polyubiquitylation of histones marks them for proteolytic breakdown, whereas monoubiquitylation may result in either gene activation or repression. The result of sumoylation is the covalent binding of small ubiquitin-like modifier molecules to histone lysines, as with ubiquitylation. Sumoylation, which may occur on any of the four core histones and blocks the acetylation or ubiquitylation of lysine residues, is responsible for gene repression [54, 55].

2.2.9 The "Histone Code": Its Writers, Readers, and Erasers

Remarkable nucleosomal variety results from covalent changes happening at various and specific places on the histones. Genetic changes in gene expression are regulated and determined by the chromatin structure, which may alter various histone modification combinations. The "Histone Code Hypothesis" describes this theory. This theory proposes that changes to histones serve as a platform for signals that may activate or silence genes by affecting the chromatin states. Histone modifications are performed by enzymes known collectively as the "writers" of the histone code. The enzymes in question include histone deacetylases (HATs), methyltransferases (HMTs), and histone kinases [56, 57]. Erasers of histone codes refer to enzymes like HDACs and KDMs that remove histone modifications. More than just changing the chromatin state, these modifications also create binding or recognition sites for a variety of effector proteins called "readers" of the histone codes, which in turn recruit additional co-regulator complexes to affect even more dramatic changes in chromatin structure and, ultimately, gene expression [58–61].

2.2.10 Noncoding RNAs

Only a tiny percentage of the human transcriptome (2–5%) was encoded for proteins, and the functions of the other transcripts were not understood before the Human Genome Project. New sequencing methods have shown that 75% of the genome is transcribed. There has been a reevaluation of the idea that most of the genome is "junk DNA" since it is not transcribed, but the argument remains heated. Classification of RNA types has undergone significant revision in recent years. It now falls into two basic categories: RNAs that code for proteins (messenger RNAs) and those that don't (noncoding RNAs) [62, 63]. Promoter-associated RNAs (PARs), Small interfering RNAs (siRNAs), PIWI-interacting RNAs (piRNAs), circular RNAs (circ RNAs), long noncoding RNAs (lncRNAs), microRNAs (miRNAs), and enhancer RNAs (eRNAs) are just some of the numerous noncoding RNAs that have been discovered in recent years. Recent years have seen much research into the activities of ncRNAs, and it has become clear that they serve several structural and regulatory functions [26, 64, 65].

What's more, noncoding RNAs have recently been recognized as a distinct epigenetic mechanism for controlling gene expression. Changes in chromatin states may be effected through cis or trans processes, and the scaffolding for chromatin-remodeling and -modifying enzyme complexes is an important function of ncRNAs. They may also facilitate the recruitment of co-repressors and other factors involved in gene silencing or transcriptional activation. These ncRNAs may offer the required sequence specificity to guide chromatin-modifying complexes to their targets. ncRNAs, whose roles are gradually revealed, are an essential component of epigenetic control [66, 67].

2.2.11 RNA Modifications

Covalent modification of RNA may be induced by a broad range of chemical additions to its sugar and nucleotide groups. However, although alterations at the nucleotide base impart new regulatory activities, RNA is largely stable because of modifications to its sugar backbone. The vast majority of coding and noncoding RNA undergo post-translational modifications; methylation of adenosine at position N(6) (m(6)A) is the most common of these modifications. The 30UTRs and stop codons are the most common places where this happens. Different stages of mRNA metabolism, including RNA polyadenylation, microRNA-mediated degradation, and pre-microRNA processing, are regulated by RNA methylation and engaged in translation and RNA degradation [68–70].

Additionally, it regulates alternative splicing for a certain group of messenger RNAs (mRNAs) and long noncoding RNAs (lncRNAs). The METTL3 RNA methyltransferase complex initiates this modification, while the FTO (fat mass and obesity-associated) RNA demethylase complex reverses it. This alteration has a critical function in development, metabolism, and fertility, and its amount is dynamically regulated by the activity of these enzymes [71, 72].

2.3 Techniques for Epigenetic Analysis

Initially, epigenetics relied on specialized nucleases (such as restriction enzymes, DNase I, and MNase); DNA- and histone-modifying enzymes were just recently discovered, yet they've already yielded a plethora of fresh data. The last several years have introduced novel methods that directly evaluate interacting proteins, chromatin modification, nucleosomal occupancy, and DNA methylation patterns [73, 74]. Understanding epigenetic processes have been sped up by the active development of new tools to examine genome-wide DNA methylation patterns and chromatin structure in the contemporary age. Following is a discussion of the many genome-wide and loci-specific methodologies built to identify epigenetic markers and evaluate their functioning [75, 76].

2.3.1 DNA Methylation

From restriction, endonuclease-based southern blotting to epigenomic microarrays and methylation-specific polymerase chain reaction to targeted or whole genome bisulfite sequencing using next-generation sequencing technology, the methods for detecting DNA methylation have progressed significantly in recent years [77, 78]. The methodology used is crucial to gain an objective response to the study topic. The sensitivity and specificity requirements, the study's goal, the volume and quality of the DNA sample, the availability of reagents and equipment, and the technique's cost-effectiveness and ease of use all play a role in determining which method will be used. Whether the information sought is genome-wide or locus-specific is another consideration when deciding on a methodology. Whether or not the candidate genes are known may also play a role in deciding which approach to take [79, 80].

2.3.2 Histone Modifications and Chromatin Remodeling

The study of histone modifications and chromatin remodeling throughout the whole genome has come a long way in the last three decades. Improvements in high-throughput sequencing in tandem with chromatin immunoprecipitation assays (ChIP) and DNA microarrays (DNA chips)—for example, ChIP-on-chip, Chromatin

Interaction Analysis by Paired-End Tag Sequencing (ChIA-PET), and ChIP-Sequencing (ChIP-Seq)—have aided in deciphering the human epigenome [81, 82].

Studying the kinetics of histone methylation is made possible by chromatin immunoprecipitation (ChIP), a potent method for analyzing protein-DNA interactions. The idea of chromatin immunoprecipitation (ChIP) is to enrich the area of DNA of interest (antigen) by immunoprecipitation, then amplify the enhanced region by polymerase chain reaction (PCR) to acquire an adequate amount of the enriched fraction. Also, southern blotting, polymerase chain reaction (PCR), and genome-wide approaches are used for the study [83, 84].

Two kinds of ChIP-on-chip methods exist, each tailored to the specific contents of microarrays: (1) promoter tiling arrays and (2) genome tiling arrays. Since the probes in promoter tiling arrays are tailored to target certain promoters and other genomic elements, some important details may be overlooked. Genome tiling arrays use probes that span the whole genome, allowing for comprehensive genomic studies. Histone alterations have been previously studied using the ChIP-on-chip technique in yeast and *Drosophila melanogaster*. Analysis of histone modifications in the human genome using ChIP-on-chip has recently proven fruitful [85, 86].

Thirdly, ChIP-seq utilizes next-generation sequencing techniques to provide an additional way for evaluating histone methylation and chromatin remodeling. To do ChIP-seq, DNA ends must be repaired and ligated to a set of adapters. The oligonucleotides are attached to the surface of the flow cell after the DNA has been amplified. These oligonucleotides are specific to adapter sequences ligated to DNA. During solid-phase PCR, the DNA sequences are read by the genome analyzer and then mapped to a reference genome to determine positions. This method helped researchers overcome the ChIP-chip technique's mediocre resolution and distracting background noise [87–89].

Mass spectrometry allows for the quantitative analysis of protein expression and the differential expression of protein modifications. Recently, chromatin affinity purification in conjunction with MS (ChAP-MS) and chromatin proteomics (ChroP)/ChIP-MS was developed to overcome the problem of MS's inability to map alteration patterns to particular promoter regions. Accordingly, it is now possible to examine histone marks and binding proteins in functionally different chromatin regions concurrently [90]. Different mass spectrometry approaches, named bottom-up, middle-down, and top-down, are distinguished by the level of the histone sequence at which they are applied. In the standard "bottom-up" approach, protease enzymes digest the target protein into shorter peptides (5e20 aa) before MS analysis. The "top-down" technique is used to characterize full-length proteins, whereas the "middle-down" method is used to analyze and characterize big peptides with less than 50 N-terminal amino acid residues of histone tails [91, 92].

2.3.2.1 Technologies for Capturing the Conformation of Chromosomes

Many methods of capturing chromosomes in their native 3D shape have been developed so that the genome's spatial organization may be analyzed.

Methods such as (3C) and (4C) chromosomal conformation capture, (6C) combined 3C-ChIP-cloning, (GCC) genome conformation capture,

(ChIA-PET) chromosome in situ hybridization, (5C) chromosome conformation capture carbon copy, etc. are all examples. Formaldehyde crosslinks the cells, a hypotonic buffer and protease inhibitors lyse the cells, and sodium dodecyl sulfate (SDS) solubilizes and digests the chromatin. Ligase is then used to relink the chromatin under diluted circumstances. This is the 3C technique. In addition, 3C libraries are produced by reverse cross-linking and purification. Higher resolution and throughput studies of the 3D structure of chromatin at a specific locus are made possible by using the 3C libraries to build 5C libraries [93, 94].

The disadvantage of the ChIP-chip technology is the development of microarrays (6, ChIA-PET). A 3C-based method called Chromatin Interaction Analysis by paired end tag ((ChIA-PET)) was developed to solve this issue. In this technique, DNA "tags" that have been immunoprecipitated are cloned into a plasmid library and then sequenced [95].

2.3.3 Methods to Analyze Methylation of RNA and ncRNA Species

Purifying RNA according to standard techniques allows for analysis of RNA modifications such as N6-methyladenosine (m6A), N1-methyladenosine (m1A), 20-OMethylation (20OMe/Nm), and 5-Methylcytosine (m5C). Multiple variables, including the number and kinds of modifications present and the RNA sequence itself, influence the analysis and characterization of RNA [96, 97]. The following are some of the methods that have been used to learn about RNA methylation.

Incorporating radioactive isotopes into RNA is one method used in radioisotope incorporation tests, which may be used to measure RNA methylation levels. Methyltransferase activity is assessed via scintillation because of the radioactivity of the methyl group added from the donor to the nucleoside. The methyl donor, S-adenosyl methionine, is tritium-labeled.

Another method for recognizing most RNA alterations is two-dimensional thin-layer chromatography. The nucleotides are spread out over the cellulose substrate in a two-dimensional RNA separation based on their charge and hydrophobicity. The disadvantage of 2D-TLC is that it offers an overall methylation status of the transcriptome. Ultraviolet light may then be used to create an image of the pattern. Site-specific cleavage, radioactive tagging, ligation-assisted extraction, and thin-layer chromatography (SCARLET) may be used to investigate stoichiometry when the sequence is known [98, 99].

Nucleotides are characterized by their mass-to-charge ratio in a conventional MS-based comparison approach. MS is comparable to chromatography-based techniques. However, it doesn't need a radioisotope or label. A key downside of MS-based techniques is the need for substantial amounts of RNA and priori-sequence information.

Sequencing using bisulfite, a thymine mutation occurs during reverse transcription. This mutation arises because sodium bisulfite converts cytosine to uracil, which is subsequently reflected in the final sequencing data set. Methylated cytosine is protected against delamination. The bisulfite sequencing method works this way, yielding information on RNA methylation at the level of a single base pair. Bisulfite sequencing has several limitations, including the need for large amounts of RNA and the resistance of surrounding changes and double-stranded areas to bisulfite treatment [100, 101].

Method number five involves antibodies; they may be purchased and used to detect methylated RNA residues like m6A, m1A, or m5C. Estimates of m5C, m1A, and m6A are also being made using antibodies directed toward the changes. "Methylated RNA immunoprecipitation sequencing" ("MeRIP-seq") is the most well-known and widely used method for mapping these modifications. The technique of N6-methyladenosine and m6A crosslinking immunoprecipitation sequencing (m6A-CLIP Seq/mi-CLIP) enables high-resolution mapping of m6A modifications on as little as 1 mg of poly(A)-selected mRNA [102, 103].

2.4 Understanding the Roles of Epigenetic Modulating Enzymes

Western blotting, enzyme-linked immunosorbent assays, chromatin immunoprecipitation (ChIP) tests, and coimmunoprecipitation are often used to investigate the expression and analysis of epigenetic modulators such as DNMTs, HDACs, and MeCPs. Co-immunoprecipitation is often employed to investigate the relationships between epigenetic regulators. In addition, imaging techniques in living organisms are being utilized to investigate the pharmacokinetics, direct binding, and activity of HDAC inhibitors [104, 105].

2.5 Single-Cell Epigenomics

Current epigenetics knowledge is based on population-level data that predominantly related epigenetics to transcriptional activation or silencing. These simplifications are a major reason why so many pressing questions in biology have yet to be resolved. Epigenetic controls must be studied at the single-cell level, where changes across cells may be investigated, to provide a fuller picture of cellular activities and dysfunctions. Recent advances in single-cell technology have shown conclusively that even seemingly similar cell populations display different degrees of gene expression, likely due to epigenetic heterogeneity. Single-cell epigenomics is a cutting-edge field of study that has tremendously helped our comprehension of gene control and associated molecular disorders. This exciting new method has the potential to advance our understanding of epigenetic regulation [106, 107]. However, concurrent analysis of genomic, transcriptomic, and epigenomic data may reveal the full potential of single-cell epigenetic research. When epigenome data is included in multi-omics measures, it may further strengthen molecular linkages between functional output and genome. In the realm of single-cell epigenetics analysis methods, the primary step involves isolating a single cell from a culture or dissociated tissue, followed by lysing the cell for subsequent epigenetic analysis. This may be done in various ways, including droplet encapsulation, fluorescence-activated cell sorting (FACS), and manual manipulations. New microfluidics technologies allow for the isolation of cells in chambers, followed by lysis and RNA-seq library preparation [108].

2.6 DNA Methylation and Other Modification

Single-cell sequencing at a single-nucleotide resolution has made it possible to investigate a variety of DNA modifications, including hydroxymethylation (5hmC), formyl cytosine (5fC), and methylation (5mC). Enrichment of CpG dense places (like CpG islands) and restriction digestion form the basis of a method called reduced representation bisulfite sequencing (scRRBS), which was first used to quantify 5mC throughout the whole genome in a single cell. Despite its ability to assess a larger proportion of CpG sites in the promotor region, it does not cover many crucial regulatory areas. The technological developments in single-cell whole-genome methylation sequencing are based on a post-bisulfite adapter-tagging (PBAT) strategy, where bisulfite modification is performed before library formation. Microfluidics has recently given rise to a new method for producing single-cell libraries. This approach has greatly improved library preparation throughput, especially when combined with cell-specific barcoding and the pooling of adapter-tagged fragments. This method can determine methylation at 50% of CpG sites in a single cell. Because of this, we have identified large differences in distal enhancer methylation across individual cells, a phenomenon often missed by scRRBS. Current technologies such as AbaSI (restriction enzyme) and TET-assisted bisulfite sequencing (TAB-seq) coupled with sequencing (Aba-seq) may be tweaked for hydroxymethylated cytosine single-cell research (5hmC) [96, 109].

2.7 Histone Modifications and Transcription Factor Binding

Chromatin immunoprecipitation followed by sequencing (ChIP-seq) is used to map histone modifications; nevertheless, ChIP-seq at the single-cell level is very difficult to conduct. Background noise from nonspecific antibody pull-down is an issue with single-cell ChIP-seq that worsens at lower target antigen concentrations. The use of barcoding and micrococcal nuclease (MNase) digestion has recently replaced this method, allowing for the efficient processing of thousands of cells. This method employs microfluidic droplet technology to simultaneously process vast swaths of cells [92]. Low-level expression of a fusion protein of *Escherichia coli* deoxyadenosine methylase (Dam) and the protein of interest in a cell line paves the way for investigating protein-DNA interactions in single cells using the DamID method. This Dam-based method employs adenine methylation adjacent to protein-binding sites, followed by cleavage of the DNA using the enzyme DpnI, sensitive to methylation, and ligation of sequencing adapters. The current state of this method prevents it from mapping transcription factor binding sites in single cells due to its low resolution (100 kb). Combining Dam fusion with targeted histone readers or modifiers might also allow for genome-wide investigation of histone changes utilizing single-cell DamID [89, 101].

2.8 Chromatin Structure and Chromosome Organization

A transposase test (ATAC-seq) measures chromatin structure in single cells. With this method, DNA is fragmented, and adaptor sequences are attached to it simultaneously using an enzyme called Tn5. ATAC-seq achieves single-cell resolution using the "combinatorial indexing" method of giving a unique barcode to each of 96 pools containing several thousand nuclei [95, 105]. Another described method for single-cell ATAC-seq involves using a piece of commercially available microfluidics equipment to carry out the transposition reaction on individual cells. By mapping an average of 70,000 reads per cell, the resolution gains from this combinatorial indexing method are substantial. DNase-seq is a method used to map DNaseI hypersensitive regions, which have been used to study open chromatin genomic regions in single cells. If the resolution of single-cell DNase-seq is improved to 300,000 mapped reads per cell, the mapping proficiency is only 2%, and the throughput is even lower [100, 109].

Recent technological advances have made it feasible to use an HiC-based approach to evaluate chromatin structure and chromosomal conformation in a single cell. An extension of chromosomal conformation capture (3C), Hi-C can detect long-range interactions throughout the whole genome. Despite the current limitations of single-cell HiC, it is possible to portray chromosome structure, including interchromosomal connections and compartmentalization [75, 78].

2.9 Epigenome Manipulation and Editing

Eukaryotic cells' epigenomes are intricate and tightly linked to essential cellular functions. If they aren't checked, it may lead to pathology and abnormal gene expression. Increased understanding of epigenetics' role in gene regulation and the ability to manipulate cell phenotype for research or therapeutic reasons are only two potential benefits of epigenome editing. Several epigenome editing platforms are available now, thanks to recent developments in genome engineering that use DNA-targeting techniques to properly adjust epigenetic modifications in a locus-specific manner [63, 66] (Table 2.1).

2.10 Epigenetic Manipulation Techniques

Classical genetic techniques can be used to simulate the perturbation of individual components of the epigenome. These methods include gene knockouts and domain deletions, point mutations, inducible expression constructs, ectopic expression of

Table 2.1 Analysis of DNA methylation using a variety of methods

Class	Method	Principle	Features	Limitations
Differentially methylated regions	Array or bead hybridization	To hybridize with a microarray, methylated DNA subsets are used	Capable of methylation-spot identification at individual loci	Potentially produces false positive results owing to cross-reactivity
	Bisulfite sequencing	The "gold standard" technique for DNA methylation studies, validated through polymerase chain reaction (PCR) amplification and Sanger sequencing, involves bisulfite treatment to facilitate the deamination of cytosine into uracil, which is then read as thymine. This treatment does not lead to uracil formation from cytosine-5 mC residues.	Resolution down to the nucleotide level	It is possible that partial uracilization of unmethylated cytosine results in bias
	Endonuclease digestion followed by sequencing	Methylation-sensitive endonucleases are used to cut off unmethylated sections of DNA in the genome, and then adapter ligation occurs	To isolate methylated CpG islands, use next-generation sequencing	It's crucial to use high-quality DNA
DNA methylation analysis	Bead array	An assay using an extension technique	Profiling of up to 384 unique CpG sites is conceivable with this method, and it is applicable to non-human animals	Costly
	COLD-PCR	The PCR's denaturation temperature is lowered such that only unmethylated fragments (those with a lower concentration of GC) are amplified and identified	Methylation levels in DNA are so low that they may be identified at the 0.1% level	Using a denaturing gradient, such as polyacrylamide, is essential
	Methylation-specific PCR	To accomplish DNA amplification, you'll need two sets of primers, one set for methylation DNA, and another set for unmethylated DNA. Two separate qPCR reactions are run on each sample, and the difference in Ct values is used to determine the level of methylation	An effective procedure that can rapidly profile a large number of samples	The methylation status of one or two CpG sites is evaluated all at once

(continued)

Table 2.1 (continued)

Class	Method	Principle	Features	Limitations
	PCR and sequencing	Amplification of region-specific DNA that has been bisulfite-treated and sequenced	Direct sequencing of PCR products is best for detecting significant methylation changes	Uses nested PCR to reduce nonspecific amplification
	PCR with high-resolution melting	DNA methylation status may be inferred from the fact that high-Tm DNA melts at a lower temperature than unmethylated DNA (low Tm). After attaching to DNA, intercalating dyes like SYBR and Eva green emit a fluorescent signal. Reduced fluorescence is a consequence of DNA dissociation as temperature increases	Depending on the quality of the PCR result, we are able to identify subtle changes in methylation (5–10%)	The need for a clean PCR product
	Pyrosequencing	Short-read pyrosequencing reaction (w100 bp) is carried out when primers are constructed, PCR products are acquired, and the reaction is initiated. Signal strengths for integrated dGTP and dATP are used to infer the methylation status of individual CpG sites within the sequenced area	Appropriate for low-throughput applications; efficient method for diverse samples; sensitive enough to pick up methylation variations as tiny as 5%	Poor proofreading and the presence of repetitive DNA make genome assembly difficult
Profiling whole genome methylation	AFLP (amplification fragment length polymorphism) and RFLP (restriction fragment length polymorphism)	Differentially methylated fragments are identified using polymerase chain reaction	Affordable, universally applicable across all species	Evaluates a negligible fraction of the whole DNA methylation landscape
	ELISA-based methods	Uses a tagged secondary antibody and colorimetric/fluorometric detection with a primary antibody produced against 5 Mc	Fast and cheap	High in error, yet adequate for providing a ballpark estimate of DNA methylation

	Global DNA methylation assay-LINE-1 (long interspersed nuclear elements-1)	Using biotinylated LINE-1 probes, DNA fragments are hybridized and then immobilized on a streptavidin-coated plate. Methylated cytosines were counted with the use of an anti-5 mC antibody, a secondary antibody coupled with HRP, and chemiluminescent detection reagents	Detects very subtle DNA methylation changes	Since pyrosequencer detection requires extensive amplification of target sequences, PCR amplification bias is a real possibility
	HPLC-UV (high-performance liquid chromatography—ultraviolet)	Reducing DNA to its nucleoside bases by hydrolysis; separating 5 mC and dC bases using chromatography; measuring fractions; and calculating the 5 mC/dC ratio	Deoxycytidine (dC) and methylcytosine (5 mC) test is the gold standard for measuring DNA content in hydrolyzed DNA samples	Requires a lot of DNA (3–10 μg) and specialist lab equipment
	LC-MS/MS (liquid chromatography coupled with tandem mass spectrometry)	MS is used for both evaluation and quantification, like HPLC, but using spiked internal standards instead	Because less hydrolyzed DNA is needed, low-quality DNA has little impact and high-throughput methylation analysis may be performed at a lower cost	It's expensive to get started, and it costs much more to buy internal standards
	LUMA (luminometric methylation assay)	DNA was cut using restriction enzymes that are sensitive to methylation, and the polymerase extension test was performed using pyrosequencing	Easily scalable, quantitative approach with excellent reproducibility	Necessitates prime DNA
	Methylated DNA immunoprecipitation (MeDIP)	Involves the use of a 5mC-specific antibody and subsequent genome-wide PCR microarray analysis of the DNA that was immunoprecipitated from the antibody	Highly repeatable and suitable for genome-wide methylation pattern study during the course of development	Requires high-grade and cross-reactive 5mC antibodies
	Reduced-representation bisulfite sequencing (RRBS)	The CpG-rich areas close to the restriction enzyme's recognition sequence are enriched, suggesting that these regions are particularly important	Cost-effective	Intergenic and distal regulatory components are not always fully covered

(continued)

Table 2.1 (continued)

Class	Method	Principle	Features	Limitations
	Whole-genome bisulfite sequencing (WGBS)	This should, in theory, include all there is to know about the C programming language. Genomic DNA is purified, then snipped into pieces and their ends mended. Sequencing is performed after a tailing (adenine base addition at 30 end), methylated adapter ligation, sodium bisulfite treatment, polymerase chain reaction, and sequencing of DNA fragments	Features genome-wide coverage and single-nucleotide resolution	Expensive and demanding careful planning

vectors, targeted knockdowns of a transcript, and various screens for gain or loss of function that alter genome structure or gene expression. These methods have been critical in laying the groundwork for our present understanding of epigenetics. However, these methods also cause widespread changes to the epigenome, which might potentially skew the findings of experiments [56, 70].

2.10.1 Small-Molecule Inhibitors

Anticancer medications and research use a class of small-molecule inhibitors that target specific epigenetic markers. The hallmarks of these medications are the histone deacetylase (HDAC) inhibitors romidepsin and suberoylanilide hydroxamic acid (SAHA), and the irreversible DNMT1 (depsipeptide or FK228) and DNMT3 inhibitors decitabine (5-aza-2 -deoxycytidine) and azacitidine (5-azacitidine). Histone-modifying enzymes are only one example of an epigenetic component that a variety of small-molecule inhibitors may target. Even though these chemicals aren't selective for any one kind of tissue or cell, they have demonstrated extraordinary effectiveness in various models when used in a narrow dosage range [63].

2.11 Targeted Epigenome Manipulations

Transcription activator-like effectors (TALEs), zinc finger proteins, and CRISPR-Cas systems are the three most essential molecular tools for targeted epigenome editing.

2.11.1 Zinc Finger Proteins

When manipulating nucleic acids with a precise focus on a certain sequence, zinc finger proteins are one of the best-studied systems. One method of editing the epigenome involves the combination of DNA-binding zinc finger proteins with effector domains that are either catalytically active or scaffolding. As a result, the chimeric proteins behave as artificial transcription factors (ATFs), altering gene expression patterns [73].

2.11.2 TALEs

TALEs' ability to recognize DNA results from its 33–35-residue-long core tandem amino acid repeat domain. TALEs' tandem repeat sequence encodes a DNA specificity that may shift depending on the context. In contrast to zinc fingers, which need triplet sequence recognition sites, TALEs could address a single nucleotide at the moment throughout its repeat variable di-residues (RVDs). This feature makes

TALEs more amenable to engineering, which has allowed the logical construction of artificial TALEs for use in epigenome editing [63].

2.11.3 CRISPR/Cas9 System

Because of its versatility and simplicity, the CRISPR/Cas9 system has been the primary focus of the new epigenome editing technologies. Several methods have been used to determine the best CRISPR/Cas9 targeting locations. Some research has successfully used innovative epigenome editing methods by simultaneously manipulating many nearby genes in the same area of the genome. In addition to the potential for additive effects, such an approach may raise the likelihood of an elevated number of off-target sites and complicate steric effects like suppression by catalytically inactive variations. Alternatively, the CRISPR-based method may identify the most effective gRNAs that modify transcription or protein expression by screening libraries containing hundreds of unique gRNAs. Cost and complexity are the only real issues with this method. Since it relies on assessing RNA or protein expression levels, which may or may not reflect epigenome alterations. CRISPR/Cas9 and other targeted epigenome editing methods provide unique platforms for the easy and flexible targeting of several epigenetically active regions of the genome [68, 74].

2.12 Conclusion

Only a tiny percentage of potential epigenetic targets have been explored, and the field of epigenome editing remains in its infancy. The development of cutting-edge technologies is expected to lead to a meteoric rise in the capacity of epigenome editing, enabling the production of novel epigenetic states. These advancements will facilitate the study of epigenetic processes responsible for altering epigenetic marks.

References

1. Angarica VE, Del Sol A. Bioinformatics tools for genome-wide epigenetic research. Adv Exp Med Biol. 2017;978:489–512.
2. Bergstedt J, Azzou SAK, Tsuo K, Jaquaniello A, Urrutia A, Rotival M, Lin DTS, MacIsaac JL, Kobor MS, Albert ML, Duffy D, Patin E, Quintana-Murci L. The immune factors driving DNA methylation variation in human blood. Nat Commun. 2022;13:5895.
3. Borchiellini M, Ummarino S, Di Ruscio A. The bright and dark side of DNA Methylation: a matter of balance. Cell. 2019;8:1243.
4. Bošković A, Rando OJ. Transgenerational epigenetic inheritance. Annu Rev Genet. 2018;52: 21–41.
5. Alharbi KS, Shaikh MAJ, Almalki WH, Kazmi I, Al-Abbasi FA, Alzarea SI, Imam SS, Alshehri S, Ghoneim MM, Singh SK, Chellappan DK, Oliver BG, Dua K, Gupta G. PI3K/Akt/mTOR pathways inhibitors with potential prospects in non-small-cell lung cancer. J Environ Pathol Toxicol Oncol. 2022;41:85–102.

6. Boulias K, Greer EL. Detection of DNA Methylation in genomic DNA by UHPLC-MS/MS. Methods Mol Biol (Clifton, N.J.). 2021;2198:79–90.
7. Brunstein J. Epigenetics: DNA methylation assays. MLO Med Lab Obs. 2015;47:28–9.
8. Anand K, Vadivalagan C, Joseph JS, Singh SK, Gulati M, Shahbaaz M, Abdellattif MH, Prasher P, Gupta G, Chellappan DK, Dua K. A novel nano therapeutic using convalescent plasma derived exosomal (CP(Exo)) for COVID-19: a combined hyperactive immune modulation and diagnostics. Chem Biol Interact. 2021;344:109497.
9. Gupta G, Al-Malki WH, Kazmi I, Thangavelu L, Gupta PK, Jha NK, Prasher P, Singh SK, Dua K. The role of HGF/MET in liver cancer. Future Med Chem. 2021;13:1829–32.
10. Burgess DJ. Epigenetics: rich pore methods for DNA methylation detection. Nat Rev Genet. 2017;18:209.
11. Cao R, Guan W. Evaluating reliability of DNA methylation measurement. Methods Mol Biol (Clifton, N.J.). 2022;2432:15–24.
12. Chao YL, Pecot CV. Targeting epigenetics in lung cancer. Cold Spring Harb Perspect Med. 2021;11:a038000.
13. Chen X, Yan F, Lin X, Shi L, Wang X, Zeng Y. DNA Methylation in chronic obstructive pulmonary disease. Adv Exp Med Biol. 2020;1255:83–98.
14. Jena R, Vishwas S, Kumar R, Kaur J, Khursheed R, Gulati M, Singh TG, Vanathi BM, Alam A, Kumar B, Chaitanya M, Gupta S, Negi P, Pandey NK, Bhatt S, Gupta G, Chellappan DK, Oliver BG, Dua K, Singh SK. Treatment strategies for HIV infection with emphasis on role of CRISPR/Cas9 gene: success so far and road ahead. Eur J Pharmacol. 2022;931:175173.
15. Pathak S, Gupta G, Gilhotra RM. The role of diazepam in epigenetics: from the molecular level to clinical implications. Adv Mind Body Med. 2021;35:25–33.
16. Chenarani N, Emamjomeh A, Allahverdi A, Mirmostafa S, Afsharinia MH, Zahiri J. Bioinformatic tools for DNA methylation and histone modification: a survey. Genomics. 2021;113:1098–113.
17. Choi BY, Han M, Kwak JW, Kim TH. Genetics and epigenetics in allergic rhinitis. Genes (Basel). 2021;12:2004.
18. Costa-Pinheiro P, Montezuma D, Henrique R, Jerónimo C. Diagnostic and prognostic epigenetic biomarkers in cancer. Epigenomics. 2015;7:1003–15.
19. Couldrey C, Cave V. Assessing DNA methylation levels in animals: choosing the right tool for the job. Anim Genet. 2014;45(Suppl 1):15–24.
20. Dabe EC, Sanford RS, Kohn AB, Bobkova Y, Moroz LL. DNA methylation in basal metazoans: insights from ctenophores. Integr Comp Biol. 2015;55:1096–110.
21. Darılmaz Yüce G, Ortaç Ersoy E. Lung cancer and epigenetic modifications. Tuberkuloz ve toraks. 2016;64:163–70.
22. Dawson MA, Kouzarides T. Cancer epigenetics: from mechanism to therapy. Cell. 2012;150: 12–27.
23. de la Rica L, Stanley JS, Branco MR. Profiling DNA methylation and hydroxymethylation at retrotransposable elements. Methods Molecular Biol (Clifton, N.J.). 2016;1400:387–401.
24. Duan R, Fu Q, Sun Y, Li Q. Epigenetic clock: a promising biomarker and practical tool in aging. Ageing Res Rev. 2022;81:101743.
25. Rohilla S, Singh M, Priya S, Almalki WH, Haniffa SM, Subramaniyan V, Fuloria S, Fuloria NK, Sekar M, Singh SK, Jha NK, Chellappan DK, Negi P, Dua K, Gupta G. Exploring the mechanical perspective of a new anti-tumor agent: melatonin. J Environ Pathol Toxicol Oncol. 2023;42:1–16.
26. Wadhwa R, Aggarwal T, Malyla V, Kumar N, Gupta G, Chellappan DK, Dureja H, Mehta M, Satija S, Gulati M, Maurya PK, Collet T, Hansbro PM, Dua K. Identification of biomarkers and genetic approaches toward chronic obstructive pulmonary disease. J Cell Physiol. 2019;234:16703–23.
27. Dueñas-Gonzalez A, Alatorre B, Gonzalez-Fierro A. The impact of DNA methylation technologies on drug toxicology. Expert Opin Drug Metab Toxicol. 2014;10:637–46.

28. Dumitrescu RG. Early epigenetic markers for precision medicine. Methods Mol Biol (Clifton, N.J.). 2018;1856:3–17.
29. Aggarwal T, Wadhwa R, Gupta R, Paudel KR, Collet T, Chellappan DK, Gupta G, Perumalsamy H, Mehta M, Satija S, Hansbro PM, Dua K, Maurya PK. MicroRNAs as biomarker for breast cancer. Endocr Metab Immune Disord Drug Targets. 2020;20:1597–610.
30. Aljabali AA, Hassan SS, Pabari RM, Shahcheraghi SH, Mishra V, Charbe NB, Chellappan DK, Dureja H, Gupta G, Almutary AG, Alnuqaydan AM, Verma SK, Panda PK, Mishra YK, Serrano-Aroca Á, Dua K, Uversky VN, Redwan EM, Bahar B, Bhatia A, Negi P, Goyal R, McCarron P, Bakshi HA, Tambuwala MM. The viral capsid as novel nanomaterials for drug delivery. Future Sci OA. 2021;7:Fso744.
31. Gupta G, Chellappan DK, de Jesus Andreoli T, Pinto PM, Hansbro M, Bebawy KD. Tumor suppressor role of miR-503. Panminerva Med. 2018;60:17–24.
32. Feng H, Jin P, Wu H. Disease prediction by cell-free DNA methylation. Brief Bioinform. 2019;20:585–97.
33. Feng L, Lou J. DNA methylation analysis. Methods Mol Biol (Clifton, N.J.). 2019;1894:181–227.
34. Awasthi R, Singh AK, Mishra G, Maurya A, Chellappan DK, Gupta G, Hansbro PM, Dua K. An overview of circular RNAs. Adv Exp Med Biol. 2018;1087:3–14.
35. Bhardwaj S, Kesari KK, Rachamalla M, Mani S, Ashraf GM, Jha SK, Kumar P, Ambasta RK, Dureja H, Devkota HP, Gupta G, Chellappan DK, Singh SK, Dua K, Ruokolainen J, Kamal MA, Ojha S, Jha NK. CRISPR/Cas9 gene editing: new hope for Alzheimer's disease therapeutics. J Adv Res. 2022;40:207–21.
36. Fischer MA, Vondriska TM. Clinical epigenomics for cardiovascular disease: diagnostics and therapies. J Mol Cell Cardiol. 2021;154:97–105.
37. Fitz-James MH, Cavalli G. Molecular mechanisms of transgenerational epigenetic inheritance. Nat Rev Genet. 2022;23:325–41.
38. Chellappan DK, Sivam NS, Teoh KX, Leong WP, Fui TZ, Chooi K, Khoo N, Yi FJ, Chellian J, Cheng LL, Dahiya R, Gupta G, Singhvi G, Nammi S, Hansbro PM, Dua K. Gene therapy and type 1 diabetes mellitus. Biomed Pharmacother. 2018;108:1188–200.
39. Chellappan DK, Yap WS, Bt Ahmad Suhaimi NA, Gupta G, Dua K. Current therapies and targets for type 2 diabetes mellitus. Panminerva Med. 2018;60:117–31.
40. Gill D, Parry A, Santos F, Okkenhaug H, Todd CD, Hernando-Herraez I, Stubbs TM, Milagre I, Reik W. Multi-omic rejuvenation of human cells by maturation phase transient reprogramming. elife. 2022;11:e71624.
41. Gjaltema RAF, Rots MG. Advances of epigenetic editing. Curr Opin Chem Biol. 2020;57:75–81.
42. Gouil Q, Keniry A. Latest techniques to study DNA methylation. Essays Biochem. 2019;63: 639–48.
43. Grunau C, Le Luyer J, Laporte M, Joly D. The epigenetics dilemma. Genes (Basel). 2019;11: 23.
44. Hübel C, Marzi SJ, Breen G, Bulik CM. Epigenetics in eating disorders: a systematic review. Mol Psychiatry. 2019;24:901–15.
45. Hüls A, Czamara D. Methodological challenges in constructing DNA methylation risk scores. Epigenetics. 2020;15:1–11.
46. Hwang JY, Aromolaran KA, Zukin RS. The emerging field of epigenetics in neurodegeneration and neuroprotection. Nat Rev Neurosci. 2017;18:347–61.
47. Jones PA, Issa JP, Baylin S. Targeting the cancer epigenome for therapy. Nat Rev Genet. 2016;17:630–41.
48. Prasher P, Sharma M, Chellappan DK, Gupta G, Jha NK, Singh SK, MacLoughlin R, Pinto TJA, Löbenberg R, Dua K. Advanced drug delivery systems targeting NF-κB in respiratory diseases. Future Med Chem. 2021;13:1087–90.
49. Shahcheraghi SH, Ayatollahi J, Aljabali AA, Shastri MD, Shukla SD, Chellappan DK, Jha NK, Anand K, Katari NK, Mehta M, Satija S, Dureja H, Mishra V, Almutary AG, Alnuqaydan

AM, Charbe N, Prasher P, Gupta G, Dua K, Lotfi M, Bakshi HA, Tambuwala MM. An overview of vaccine development for COVID-19. Ther Deliv. 2021;12:235–44.
50. Jones PA, Ohtani H, Chakravarthy A, De Carvalho DD. Epigenetic therapy in immune-oncology. Nat Rev Cancer. 2019;19:151–61.
51. Kalish JM, Jiang C, Bartolomei MS. Epigenetics and imprinting in human disease. Int J Dev Biol. 2014;58:291–8.
52. Singhvi G, Manchanda P, Krishna Rapalli V, Kumar Dubey S, Gupta G, Dua K. MicroRNAs as biological regulators in skin disorders. Biomed Pharmacother. 2018;108:996–1004.
53. Sunkara KP, Gupta G, Hansbro PM, Dua K, Bebawy M. Functional relevance of SATB1 in immune regulation and tumorigenesis. Biomed Pharmacother. 2018;104:87–93.
54. King SE, Skinner MK. Epigenetic transgenerational inheritance of obesity susceptibility. Trends Endocrinol Metab. 2020;31:478–94.
55. Klein CB. Emerging confluences of epigenetics and DNA repair in cancer and disease. Mutat Res Rev Mutat Res. 2019;780:11–4.
56. Klemp I, Hoffmann A, Müller L, Hagemann T, Horn K, Rohde-Zimmermann K, Tönjes A, Thiery J, Löffler M, Burkhardt R, Böttcher Y, Stumvoll M, Blüher M, Krohn K, Scholz M, Baber R, Franks PW, Kovacs P, Keller M. DNA methylation patterns reflect individual's lifestyle independent of obesity. Clin Transl Med. 2022;12:e851.
57. Komaki S, Ohmomo H, Hachiya T, Sutoh Y, Ono K, Furukawa R, Umekage S, Otsuka-Yamasaki Y, Tanno K, Sasaki M, Shimizu A. Longitudinal DNA methylation dynamics as a practical indicator in clinical epigenetics. Clin Epigenetics. 2021;13:219.
58. Lemaître JF, Rey B, Gaillard JM, Régis C, Gilot-Fromont E, Débias F, Duhayer J, Pardonnet S, Pellerin M, Haghani A, Zoller JA, Li CZ, Horvath S. DNA methylation as a tool to explore ageing in wild roe deer populations. Mol Ecol Resour. 2022;22:1002–15.
59. Li L, Zhang C, Liu S, Guan H, Zhang Y. Age prediction by DNA methylation in neural networks. IEEE/ACM Trans Comput Biol Bioinform. 2022;19:1393–402.
60. Tiwari J, Gupta G, de Jesus Andreoli T, Pinto R, Sharma K, Pabreja Y, Matta N, Arora A, Mishra R, Sharma KD. Role of microRNAs (miRNAs) in the pathophysiology of diabetes mellitus. Panminerva Med. 2018;60:25–8.
61. Vadivalagan C, Shitut A, Kamalakannan S, Chen RM, Serrano-Aroca Á, Mishra V, Aljabali AAA, Singh SK, Chellappan DK, Gupta G, Dua K, El-Tanani M, Tambuwala MM, Krishnan A. Exosomal mediated signal transduction through artificial microRNA (amiRNA): a potential target for inhibition of SARS-CoV-2. Cell Signal. 2022;95:110334.
62. Li S, Tollefsbol TO. DNA methylation methods: global DNA methylation and methylomic analyses. Methods (San Diego, Calif). 2021;187:28–43.
63. Li Y. Modern epigenetics methods in biological research. Methods (San Diego Calif). 2021;187:104–13.
64. Liberman N, Wang SY, Greer EL. Transgenerational epigenetic inheritance: from phenomena to molecular mechanisms. Curr Opin Neurobiol. 2019;59:189–206.
65. Lin W, Hu S, Wu Z, Xu Z, Zhong Y, Lv Z, Qiu W, Xiao X. iCancer-Pred: a tool for identifying cancer and its type using DNA methylation. Genomics. 2022;114:110486.
66. Mancia A. Genome-wide DNA Methylation protocol for epigenetics studies. Methods Mol Biol (Clifton, N.J.). 2022;2498:19–41.
67. Meng H, Cao Y, Qin J, Song X, Zhang Q, Shi Y, Cao L. DNA methylation, its mediators and genome integrity. Int J Biol Sci. 2015;11:604–17.
68. Michels KB, Binder AM. Considerations for design and analysis of DNA Methylation studies. Methods Mol Biol (Clifton, N.J.). 2018;1708:31–46.
69. Ming X, Zhu B, Li Y. Mitotic inheritance of DNA methylation: more than just copy and paste. J Genet Genomics. 2021;48:1–13.
70. Morris MJ, Monteggia LM. Role of DNA methylation and the DNA methyltransferases in learning and memory. Dialogues Clin Neurosci. 2014;16:359–71.

71. Nannini DR, Zheng Y, Joyce BT, Gao T, Liu L, Jacobs DR Jr, Schreiner P, Liu C, Horvath S, Lu AT, Yaffe K, Sidney S, Greenland P, Lloyd-Jones DM, Hou L. Marijuana use and DNA methylation-based biological age in young adults. Clin Epigenetics. 2022;14:134.
72. Nichols RV, O'Connell BL, Mulqueen RM, Thomas J, Woodfin AR, Acharya S, Mandel G, Pokholok D, Steemers FJ, Adey AC. High-throughput robust single-cell DNA methylation profiling with sciMETv2. Nat Commun. 2022;13:7627.
73. Noehammer C, Pulverer W, Hassler MR, Hofner M, Wielscher M, Vierlinger K, Liloglou T, McCarthy D, Jensen TJ, Nygren A, Gohlke H, Trooskens G, Braspenning M, Van Criekinge W, Egger G, Weinhaeusel A. Strategies for validation and testing of DNA methylation biomarkers. Epigenomics. 2014;6:603–22.
74. Ochoa E, Zuber V, Bottolo L. Accurate measurement of DNA Methylation: challenges and bias correction. Methods Mol Biol (Clifton, NJ). 2022;2432:25–47.
75. Pan H, Elemento O. Analyzing DNA Methylation patterns during tumor evolution. Methods Mol Biol (Clifton, N.J.). 2018;1711:27–53.
76. Papareddy RK, Nodine MD. Plant epigenetics: propelling DNA methylation variation across the cell cycle. Curr Biol. 2021;31:R129–r131.
77. Peixoto P, Cartron PF, Serandour AA, Hervouet E. From 1957 to nowadays: a brief history of epigenetics. Int J Mol Sci. 2020;21:7571.
78. Reale A, Tagliatesta S, Zardo G, Zampieri M. Counteracting aged DNA methylation states to combat ageing and age-related diseases. Mech Ageing Dev. 2022;206:111695.
79. Renaudineau Y, Ballestar E. Epigenetics: DNA methylation signatures in Sjögren syndrome. Nat Rev Rheumatol. 2016;12:565–6.
80. Richardson BC, Patel DR. Epigenetics in 2013. DNA methylation and miRNA: key roles in systemic autoimmunity. Nat Rev Rheumatol. 2014;10:72–4.
81. Rooney K, Sadikovic B. DNA Methylation episignatures in neurodevelopmental disorders associated with large structural copy number variants: clinical implications. Int J Mol Sci. 2022;23:7862.
82. Rosen ED, Kaestner KH, Natarajan R, Patti ME, Sallari R, Sander M, Susztak K. Epigenetics and epigenomics: implications for diabetes and obesity. Diabetes. 2018;67:1923–31.
83. Roy R, Ramamoorthy S, Shapiro BD, Kaileh M, Hernandez D, Sarantopoulou D, Arepalli S, Boller S, Singh A, Bektas A, Kim J, Moore AZ, Tanaka T, McKelvey J, Zukley L, Nguyen C, Wallace T, Dunn C, Wersto R, Wood W, Piao Y, Becker KG, Coletta C, De S, Sen JM, Battle A, Weng NP, Grosschedl R, Ferrucci L, Sen R. DNA methylation signatures reveal that distinct combinations of transcription factors specify human immune cell epigenetic identity. Immunity. 2021;54:2465–80.e2465
84. Rustad SR, Papale LA, Alisch RS. DNA methylation and hydroxymethylation and behavior. Curr Top Behav Neurosci. 2019;42:51–82.
85. Saddiki H, Colicino E, Lesseur C. Assessing differential variability of high-throughput DNA methylation data. Curr Environ Health Rep. 2022;9:625–30.
86. Sarkies P. Encyclopaedia of eukaryotic DNA methylation: from patterns to mechanisms and functions. Biochem Soc Trans. 2022;50:1179–90.
87. Scott M, De Sario A. DNA methylation changes in cystic fibrosis: cause or consequence? Clin Genet. 2020;98:3–9.
88. Shao S, Gudjonsson JE. Epigenetics of psoriasis. Adv Exp Med Biol. 2020;1253:209–21.
89. Sinčić N, Herceg Z. DNA methylation and cancer: ghosts and angels above the genes. Curr Opin Oncol. 2011;23:69–76.
90. Skvortsova K, Stirzaker C, Taberlay P. The DNA methylation landscape in cancer. Essays Biochem. 2019;63:797–811.
91. Smith J, Banerjee R, Weeks RJ, Chatterjee A. Editing of DNA methylation patterns using CRISPR-based tools. Methods Mol Biol (Clifton, N.J.). 2022;2458:63–74.
92. Stoccoro A, Coppedè F. Role of epigenetics in Alzheimer's disease pathogenesis. Neurodegener Dis Manag. 2018;8:181–93.

93. Svoboda LK, Neier K, Wang K, Cavalcante RG, Rygiel CA, Tsai Z, Jones TR, Liu S, Goodrich JM, Lalancette C, Colacino JA, Sartor MA, Dolinoy DC. Tissue and sex-specific programming of DNA methylation by perinatal lead exposure: implications for environmental epigenetics studies. Epigenetics. 2021;16:1102–22.
94. Szyf M. The epigenetics of perinatal stress. Dialogues Clin Neurosci. 2019;21:369–78.
95. Tang J, Fang F, Miller DF, Pilrose JM, Matei D, Huang TH, Nephew KP. Global DNA methylation profiling technologies and the ovarian cancer methylome. Methods Mol Biol (Clifton, N.J.). 2015;1238:653–75.
96. Tirnaz S, Batley J, Methylation DNA. Toward crop disease resistance improvement. Trends Plant Sci. 2019;24:1137–50.
97. Umer M, Herceg Z. Deciphering the epigenetic code: an overview of DNA methylation analysis methods. Antioxid Redox Signal. 2013;18:1972–86.
98. Unnikrishnan A, Freeman WM, Jackson J, Wren JD, Porter H, Richardson A. The role of DNA methylation in epigenetics of aging. Pharmacol Ther. 2019;195:172–85.
99. Velagacherla V, Mehta CH, Nayak Y, Nayak UY. Molecular pathways and role of epigenetics in the idiopathic pulmonary fibrosis. Life Sci. 2022;291:120283.
100. Visconti VV, Cariati I, Fittipaldi S, Iundusi R, Gasbarra E, Tarantino U, Botta A. DNA Methylation signatures of bone metabolism in osteoporosis and osteoarthritis aging-related diseases: an updated review. Int J Mol Sci. 2021;22:4244.
101. von Meyenn F. Profiling DNA Methylation in human Naïve pluripotent stem cells. Methods Mol Biol (Clifton, N.J.). 2022;2416:157–80.
102. Werner RJ, Kelly AD, Issa JJ. Epigenetics and precision oncology. Cancer J (Sudbury, Mass). 2017;23:262–9.
103. Zhang H, Hou L, Liu L. A review of high-dimensional mediation analyses in DNA Methylation studies. Methods Mol Biol (Clifton, N.J.). 2022;2432:123–35.
104. Zhang L, Lu Q, Chang C. Epigenetics in health and disease. Adv Exp Med Biol. 2020;1253:3–55.
105. Zhang W, Wu H, Li Z. Complete deconvolution of DNA methylation signals from complex tissues: a geometric approach. Bioinformatics (Oxford, England). 2021;37:1052–9.
106. Zhou S, Wang X, Gao H, Zeng Y. DNA methylation in pulmonary fibrosis. Adv Exp Med Biol. 2020;1255:51–62.
107. Zlotorynski E. Epigenetics: DNA methylation prevents intragenic transcription. Nat Rev Mol Cell Biol. 2017;18:212–3.
108. Zlotorynski E. Am-bivalency towards DNA methylation. Nat Rev Mol Cell Biol. 2020;21:497.
109. Zuo T, Tycko B, Liu TM, Lin JJ, Huang TH. Methods in DNA methylation profiling. Epigenomics. 2009;1:331–45.

Epigenetic Mechanisms in Inflammation 3

Rajiv Dahiya, Riya Thapa, Narender Kumar Kumawat, Manisha Singh, Shikha Jakhotiya, Deepika Deopa, Yogendra Singh, Neelam Singla, and Gaurav Gupta

Abstract

Epigenetic modifications are triggered by environmental changes and play a critical role in how genes are expressed in response to stimuli. The common epigenetic processes include histone, DNA methylation, and regulatory factor methylation, acetylation, and the short non-coding RNAs appearance. Epigenetic modifications cause susceptibility to disease and are modified by environmental factors like nutrition, pollution, and infection. Despite the rising form of knowledge on gene expression that is affected by the environment, other than that a very little is known about the epigenetic processes involved in the regulation of inflammatory and anti-inflammatory genes. Herein, we provide the present update on the mechanism of epigenetics during inflammation.

R. Dahiya
School of Pharmacy, Faculty of Medical Sciences, The University of the West Indies, St. Augustine, Trinidad and Tobago

R. Thapa · N. Singla · G. Gupta (✉)
School of Pharmacy, Suresh Gyan Vihar University, Jaipur, India

N. K. Kumawat · Y. Singh
Department of Pharmacology, Maharishi Arvind College of Pharmacy, Ambabari, Jaipur, India

M. Singh
Department of Pharmaceutical Chemistry, Maharishi Arvind College of Pharmacy, Ambabari, Jaipur, India

S. Jakhotiya
Department of Pharmaceutics, Maharishi Arvind College of Pharmacy, Ambabari, Jaipur, India

D. Deopa
School of Pharmacy, Suresh Gyan Vihar University, Jaipur, India

Six sigma institute of technology and science, Rudrapur, Uttarakhand, India

G. Gupta et al. (eds.), *Targeting Epigenetics in Inflammatory Lung Diseases*,
https://doi.org/10.1007/978-981-99-4780-5_3

3.1 Introduction

Epigenetics is the field that examines how alterations to DNA may modulate gene expression even if they are not part of the genetic code. Many different chemical mechanisms, including DNA methylation, the histone variants positions, histone modifications, and the control of RNAs genes, i.e. non-coding, all contribute to the complicated molecular underpinnings of epigenetic processes [1, 2]. Since epigenetic changes may be undone, studying them in depth might provide new drug aims for treating disorders. The epigenome is very important by way of the genome for healthy growth and development [3]. Environmental factors including infections, pollutants, diet, and hypoxia may all drastically affect the epigenetic signature, which may explain why they can all play a role in disease susceptibility [4]. For instance, recent studies have shown that environmental variables present during pregnancy may modify the epigenome, with far-reaching implications on gene regulation and the onset of age-related diseases. There may be a link between a placental-foetal exposure and an inflammatory response in the foetus [5] (Fig. 3.1).

3.2 Inflammation

Inflammation is one of many complex physiological reactions triggered by an organism in response to a perceived danger, such as a damaged cells, pathogen, or an irritant. The early reaction of the body to stimuli in acute inflammation is the

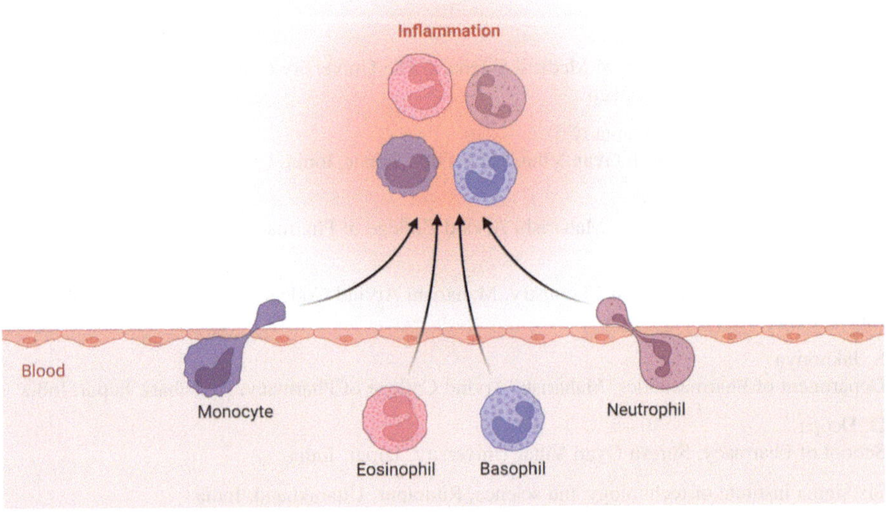

Fig. 3.1 Immune cell recruitment

recruitment of additional plasma and leukocytes in the bloodstream to the wounded areas. Inflammation that persists over time is said to be chronic [6, 7]. In contrast to the more intense symptoms associated with acute inflammation, the disorder of chronic inflammation lasts longer. Chronic inflammation has been related to a wide variety of disorders, like diabetes mellitus and periodontal disease [8]. The inflammatory response, being so multifaceted, demands for the expansion of a complex monitoring system to transmit out process at the gene and signal levels. As part of this network, genes involved in tissue repair, immune response, and tissue remodelling are triggered [9]. Current data analysis supports the idea that chromatin changes cause an important mechanistic part in the maturation of the macrophage phenotype. Macrophages are involved in the development of several chronic disorders, including cancer and allergic responses [10]. In addition to transcription factors from the IRF, FOXP3, NF-kB, and STAT families, epigenetic mechanisms including methylation of DNA and covalent modifications of histone have been playing a substantial role in the regulation of inflammatory genes. Not only that, but a variety of critical regulatory parameters are precise by epigenetic mechanisms in monocytes and T-cells [11, 12] (Fig. 3.2).

3.3 Histone Modifications

The nucleosome is the smallest functional element of chromatin and is made up of two copies of the histone proteins H3, H2B, H2A, and H4. Structured chromatin is the result of this arrangement [13]. The modification of covalent histones is a key epigenetic method for regulating genes. Post-translational modifications, such as methylation of arginine and acetylation, SUMOylation, phosphorylation, ADP-ribosylation, and ubiquitination of lysine residues, are predominantly observed in the N-terminal tails of essential histones [14, 15].

The histone acetylation is associated with an "open" chromatin shape that aids transcription. In order to recruit the common transcription machinery, chromatin must adopt a more relaxed shape, and the acetylated N-terminal tails extending from the nucleosome core should exhibit reduced affinity for the DNA [16]. Many acetylated marks of histone, including H3K39ac and H3K4ac, have been associated with transcriptional activity. CREB-binding protein (CBP) and its nearby p300 homolog are examples of histone acetyltransferases (HATs) that accomplish these modifications [17]. Histone deacetylases (HDACs) are enzymes that counteract HAT activity by increasing chromatin condensation and repressing genes. Nevertheless, histone methylation may contribute to the maintenance of chromatin in either an active or inactive state [18]. H3K4me3 and H3K36me3, a tri-methylation of lysine 4 and 36 on histone H3, open the chromatin and allow transcription to take place [19]. Compacted chromatin and gene repression are often associated with modifications to histone lysine's 9 and 27 (H3K9me3 and H3K27me3). Genes with double alteration (H3K27me3 and H3K4me3) are frequently significant evolving regulators that are silenced but ready to be activated as differentiation continues in stem cells or progenitor cells. Increased binding of acetyltransferase co-factors

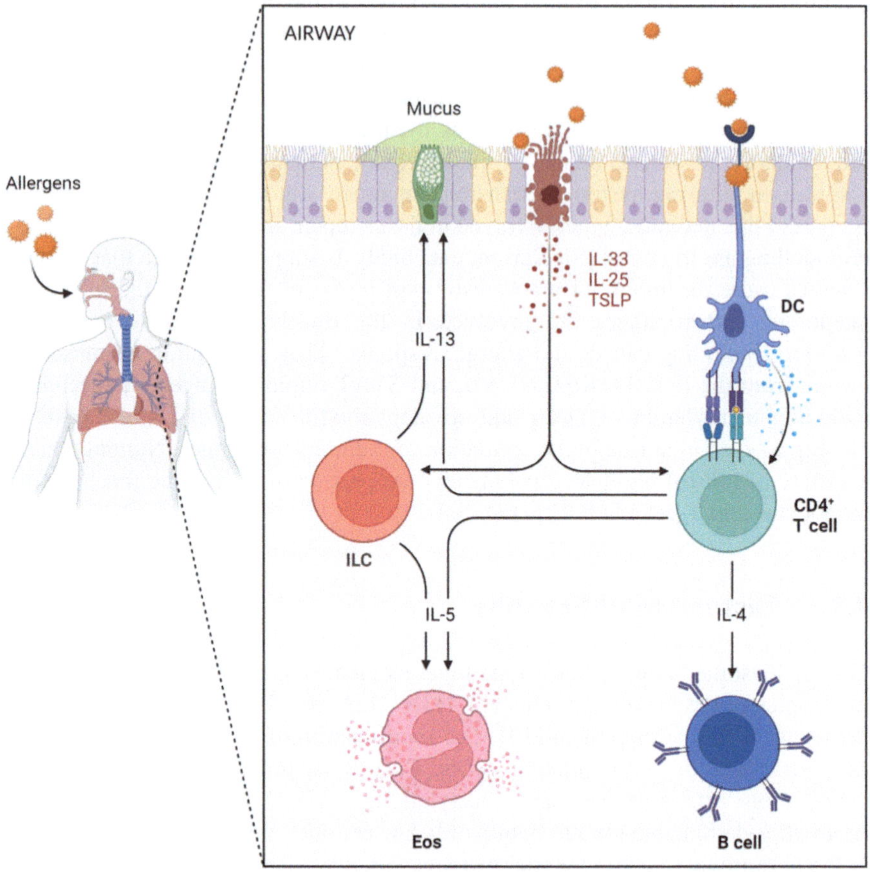

Fig. 3.2 Allergic airway inflammation

CBP/p300 of histones and histone mono-methylation of H3 at lysine 4 define enhancers, which control gene expression in certain tissues (H3K4me1). Certain enzymes, known as HMT and HDM, may add or remove methyl groups from histones [20–22].

3.4 Histone Methyltransferases

Polycomb Group (PcG) proteins show a vital role in cell fate decisions throughout embryonic and foetal development. There are two forms of complexes in mammals, and they both function to silence genes (PRC1 and PRC2). PRC1 is a protein complex composed of four individual proteins: PHC, the CBX family, RING1, and MEL18/BMI1. The proteins EED, RBAP48/46, EZH2/EZH1, and SUZ12 are essential components of PRC2 [23, 24]. PRC2 repression regulates transcription by catalyzing the tri- and di-methylation of lysine 27 on histone H3. Current mapping of the supervisory domains of some evolving imperative genes, like as those in the Sox, Gata, Dlx, and Tbx families, has shown the co-localization of PRC2, PRC1, and H3K27me3 [25]. Most of the time, PRC2 must be running for PRC1 to bind to DNA. On the other hand, PRC2 might practice different conscription strategies to zero in on diverse targets in the gene regulatory circuit [26]. Related to the Jumonji C (JmjC) and domain protein of ARID families, the transcription repressor JARID2 forms a stable complex with PRC2 [27]. In order to attach to the PcG-responsive areas, JARID2 utilises its DNA-binding domain ARID. Non-coding RNAs (ncRNAs) may also play a role in directing PRC2 to certain promoters [28].

3.5 DNA Methylation

Methylation of DNA involves a covalent transmission of a methyl group from S-adenosyl-L-methionine to cytosines in dinucleotides CpG. Across the mammalian genome are CpG islands, areas with a high concentration of CpG sites that are not methylated and are located in close proximity to the promoters of the vast majority of genes [29]. CpG methylation sites are frequently found in specific regions of various genes within eukaryotic genomes, including those associated with repetitive sequences, development, and imprinted genes [30]. Human cancers generally exhibit hypomethylation and other changes in DNA methylation, along with chromosome instability and transposable element activity. Groups of enzymes belonging to the same family catalyse the process of DNA methylation (DNMT3a, DNMT1, and DNMT3b) [31]. The most common kind of DNA methyltransferase in mammals is DNMT1, which functions as a maintenance methyltransferase. The process adds methylation to the strand of DNA that is already partially methylated (hemi-methylated). Proteins that possess methyl-CpG-binding domains include MeCP2 and Kaiso [32, 33]. Importantly, methylation of DNA and methylation of histones are both firmly precise processes in eukaryotes. N-terminal of Histone H3 tails containing lysine 4 unmethylated are mandatory for DNA methylation (H3K4) [34]. Histone H3 tri-methylation at lysine 9 is an additional step in DNMT1-mediated DNA methylation. The NP95 protein links DNA with histone H3 methylation via its SET-, Ring-, and Tudor domains. Hence, the two main epigenetic processes responsible for silencing are the coordination of NP95 and DNA methylation, along with histone methylation [35].

3.6 Epigenetic Events in Inflammation

3.6.1 Histone Methylation and Inflammation

Histone alterations may be eliminated by inducible Jmjd3 from the Jumonji family. Recent research has linked inflammation to epigenome reprogramming by identifying a protein that controls macrophage growth and cell identity [36]. To regulate the amounts of H3K27me3 and the transcriptional activity of the PcG target genes, macrophages exposed to contaminated things and inflammatory cytokines produce more Jmjd3, which fixes the PcG target genes [37]. Continued treatment with IL-4 initiates Jmjd3 activity, resulting in the subsequent removal of repressive H3K27me3 marks from the promoter of STAT6. The binding of active promoter to STAT6 of the Jmjd3 gene results in positive regulation of that gene's expression [38]. Methylation marks on histone H3 (H3K27) govern the expression of inflammatory genes; Jmjd3 eliminates these marks [39]. However, Jmjd3 may also function independently of demethylation H3K27. The study demonstrates that the marker activation of H3K4me3 and the Polymerase II RNA complex are required for Jmjd3 to be recruited to transcription start sites. These findings suggest that Jmjd3 binding to target genes does not coincide with H3K27 demethylation [40]. These discoveries suggested that the switching around of methyl groups on H3K4 and H3K27 is an important epigenetic process in gene regulation. Many genes in the mammalian genome have been identified as PcG target genes. A possible link between chronic inflammation and aberrant DNA methylation at these locations has just recently been discovered. Research has unveiled that PcG proteins bind to target genes and facilitate the recruitment of DNMTs, thereby enhancing gene repression [41, 42]. More recently, it has been discovered that an alternate mechanism leading to gene suppression in SSI triggered by acute inflammatory events is NF-kB/RelB-dependent silence. The induction endotoxin of RelB by activation is necessary and sufficient for the acute pro-inflammatory gene repression [43]. H3K9 methyltransferase G9a forms heterochromatin in response to RelB's direct interaction with it to silence these genes during SSI [44]. This interaction leads to H3K9me3 and the conscription of heterochromatin protein 1. HP1 and G9a form a restrictive intricate at the promoters of dependent genes of RelB, leading to the recruitment of CpG methylation. In the case of the promoter TNF in blood leukocytes, it was recently shown that SSI creates a supportive interface between histone methylation and DNA methylation [45].

3.6.2 Histone Acetylation and Inflammation

Upregulation activity of histone deacetylases (HDACs) cause inhibition of inflammatory genes, whereas HAT acetylation stimulates them. In COPD airway biopsies and alveolar macrophages, there is an observed increase in histone acetylation in the promoter region of inflammatory genes, which is mediated by NF-kB. Reduced activity of the enzyme histone deacetylase (HDAC) underlies the observed upregulation of histone acetylation. Pro-inflammatory cytokine promoters (IL-8, IL-2, IL-1, and IL-12) are quickly CBP/p300 acetylated, resulting in activation of transcriptional factors and decreased activity of HDAC [46, 47]. Histone deacetylation and gene suppression occur instead when HDACs are recruited. To do this, HDACs recruit complex corepressor and factors of transcription like STATs, ZEB1, GATAs, FOXP3, and NF-kB to gene promoters, where they control pro- and anti-inflammatory cytokine transcriptions, respectively [48]. In response to cytokine stimulation, NF-kB activation is strictly regulated by the kinase of IB. IκB kinase (IKK)-binds to dependent NF-kB promoters and histone H3 acetylates at Lys9 and phosphorylates histone H3 at Ser10, with the help from CBP and polymerase II complex [49]. This phosphorylation that is cytokine-induced is essential for the following histone acetylation H3 at Lys14 through the CBP pathway. After inflammation, an increased recruitment of NF-kB to the promoters of cytokines and chemokines is observed, coinciding with histone H3 acetylation. The glucocorticoid receptor (GR) cooperates with histone deacetylase 2 to possibly trigger NF-kB-dependent inflammatory genes repression [50, 51].

3.6.3 DNA Methylation and Inflammation

Methylation of DNA has been shown to have a crucial role in the inflammation for the regulation of gene expression. Hypomethylation of the promoter of the TLR2 gene has been associated to an exacerbated response of pro-inflammatory peptidoglycan of bacteria in CF epithelial bronchial cells [52]. Methylation of DNA and histone acetylation are two mechanisms that regulate TLR4 expression in intestinal epithelial cells. Methylation of DNA and histone alteration play a significant role in establishing the TNF locus' epigenetic environment [53]. Environmental factors, like bacterial infection, have been discovered to alter the genome's epigenetic state. DNA methylation has a major impact on the expression of imprinted genes throughout embryonic development. For example, bacterial infection in mice causes hypermethylation of the Igf2 gene's promoter. Many imprinted genes and other crucial development genes were shown to have altered expression in the placenta after maternal infection, according to expression microarray studies in mice [54]. A persistent infection with *Helicobacter pylori* (HP) in the human stomach causes gastric mucosa inflammation and activates many carcinogenic pathways [55]. The DNA methylation pattern of cells lining the stomach was changed by HP. DNA methylation induced by HP at the locus of Runx3 was demonstrated to be responsible for the expression loss of the gene in epithelial gastric cells [56].

3.6.4 MicroRNAs and Inflammation

MicroRNAs (miRNAs) function as post-transcriptional regulators and are a kind of tiny, non-coding RNA. Pre-miRNA transcripts are first cleaved by the Pasha and Drosha complexes in the nucleus before being shuttled to the cytoplasmic Dicer for further processing into mature miRNA duplexes of 18–24 base pairs in length [57]. The RISC (RNA-induced silencing complex) incorporates these short RNA duplexes and binds to their 3′ untranslated region (3′UTR) to selectively target specific messenger RNAs for degradation or translational inhibition. Recent studies have shown that miRNAs have a significant regulatory role in an extensive variety of developmental, differentiation, and pathological processes [58]. In contrast to miR-155, which inhibits macrophage activation by targeting the lipid phosphatase

SHIP1, miR-146a inhibits TRAF6-mediated signalling from Toll-like receptors. Acetylcholine (ACH) mRNA is a target of miR-132, which activates a key regulator of inflammation in the periphery. MiR-155 is activated when macrophages are exposed to lipopolysaccharide (LPS) [59]. MiR-155 regulates CCAAT/enhancer expression which binding protein Beta (C/EBP Beta) during activation of macrophage mRNA and the response of acute phase to control the production of pro-inflammatory cytokines. Liu et al. showed that a negative-feedback loop mechanism is responsible for TLR's activation of miR-147 to avoid an overly reactive inflammatory response. Both NF-kB and IRF3 activation are necessary for miR-147 induction by TLR stimulation [60]. In addition, macrophage inflammation triggered by TLRs is dampened by miR-147. Another research found that miR-105 might control the TLR-2 translation in gingival human of keratinocytes. Recently, it has been hypothesised that there are two separate levels of regulation at play in the reprogramming of inflammatory genes that is TLR4-dependent [61]. Epigenetic alterations facilitate transcriptional control, and differential expression of miRNAs, which is TLR4-dependent, regulates a higher level of regulation (miR579, miR-221, and miR-125b). It is believed that around 30% of all human genes may act as targets for miRNAs [62]. Epigenetic indicators such as EZH2, DNMT3b, DNMT3a, and HDACs are encoded by a number of these genes. Both miR-29 and miR-143 target DNMT3a and DNMT3b, and miR-29 can correct aberrant methylation in lung cancer whereas miR-143 controls DNMT3a in colorectal cancer [63]. By inhibiting HDAC5, miR-2861 regulates osteoblast development while miR-29 promotes osteogenesis via targeting HDAC4. MiR-140 controls HDAC4 in cartilage, whereas in prostate cancer, MiR-449a targets HDAC1 [64, 65]. However, microRNA's effect on epigenetic markers may be indirect. Retinoblastoma (Rb)-like protein 2 (Rb12) is a virally encoded microRNA known to control 3b mRNA and DNMT1, 3a, levels. Expression of EZH2 may be repressed by miR-26a, miR-101, and miR-214 [66]. In the second case, PcG-induced changes in gene expression are controlled by miR-214, which in turn is regulated by EZH2, the catalytic component of PRC2. Expression of PcG proteins downregulates miR-214 transcription in non-differentiated skeletal muscle cells. During the process of differentiation, PcG disengagement and miR-214 transcriptional activation take place [67]. Similarly, miR-214 aims for EZH2. Finally, DNA methylation, histone modifications, and microRNA targeting were examined as three separate epigenetic mechanisms of gene suppression [68]. A newly discovered epigenetic relay channel has been suggested to explain the downregulation of PPAR transcription. There is a highly controlled mechanism required for this route, and it includes EZH2, MeCP2, and miR-132. In this study, the scientists showed that MeCP2 translation was unblocked when miR-132 was downregulated. As a result, MeCP2 brings in HP1/G9a and PRC1/2 complexes, which methylate histone H3 at Lys9 and Lys27 in the PPAR promoter (H3K9 and H3K27) [69].

3.6.5 Inflammation and Cancer

Inflammation may have a role in the growth and spread of cancer. Up until recently, the mechanical basis of this phenomenon was unknown. When Iliopoulos et al. proposed a link between inflammation and neoplastic transformation, they revolutionised our understanding of the two processes [52, 70, 71]. The cytokine IL-6, the microRNA let-7, and the RNA-binding protein Lin-28 are all essential players in this paradigm. During the cellular transformation process, the activation of oncoprotein, particularly triggered by Src, initiates an inflammatory response dependent on the transcription factor NF-kB. This response induces the transcription of lin-28 and suppresses let-7 miRNA [72]. Let-7 inhibited transformation of cells by blocking the effects of IL-6. For cells to undergo metamorphosis, an uptick in IL-6 is needed, which was brought about by the activation of NF-kB and the subsequent let-7 repression. IL-6 has been proven to trigger oncogenic transformation by targeting STAT3 [73, 74].

3.6.6 Epigenetics in Human Diseases and Ageing

Rheumatoid arthritis (RA) is an autoimmune disease classified as a chronic inflammatory disorder that leads to gradual deterioration of the cartilage in the joints and bones. Many studies have demonstrated that epigenetic changes such as methylation of DNA, acetylation of histone, and microRNAs have a role in RA pathogenesis [75, 76]. Synovial fibroblasts from patients with rheumatoid arthritis were observed in one research to have a significantly reduced global DNA methylation pattern compared to control cells. Multiple studies have linked chronic inflammation to an increased risk of both COPD and lung cancer [77]. Recent studies suggest that epigenetic changes may play a role in the growth of acceptance in T-cell and macrophages activity. HDAC2 appearance and action are reduced in macrophages of lung, biopsies of lung, and blood cells of people with severe asthma, COPD, and smoking-induced asthma [78]. Our immune systems change as we age, making us more vulnerable to infections, inflammation, and hyperreactivity to self-antigens, all of which may increase our chance of developing cancer. DNA hypomethylation as a result of ageing has been associated with both chronic inflammation and malignancy [79]. El Mezayen et al. showed that dendritic cells' IL-23p19 gene expression was upregulated in tandem with H3K4 methylation. Periodontitis is a complicated disorder characterised by inflammation and loss of the tissues that keep teeth in place. Prostaglandin-endoperoxide synthase-2 (PTGS2) and prostaglandin E levels upregulated with the progression lesions of periodontal, but downregulated in chronic sickness. Hypermethylation of the PTGS2 promoter has been allied to decreased expression in patients with chronic periodontitis [80, 81].

3.7 Conclusions

Changes in the modifications of histone, methylation of DNA, and regulation of microRNA are key pieces of evidence needed to decipher the molecular basis of chronic inflammatory illnesses. There is hope for the development of highly specific medications thanks to recent discoveries about epigenetic modifications that occur during the inflammatory response. Medications being examined for therapy of epigenetics include deacetylase of histone inhibitors and demethylating agents, both target chromatin in rapidly reproducing tumour cells. Recent technological developments in whole-genome expression of microarray profiling and chromatin immunoprecipitation-based sequencing (ChIP-seq) methods may allow for better epigenetic drug development.

References

1. Ahmad S, et al. Epigenetic underpinnings of inflammation: connecting the dots between pulmonary diseases, lung cancer and COVID-19. Semin Cancer Biol. 2022;83:384–98.
2. Alam R, Abdolmaleky HM, Zhou JR. Microbiome, inflammation, epigenetic alterations, and mental diseases. Am J Med Genet B Neuropsychiatr Genet. 2017;174(6):651–60.
3. Aleksandrova K, Romero-Mosquera B, Hernandez V. Diet, gut microbiome and epigenetics: emerging links with inflammatory bowel diseases and prospects for management and prevention. Nutrients. 2017;9:9.
4. Badii M, et al. Trained immunity and inflammation in rheumatic diseases. Joint Bone Spine. 2022;89(4):105364.
5. Barnes PJ. Pathophysiology of allergic inflammation. Immunol Rev. 2011;242(1):31–50.
6. Binnie A, et al. Epigenetics of sepsis. Crit Care Med. 2020;48(5):745–56.
7. Buch MH, Eyre S, McGonagle D. Persistent inflammatory and non-inflammatory mechanisms in refractory rheumatoid arthritis. Nat Rev Rheumatol. 2021;17(1):17–33.
8. Charras A, Hedrich CM. The role of epigenetics in paediatric rheumatic disease. Curr Opin Rheumatol. 2019;31(5):450–63.
9. Crujeiras AB, et al. Molecular basis of the inflammation related to obesity. Oxidative Med Cell Longev. 2019;2019:5250816.
10. Czimmerer Z, et al. The epigenetic state of IL-4-polarized macrophages enables inflammatory cistromic expansion and extended synergistic response to TLR ligands. Immunity. 2022;55(11):2006–26.e6
11. Dąbek A, et al. Modulation of cellular biochemistry, epigenetics and metabolomics by ketone bodies. Implications of the ketogenic diet in the physiology of the organism and pathological states. Nutrients. 2020;12:3.
12. Dai Y, et al. Classical HDACs in the regulation of neuroinflammation. Neurochem Int. 2021;150:105182.
13. Davis FM, et al. Epigenetic regulation of the PGE2 pathway modulates macrophage phenotype in normal and pathologic wound repair. JCI Insight. 2020;5:17.
14. Ding Q, et al. Inflammation-related epigenetic modification: the bridge between immune and metabolism in type 2 diabetes. Front Immunol. 2022;13:883410.
15. Ebersole JL, et al. Aging, inflammation, immunity and periodontal disease. Periodontol 2000. 2016;72(1):54–75.
16. Fernández-Sanlés A, et al. DNA methylation biomarkers of myocardial infarction and cardiovascular disease. Clin Epigenetics. 2021;13(1):86.

17. Gao P, et al. Salt-induced hepatic inflammatory memory contributes to cardiovascular damage through epigenetic modulation of SIRT3. Circulation. 2022;145(5):375–91.
18. Garagnani P, Pirazzini C, Franceschi C. Colorectal cancer microenvironment: among nutrition, gut microbiota, inflammation and epigenetics. Curr Pharm Des. 2013;19(4):765–78.
19. Gasmi A, et al. Obesity and insulin resistance: associations with chronic inflammation, genetic and epigenetic factors. Curr Med Chem. 2021;28(4):800–26.
20. Giallongo S, et al. The role of epigenetics in neuroinflammatory-driven diseases. Int J Mol Sci. 2022;23:23.
21. Gibson F, et al. Epigenetic dysregulation in autoimmune and inflammatory skin diseases. Clin Rev Allergy Immunol. 2022;63(3):447–71.
22. Gomez JL. Epigenetics in asthma. Curr Allergy Asthma Rep. 2019;19(12):56.
23. Grolleau-Julius A, Ray D, Yung RL. The role of epigenetics in aging and autoimmunity. Clin Rev Allergy Immunol. 2010;39(1):42–50.
24. Guzik TJ, Cosentino F. Epigenetics and immunometabolism in diabetes and aging. Antioxid Redox Signal. 2018;29(3):257–74.
25. Hey J, et al. Epigenetic reprogramming of airway macrophages promotes polarization and inflammation in muco-obstructive lung disease. Nat Commun. 2021;12(1):6520.
26. Hull EE, Montgomery MR, Leyva KJ. HDAC inhibitors as epigenetic regulators of the immune system: impacts on cancer therapy and inflammatory diseases. Biomed Res Int. 2016;2016: 8797206.
27. Iyer SS, Cheng G. Role of interleukin 10 transcriptional regulation in inflammation and autoimmune disease. Crit Rev Immunol. 2012;32(1):23–63.
28. Jenny NS. Inflammation in aging: cause, effect, or both? Discov Med. 2012;13(73):451–60.
29. Joseph PV, Abey SK, Henderson WA. Emerging role of nutri-epigenetics in inflammation and cancer. Oncol Nurs Forum. 2016;43(6):784–8.
30. Jurdziński KT, Potempa J, Grabiec AM. Epigenetic regulation of inflammation in periodontitis: cellular mechanisms and therapeutic potential. Clin Epigenetics. 2020;12(1):186.
31. Kanasi E, Ayilavarapu S, Jones J. The aging population: demographics and the biology of aging. Periodontol. 2016;72(1):13–8.
32. Karin M, Shalapour S. Regulation of antitumor immunity by inflammation-induced epigenetic alterations. Cell Mol Immunol. 2022;19(1):59–66.
33. Katoh M. Multi-layered prevention and treatment of chronic inflammation, organ fibrosis and cancer associated with canonical WNT/β-catenin signaling activation (review). Int J Mol Med. 2018;42(2):713–25.
34. Kessler C. Pathophysiology of obesity. Nurs Clin North Am. 2021;56(4):465–78.
35. Khyzha N, et al. Epigenetics of atherosclerosis: emerging mechanisms and methods. Trends Mol Med. 2017;23(4):332–47.
36. Kirsch-Volders M, et al. Micronuclei, inflammation and auto-immune disease. Mutat Res Rev Mutat Res. 2020;786:108335.
37. Kowluru RA. Cross talks between oxidative stress, inflammation and epigenetics in diabetic retinopathy. Cell. 2023;12:2.
38. Kuzet SE, Gaggioli C. Fibroblast activation in cancer: when seed fertilizes soil. Cell Tissue Res. 2016;365(3):607–19.
39. Lee HN, Na HK, Surh YJ. Resolution of inflammation as a novel chemopreventive strategy. Semin Immunopathol. 2013;35(2):151–61.
40. Li X, et al. Epigenetics in the pathogenesis of diabetic nephropathy. Acta Biochim Biophys Sin Shanghai. 2022;54(2):163–72.
41. Li X, et al. The methyltransferase METTL3 negatively regulates nonalcoholic steatohepatitis (NASH) progression. Nat Commun. 2021;12(1):7213.
42. Liu D, et al. TRIM14 inhibits OPTN-mediated autophagic degradation of KDM4D to epigenetically regulate inflammation. Proc Natl Acad Sci U S A. 2022;119:7.
43. Liu YJ, et al. Editorial: epigenetics of the immune component of inflammation. Front Immunol. 2022;13:1000836.

44. Locati M, Curtale G, Mantovani A. Diversity, mechanisms, and significance of macrophage plasticity. Annu Rev Pathol. 2020;15:123–47.
45. Lopez Krol A, et al. Lactate induces metabolic and epigenetic reprogramming of pro-inflammatory Th17 cells. EMBO Rep. 2022;23(12):e54685.
46. Lund G, Zaina S. Atherosclerosis, lipids, inflammation and epigenetics. Curr Opin Lipidol. 2007;18(6):699–701.
47. Ma Y, et al. The regulation of miRNAs in inflammation-related carcinogenesis. Curr Pharm Des. 2015;21(21):3023–31.
48. Maiuri AR, O'Hagan HM. Interplay between inflammation and epigenetic changes in cancer. Prog Mol Biol Transl Sci. 2016;144:69–117.
49. Mayo S, et al. Recent evidence in Epigenomics and proteomics biomarkers for early and minimally invasive diagnosis of Alzheimer's and Parkinson's diseases. Curr Neuropharmacol. 2021;19(8):1273–303.
50. McCall CE, et al. Epigenetics, bioenergetics, and microRNA coordinate gene-specific reprogramming during acute systemic inflammation. J Leukoc Biol. 2011;90(3):439–46.
51. Mengozzi A, et al. Microvascular inflammation and cardiovascular prevention: the role of microcirculation as earlier determinant of cardiovascular risk. High Blood Press Cardiovasc Prev. 2022;29(1):41–8.
52. Murata M. Inflammation and cancer. Environ Health Prev Med. 2018;23(1):50.
53. Nair J, Maheshwari A. Epigenetics in necrotizing enterocolitis. Curr Pediatr Rev. 2021;17(3): 172–84.
54. Novakovic B, et al. β-Glucan reverses the epigenetic state of LPS-induced immunological tolerance. Cell. 2016;167(5):1354–68.e14
55. Oldenburg KS, O'Shea TM, Fry RC. Genetic and epigenetic factors and early life inflammation as predictors of neurodevelopmental outcomes. Semin Fetal Neonatal Med. 2020;25(3):101115.
56. Onodera A, et al. Epigenetic regulation of inflammation by CxxC domain-containing proteins. Immunol Rev. 2022;305(1):137–51.
57. Pacini G, et al. Epigenetics, pregnancy and autoimmune rheumatic diseases. Autoimmun Rev. 2020;19(12):102685.
58. Pandareesh MD, Kameshwar VH, Byrappa K. Prostate carcinogenesis: insights in relation to epigenetics and inflammation. Endocr Metab Immune Disord Drug Targets. 2021;21(2): 253–67.
59. Payne JL, Maguire J. Pathophysiological mechanisms implicated in postpartum depression. Front Neuroendocrinol. 2019;52:165–80.
60. Pearce EL, Shen H. Making sense of inflammation, epigenetics, and memory CD8+ T-cell differentiation in the context of infection. Immunol Rev. 2006;211:197–202.
61. Pietropaolo V, Prezioso C, Moens U. Role of virus-induced host cell epigenetic changes in cancer. Int J Mol Sci. 2021;22:15.
62. Potaczek DP, et al. Epigenetics and allergy: from basic mechanisms to clinical applications. Epigenomics. 2017;9(4):539–71.
63. Raghuraman S, et al. The emerging role of epigenetics in inflammation and immunometabolism. Trends Endocrinol Metab. 2016;27(11):782–95.
64. Ray D, Yung R. Immune senescence, epigenetics and autoimmunity. Clin Immunol. 2018;196: 59–63.
65. Regan EA, et al. Omics and the search for blood biomarkers in chronic obstructive pulmonary disease. Insights from COPDGene. Am J Respir Cell Mol Biol. 2019;61(2):143–9.
66. Roger L, Tomas F, Gire V. Mechanisms and regulation of cellular senescence. Int J Mol Sci. 2021;22:23.
67. Romagnolo DF, et al. N-6 linoleic acid induces epigenetics alterations associated with colonic inflammation and cancer. Nutrients. 2019;11:1.
68. Rossi JF, et al. Dynamic immune/inflammation precision medicine: the good and the bad inflammation in infection and cancer. Front Immunol. 2021;12:595722.
69. Sapienza C, Issa JP. Diet, nutrition, and cancer epigenetics. Annu Rev Nutr. 2016;36:665–81.

70. Saul D, Kosinsky RL. Epigenetics of aging and aging-associated diseases. Int J Mol Sci. 2021;22(1):401.
71. Shanmugam MK, Sethi G. Role of epigenetics in inflammation-associated diseases. Subcell Biochem. 2013;61:627–57.
72. Shao S, Gudjonsson JE. Epigenetics of psoriasis. Adv Exp Med Biol. 2020;1253:209–21.
73. Shen J, et al. Inflammation and epigenetic regulation in osteoarthritis. Connect Tissue Res. 2017;58(1):49–63.
74. Shigdar S, et al. Inflammation and cancer stem cells. Cancer Lett. 2014;345(2):271–8.
75. Smolen JS, et al. Rheumatoid arthritis. Nat Rev Dis Primers. 2018;4:18001.
76. Sommese L, et al. Clinical relevance of epigenetics in the onset and management of type 2 diabetes mellitus. Epigenetics. 2017;12(6):401–15.
77. Storino Farina M, et al. Statins and atherosclerosis: the role of epigenetics. Medwave. 2015;15 (10):e6324.
78. Surace AEA, Hedrich CM. The role of epigenetics in autoimmune/inflammatory disease. Front Immunol. 2019;10:1525.
79. Takeshima H, et al. TET repression and increased DNMT activity synergistically induce aberrant DNA methylation. J Clin Invest. 2020;130(10):5370–9.
80. Tsou PS, Varga J, O'Reilly S. Advances in epigenetics in systemic sclerosis: molecular mechanisms and therapeutic potential. Nat Rev Rheumatol. 2021;17(10):596–607.
81. Turpin W, et al. Determinants of IBD heritability: genes, bugs, and more. Inflamm Bowel Dis. 2018;24(6):1133–48.

Epigenetic Regulators of Inflammatory Gene Expression

4

Satinder Kaur, Jayapriya Mishra, Abhishek Sehrawat, Gurjit Kaur Bhatti, Umashanker Navik, P. Hemachandra Reddy, and Jasvinder Singh Bhatti

Abstract

Epigenetics is defined as changes in the expression of genes whose core is confined within a non-expressed portion of DNA. All the mechanisms of epigenetics including DNA methylation, acetylation of histones and micro-RNAs and all these factors are under the direct control of diet, age, maternal

S. Kaur · J. Mishra · A. Sehrawat · J. S. Bhatti (✉)
Laboratory of Translational Medicine and Nanotherapeutics, Department of Human Genetics and Molecular Medicine, School of Health Sciences, Central University of Punjab, Bathinda, India
e-mail: jasvinder.bhatti@cup.edu.in

G. K. Bhatti
Department of Medical Lab Technology, University Institute of Applied Health Sciences, Chandigarh University, Mohali, India

U. Navik
Department of Pharmacology, School of Health Sciences, Central University of Punjab, Bathinda, India

P. H. Reddy
Department of Internal Medicine, Texas Tech University Health Sciences Center, Lubbock, TX, USA

Department of Pharmacology and Neuroscience and Garrison Institute on Aging, Texas Tech University Health Sciences Center, Lubbock, TX, USA

Department of Public Health, Graduate School of Biomedical Sciences, Texas Tech University Health Sciences Center, Lubbock, TX, USA

Department of Neurology, Texas Tech University Health Sciences Center, Lubbock, TX, USA

Department of Speech, Language, and Hearing Sciences, Texas Tech University Health Sciences Center, Lubbock, TX, USA

Nutritional Sciences Department, College of Human Sciences, Texas Tech University, Lubbock, TX, USA

G. Gupta et al. (eds.), *Targeting Epigenetics in Inflammatory Lung Diseases*,
https://doi.org/10.1007/978-981-99-4780-5_4

mental health and environmental factors. There is an increasing suggestion of epigenetics in the expansion of several lung diseases. Epigenetics also exerts its influence over inflammation which is a common feature of many lungs inflammatory diseases including COPD, asthma and IPF. All the immune cells which play a vital part in the growth of inflammation also fall under the control of epigenetic changes. Epigenetics influences both adaptive and innate branches of protection to bring inflammation at a heightened rate in inflammatory diseases by controlling their activity and differentiation. Immune cells are generally associated with protective responses to harm in lung inflammatory diseases. Hence, therapeutics targeting epigenetics have emerged as attractive candidates that in addition to controlling epigenetic mechanisms control gene activity, thus providing a suitable alternative in place of already available drugs.

Keywords

Epigenetics · Lung diseases · Inflammation · Gene regulation · Therapeutics

4.1 Introduction

Several lung diseases involve inflammation as a main culprit in afflicting lung damage. Inflammation generally involves immune cells as an integral part to contribute to pathogenesis. Death of around three million persons is caused by inflammatory disease of the respiratory passage. There is a dynamic interplay operating between inflammation and immune cells in most lung inflammatory diseases. Pulmonary diseases are one of the central reasons for demise. Pulmonary diseases include asthma, pulmonary fibrosis and COPD. Lung disease that involves inflammation as a predominant manifestation includes pulmonary fibrosis, asthma and COPD. Data obtained from WHO shows that the mean age of death rate in the person suffering from IPF is 3.75 per 10,000 males and 1.50 per 100,000 females. Similarly, asthma is another lung inflammatory disease that is generally found in adults and children. Adults suffering from asthma show a higher death rate, but the cases of asthma are more in children [1]. Females generally are more prone to the development of asthma than males [2]. IPF is a kind of interstitial lung fibrotic disease. There are about 40,000 new patients of IPF that are found each year in Europe [3]. Out of a total of 1000 persons, only five persons suffer from IPF thus making it an uncommon respiratory disease. The occurrence of IPF varies geographically with Japan showing a higher number of sufferers. About 10.5 persons out of 100,000 persons in Japan are suffering from IPF. According to data obtained in 2019, there were about 339 million people who are diagnosed with asthma out of which 25,000 people died in 2010. About three million patients died from COPD in 2010 out of 384 million people diagnosed with COPD [4]. There is a lesser chance of IPF happening in people with age below 50 years [5]. There is lethal nexus that involves genes, environmental factors and epigenetics as key participants in the progression of IPF that primes to the devastation of alveolar epithelial cells, ageing and manufacture of certain pro-fibrotic mediators that in turn brings fibroblast into action and causes their conversion into myofibroblast.

Earlier it was thought that few genes are responsible for causing lung inflammatory disease. Recently the role of non-coding portion of DNA including miRNAs has also come to light. There are certain enzymes called DNA methyltransferases and DNA demethyltransferases that change heterochromatin to euchromatin thus bringing an inaccessible or unexpressed portion of the genome into an accessible or expressed portion. Many factors including diet, pollutants in the environment, mental health and age offer their share in manipulating changes happening in epigenetics to kickstart events linked to the development of the disease as originated in several lung diseases namely COPD, asthma and lung fibrosis. At the molecular level, epigenetics is regulated by conformational changes happening in euchromatin by the involvement of DNA methyltransferases (DNMT) and DNA demethyltransferases. Similarly, acetylation and deacetylation of histone residues present inside nucleosome help to fold or unfold the euchromatin. Generally, transcription is positively stimulated by demethylation and acetylation of histones as they open up chromatin and make it readily available to RNA-polymerase-II (RNA poly-II). Similarly, the deacetylation and methylation of histones present over nucleosomes help to open up DNA thus causing repression in gene expression. Another fraction is miRNAs that control the expression of genes by degrading them or stop them from expression by binding them at 5′UTR. Besides molecular levels, epigenetics also translates its effects by bringing environmental influences like inhalation of smoke, presence of pollutants in surroundings, maternal diet and mental health and ageing that are predominant in regulating development of this disease.

4.2 Basics of Epigenetics

The word epigenetics is formed by the union of 'epi' plus 'genetics' which means above genetics or beyond genetics. Epigenetics comprises all those changes that are not present in coding sequences of DNA. Generally, epigenetic changes operate in three major ways including modifications of DNA, modifications of histone and non-coding RNAs. All these three changes fall under the control of circumstances prevailing in environment, food taken by mother, disease and age of the individual. There is a significant amount of evidence that enlightens the concealed part of epigenetic besides genetics in the growth of lung inflammatory illness. Gene methylation is performed by methyltransferases of DNA that add a methyl group specifically to cytosine thus forming 5-methylcytosine. Generally, gene promoters exhibiting hypermethylation in gene promoters are characterised by silent gene expression; meanwhile, hypomethylation in gene promoters tends to free gene promoters from silencing hence activating gene expression. CpG islands are certain regions within gene promoters that exhibit methylation in CpG nucleotides and thus exert its effect over gene expression by concealing expression of gene utilising tight packing of nucleosomes. However, contrasting roles of DNA methylation have been realised recently as Ten-eleven translocase protein (TERT) promoters bearing methylation in histones belonging to nucleosomes that positively stimulate gene expression and enhance telomerase activity. Histone tails lying within nucleosomes are altered by several modifications including ubiquitination, phosphorylation, methylation and acetylation. Phosphorylation produces its effect on the conformation of nucleosomes by tightening or loosening it. The tight conformation of nucleosomes generally conceals or hides the gene expression by hindering RNA poly-II from reaching the promoters. The loosened conformation of nucleosomes unfolds the DNA inside and thus makes all the DNA present within nucleosomes readily accessible to RNA poly-II.

Generally, histones, composed of octamer H3, H2B, H2A and H4, wrap DNA into nucleosomes. Lysine and arginine are the most active amino acid residing within histones that undergo modifications readily. These modifications are responsible for changing euchromatin into heterochromatin by controlling the packing conformation of histones. The trimethylation of histone H3K4 positively influences gene transcription and H3 K27 methylation is usually linked with the hindrance expression of gene. Similarly, histone acetyltransferases are a group of enzymes that adds acetyl residues on histone residues and unlocks gene expression while histone deacetyltransferases are a contrasting enzymes class that eliminates groups of acetyl and hence helps to lock the gene expression. Normally RNA is destructed or stopped from participating in translation by miRNAs. miRNAs comprise a non-expressible and non-coding part of the genome that exerts its control over the genetic portion by binding with 3′UTR region of RNAs to destruct or stop the protein from expressing. Drosha is an enzyme that is responsible for creating pre-miRNA from primary miRNA. An enzyme called Dicer breaks mature MiRNAs from hairpin precursor RNA also known as pre-mRNA and is made of around 70 nucleotides. Around 1000 miRNAs are housed inside the genome of human beings. Epigenetics is controlled by several factors including food ingested, surroundings and the mental health of mother [6]. Epigenetics pertains to all changes happening in unexpressed portion of DNA. Generally, all methylases and deacetylases are responsible for hiding DNA, hence suppressing gene expression, while all acetylases and demethylases serve a contrasting role by unfolding the wrapped DNA and thus making it available for occupation by RNA polymerase-II. Those enzymes that remove acetyl group are known as histone deacetyltransferase (HDAC) while those that add methyl group are known as HAT.

4.3 Epigenetic Regulation of Immune Cell Development and Function

4.3.1 Innate Immunity

The binding of transcription factors depends upon the conformation of chromatin found within genes of immune cells as open conformation of chromatin easily allows factors of transcription to occupy the organizer areas of the genes. Accessibility of chromatin to transcription factors falls under the regulation of enhancers, promoters and silencers. Generally, when a tissue is subjected to damage or injury, the native immune cells housed within that tissue are the first respondents to wound and evoke inflammation by creating a proinflammatory signal-rich environment calling other immune cells present in circulation, namely neutrophils and monocytes, at the site of damage. The innate immune organization is composed of dendritic cells, neutrophils, macrophages and natural killer cells. Antigens are mostly presented by antigen-presenting cells including dendritic cells and macrophages which engulf the antigen and present their components to lymphocytes. There are specific histone markers present inside specific enhancers of gene which are altered epigenetically to favour pro-inflammatory or anti-inflammatory fate of macrophages [7]. Macrophages found a vital portion of immune system that exerts its control over inflammation and safeguards organisms against infection by ensuring the finalization of functions belonging to a particular tissue. Macrophages are potent eaters that engulf the pathogens and then presents its fragment to T-cells. Gene expression of macrophages falls under control of transcription factors and epigenetic modifications. Macrophages are known to unleash cytokines after receiving stimulus that bridges the innate and adaptive arm of immunity to bring inflammation into action. Several macrophages are found inside different tissues and thus have vast distinctive population. As each tissue has its own collection of macrophages, hence tissue-specific macrophages are important for clearing off pathogen and maintaining homeostasis of tissues. About more than 12,000 enhancers occupying macrophages are reprogrammed upon transplantation into a new ambience as evidenced by studies conducted in mouse [8]. Epigenetic control of macrophages operates at two levels including lineage-specific transcription factor and tissue-specific transcription factors. Chromatin found inside macrophages is directly made accessible to transcription factors by means of lineage-determining transcription factors (LTDs). PU-1 is a lineage-determining transcription factor (LTD) that is present at enhancers of genes and helps to maintain basal activation state and H3K4Me3 of many promoters. At enhancers, Pu-1 sustains H3K4Me1 and thus helps in choosing enhancer elements according to tissue, hence pointing out that PU-1 is an essential requirement in genomic locations to carry its functioning as an enhancer. In both activated and dormant cells PU-1 is the master transcription factor that calls its other companion transcription factor (TFs) by making chromatin more readily accessible by opening chromatin. C/EBP family members, IRF, NF-κβ and AP-1, are other transcription factors that co-operate with enhancers chosen by PU-1 TFs. PU-1 is extremely important as its absence makes macrophages deficient in accepting genomic cues. The absence of PU-1 keeps chromatin in a closed conformation by utilising H3K27Me3 and the occupation of co-repressors. Lineage commitment of macrophages is associated with the binding of PU-1 TF that helps to unfold the closed chromatin. Environmental influences like the presence of Interlekin-4 play a vital role in allowing efficient binding of NF-κβ at macrophage enhancers. This binding is consequently followed by the elimination of H3K27Me3 and H3K27 Ac spot on enhancers. All these events lead to the activation expression of gene in macrophages. H3K4 monomethylating at many genes residing within macrophages is stimulated by PU-1 binding at histones [9]. Expression of Pu-1 having H3K4Me2and H3K27Ac in all macrophage pools is regulated by enhancer of Spil [10]. Therefore, a specific combination of transcription factors found in various tissues in combination with distinct envirnomental condition governs the epigenetics in tissue-specific manner. This tissue specific variant effects manifests its consequences by controlling gene expression, function and the environment in which these macrophages reside.

Various signalling pathways diverge from signal promoters depending upon its ability to bind different cocktails of transcription factors. Hence, response of macrophages is always elicited following their immediate need, and this response is adjusted by enhancers that regulate gene expression according to the need and situation faced by macrophages. Regulatory regions lying within specific macrophages belonging to a particular tissue and gene expression from specific macrophages fall under the order of cues coming from the environment. Macrophages harbouring active promoters contain H3K4Me and H3K27Ac while inactivated promoters within macrophages depict H3K4Me1 and H3K27Ac [11]. There is a wide variety of macrophages found in different tissues including intestinal, brain, splenic, peritoneal, liver and alveolar macrophages. The intestine is the largest reservoir of macrophage population. Liver monocytes located inside the foetal liver under influence of GM-CSF instigate PPAR-γ signalling to develop into alveolar macrophages. The immune response of genes in alveolar macrophages during bacterial infections is controlled by long non-coding RNA (lncRNA) MEG-3,4 that effects the transcription factors of alveolar macrophages to control the inflammatory response in lung tissue. The inflammatory response in the lung is associated with the lncRNA MEG-3 and microRNA MiR-38 in companionship with enhanced levels of IL-1β. The rise in levels of IL-1β is followed by increased binding between maternally expressed gene-3,4 (MEG-3,4) and mRNA of pro-inflammatory cytokine IL-1β [12]. The immune system is endowed with an inherent capability of fighting against external environmental insults and attacks. Our body is inevitably endlessly unprotected to a diversity of environmental toxicants, pathogens and other foreign particles. The immune system acts as a safeguard and shields our body against these insults.

There are two arms of immune system that helps to prevent our body and organs against external invaders. IFN-γ and microbial goods tend to stimulate pro-inflammatory M1 macrophages while IL-10, IL-4 and helminthic stimulate anti-inflammatory macrophages M2. M1 macrophages are known for shielding the host while M2 macrophages are known for their role in healing tissue damage. When human monocyte-derived dendritic cells are cultured and exposed to bacterial endotoxin lipopolysaccharide (LPS), then it underwent a loss of H3K9 methylation at organizer areas which subsequently led to stimulation of pro-inflammatory cytokines like IL-8 and MIP-1α [13]. Efficient presentation of antigens is performed by M1 macrophages that have high bacterial killing capacity owing to the making of pro-inflammatory reactive oxygen and cytokines and nitrogen species. M2 macrophages are activated alternately and decrease the making of pro-inflammatory cytokines and are less efficient at presenting antigens. Efficient and accurate eating of macrophages is dependent on the structure of macrophages which is composed of an actin cytoskeleton, and this actin cytoskeleton is modulated by DNA demethylation. The tolerant state of macrophages rests on the shoulders of inhibitory mechanisms emerging from NF-κβ. A suppressive complex formed by tri-union of nuclear receptor corepressor-HDAC-3-P50 (NCOR-HDAC-3-P50) is called by NF-κβ into aimed genes [14]. H3K9 methylation in promoters of macrophages is performed by histone methyltransferase named G9a under command received by NF-κβ. This H3K9 methylation helps AP-1 to partner with this methylated spot under the direction of NF-κβ [15]. TLR signalling tolerance is generated in macrophages by mi-RNA 146a following stimulation of the MYD88-dependent TLR pathway of inflammation by controlling *BRG-1*. miRNA-221 and mi-RNA 222 revamp functional aspects of macrophages under stimulus provided by LPS tolerization [16]. The innate immune system is quite non-specific and unleashes attack solely on recognising the nucleic acid released from pathogen or lipids, noxious substances and venom but the adaptive arm is quite specific and showers attack on the pathogen in a highly specific form by modulating gene rearrangements with B- and T-cells that help in the creation of diverse kinds of antibodies to attack diverse pathogens. Some recent studies strongly support that even innate immune cells keep the memory of previous encounters with pathogens which helps them to counteract the attacks of the same pathogens when that pathogen attacks again in large numbers. Epigenetics has been shown to have substantial involvement in controlling the memory of immune cells to exert its influence on the transcriptional level. This fact is supported by studies carried out in monocytes that are stimulated on coming in contact with β-glucan releasing from *Candida* that in turn leads to trimethylation of H3K4 and acetylation of H3K27 which produces its effects by uplifting levels of IL-6, TNF cytokines, inflammation and trained immunity. Epigenetics also offers its share in stimulating the difference of human monocytes into dendritic cells under the effect of external factors. Differentiation of human monocytes into dendritic cells is accompanied by acetylation of histone H3K9Ac and removal of H3K9me3 along with methylation of DNA at its organiser. These changes in histones enhance the expression of cluster differentiation-209 (CD-209) on dendritic cells [17]. The memory found within macrophages is often regarded as trained immunity and this trained immunity is nurtured by epigenetic memory. Macrophages acquire this memory after combating a pathogen. Promoters of these macrophages harbouring trained immunity are characterised by enhanced levels of H3K4Me3 at agents of genetic factor intricate in the manufacture of Myd88 and pro-inflammatory cytokines named IL-6, TNF-α and IL-18 [18]. A characteristic partnership is established between lncRNA of inflammatory chemokine secretion and ELR + CXCL chemokines like CXCL-2, CXCL-1, IL-8 and CXCL-3. The H3K4me3 imprint is established in macrophages under the direction provided by (WD-repeat-containing protein-5/mixed-lineage leukaemia-11) WDR5/MLL-11 complex at CXCL promoters.

4.3.2 Adaptive Immunity

Adaptative immunity involves T-cells and B-cells as their key participants that causes the emergence of a specific form of attack against pathogens by keeping a memory of the previous encounter. B-cells are unique in the fact that they produce specific antibodies against invaders entering our body by using VDJ recombination and somatic hypermutation. The functioning of immune cells is compromised by ageing and is often termed immune senescence. This drop in the functioning of adaptive immunity with age generally finds its roots in epigenetic changes.

4.4 T-Lymphocytes

T-cells diverge into several fates including effector T-cells and memory T-cells. TH1, TH17, TH9, TH25, TH22, TH2, TH9, T-regulator cells, natural killer T-cells, T-helper cells and cytotoxic T-cells are some of the effector T-cells [19]. T-cells exaggerate immune response to inflammation and heighten it by secreting cytokines and chemokines. Naïve T-cells differentiate into TH1, TH2, TH17 cell fate or T-regulatory cell fate. Imprints of histone signatures are established in the IFN-γ agent of TH1 and TH2 cell lines by utilising transcription factors STAT-4 and T-bet in the TH1 subtype while GATA-3 and STAT-6 in TH2 subtype. Main transcription factors that control the fate of binding of menin at Bach-2 locus inside T-cells maintain acetylation of Bach-2 locus, thus controlling the ageing of T-cells and cytokines releasing from T-cells. The thymus is a central location where most T-lymphocytes are matured and decrease in size with age. Thymus also falls victim at hands of increased ageing effects that in turn destabilise the stability of genes of the thymus by removing heterochromatin signs at H3K9me3 associated with loss of histone-lysine-N-methyltransferase (SUV39HI) expression [20]. A peculiar kind of T-cells named γδ T-cells are the major army of T-cells present inside mucosal tissue and skin. Certain specific γδ T-cells manufacture IL-22 to stop fibrosis. The choice to secrete IL-17 or IFN-γ is prominently concluded during their development in the thymus. Chromatin structure is principally responsible for producing IL-17A, IL-22 IL-17F or IFN-γ. The excessive infiltration of inflammatory cells at the damaged site tarnishes tissue healing. This excessive inflammation is counteracted by T-regulatory cells. The increased drop in function with ageing is correlated with the decrease in methylation of aged cells [21].

The innate immune system functions in coordination with the adaptive immune system, since cells from the innate immune system presents the antigens to B- and T-cells, allowing them to differentiate into effector cells. There is a lack of IL-2 expression in both naïve and $CD4^+$ T-cells, but this cytokine is articulated in T-cells when these cells come in contact with a pathogen. Expression of IL-2 is induced on the surface of $CD4^+$ T-cells by utilising demethylation at a single specific CpG site in the enhancer region of these T-cells; the demethylation at CpG enhancer region works to increase transcription of IL-2 and forms a memory imprint on T-cell that has encountered the pathogen [22]. $CD4^+$ T-cells exhibit MHC class-I particles to recognise antigen while $CD8^+$ T-cells exhibit MHC class II molecules to recognise the antigen. $CD4^+$ T-cells differ from CD 8^+ T-cells that directly destruct the infected cells. $CD4^+$ T-cells depend on antigen-presenting cells to recognise antigens and destroy them. Memory $CD8^+$ T-cells maintain enhanced acetylation in H3Ac at IFN-γ promoter and enhancer region of stimulated $CD8^+$ T-cells that help to build a stronger cytotoxic response under the effect of excessive stimulation provided by large number of pathogens [23]. Expression of MHC-II on the surface of $CD4^+$ T-cells is controlled by a class-IIA transactivator (CIITA) by exerting its influence over epigenetics [24]. Generally, $CD4^+$ T-cells are differentiated into TH1 type or TH2 type. Cytokines like IFN-γ, IL-12 or IL-15 help to direct the differentiation of $CD4^+$ T-cells into TH1 type while IL-4, IL-10 and IL-13 help to direct the differentiation of $CD4^+$ T-cells into TH2 subtype. Studies suggest that in addition to cytokines, epigenetic changes also dictate the transition of $CD4^+$ T-cells into TH1 or TH2 subtype. Differentiation of naïve T-cells into TH1 cells is associated with changes in the methylation pattern of IFN-γ promoter as this promoter is hypermethylated in naïve state but becomes demethylated in TH1 cells [25]. In addition to TH1 and TH2 cells, a third kind of cells named TH17 cells are known for exhibiting IL-17 cytokine and RAR-related orphan receptor-c (ROR-C) transcription factor. There is a concordance between gene expressed in T-cells and the demethylation of transcription factors like IL-17A and ROR-C. IL-17 locus of TH17 cells harbours histone acetylation at spots H3Ac and H3K4me3 [26, 27]. Stable expression of regulatory T-cells is sustained by demethylation and hyperacetylation found in forkhead box protein-3 (FOXP-3) [28, 29]. An intermediate phenotype of TH1/TH17 cells emerges during chronic inflammation as TH17 cells exhibit IFN-γ and IL-FOXP3 and 17 jointly. Epigenetic signatures of these intermediate cells resemble TH17 cells that harbour demethylated IFNG gene and lose the suppressive imprint of H3K27Me3 found in the bivalent domain of TBX21 gene that triggers expression of IFN-γ [26].

4.5 B-Lymphocytes

B-lymphocytes differentiate into plasma cells that secrete antibodies after receiving stimulation from T-cells. Diverse antibodies generated in such a way are highly specific and recognise a variety of pathogens that helps to improve the elimination of pathogen by bringing the complement system also into play. B-cell differentiation and function is controlled by monoallelic VDJ genes that helps to generate a diverse and specific repertoire of antibodies. Pax-5 is a transcription factor that helps to direct differentiation of B-cells into plasma cells. Progenitor B-cells demonstrate hypermethylation in CD79a gene promoter, followed by demethylation of CD79a promoter in the early stages of B-cell differentiation. Further, it stimulates the Pax-5 to activate histone acetyltransferase, thus allowing the genes of B-cells to be expressed [30]. Different kinds of antibodies are generated by utilising V(D)J rearrangement and AID-1 found in B-cells at specific steps. Genes are not expressed in naïve B-cells owing to hypermethylation of AID gene promoter. Stimulation of

B-cells is accompanied by demethylation of AID gene promoter and obtain enhanced levels of histone H3Ac [31]. Differentiation of B-cells into plasma cells is assisted by characteristic histone signatures in active promoters and enhancers of these cells [31]. The characteristic identity of B-cells is sustained through Blimp-1 which is a transcriptional repressor that supresses the mature B-cell differentiation and helps to maintain B-cells in their final fate. Memory B-cells are more efficient at disposing of pathogens as they carry memories of previous encounter with pathogens that help them to clear pathogen more readily and efficiently. Several epigenetic enzymes like enhancer of Zeste homoluge-2, histone acetyltransferase monocytic leukaemia zinc finger protein and DNAMT3a found in naïve and stimulated B-cells point out that these modifications provide a push to memory B-cells to give a robust and tough fight to pathogens [32–34].

Epigenetic alterations in transcription factors like T-bet and eomesodermin and cytokines like IL-2 and IFN-γ and other molecules including CD40L, CD70, integrin-L-α-chain (ITGAL), PRF and CCR6 are known to control transcription and function of memory T-cells [35]. $CD4^+$ T-cells have an essential need to migrate towards renal proximal tubular epithelial cells in order to exert their function properly. Demethylation of *CCR6* gene in memory $CD4^+$ T-cells helps to stabilise expression of chemokine that in turn provide stimulus to $CD4^+$T-cells to migrate towards epithelial cells residing in renal tubular portion [36]. A stronger cytotoxic attack is generated by means of hyperacetylating histone 3 at ninth lysine residue (H3K9Ac) in effector molecules including Eomes, PRF and GRZ as antigen is again encountered. There are four epigenetic forms, namely active, poised, bivalent and repressed, that shed light on functioning of different genes [37]. Transition of hematopoietic stem cell into antibody manufacturing B-cells require changes at the level of chromatin by utilising epigenetic mechanism. A correct functional pattern and variable diversity joining (VDJ) recombination in B-cells is built by employing demethylation of transcription factors like Pax-5 and PU-1 and their aimed genes including CD79, CD19 and IRF-4. Somatic hypermutation recombination and class switch recombination are essential for generating variety of antibodies with high affinity. This process is controlled by activation induced cytidine deaminase which is involved in DNA demethylation [38]. Natural killer (NK) cells are normally linked with inflammatory and regulatory functions. Epigenetic changes happening in NK cell receptors and cytotoxic molecules help to alter activity of NK cells [39]. Generally, T-lymphoid cells are preferred for studying epigenetics. Both B- and T-cells arise from common lymphoid progenitors. B-cells comes into existence from common lymphoid progenitor cells by starting B-cell development in bone marrow by controlling gene rearrangements in heavy immunoglobulin chain (IgH) and light immunoglobulin genes (IgL) in B-cells. B-cells reacting with self MHC molecules are generally eliminated.

4.6 Inflammatory Lung Diseases

4.6.1 Asthma

Asthma is a kind of hypersensitive reaction that ultimately results in inflammation of airway and hence provide hindrance to air flowing inside lungs by blocking airway. This blockage manifests its affects by causing shortness of breath which is a characteristic feature of asthma. Major hallmarks of asthma comprise insult of epithelial cells and intrusion of many kinds of immune cells including lymphocytes, eosinophils, mast cells and phagocytes. Asthma is caused by a teamwork of innate immune cells, adaptive immune cells and respiratory epithelium releasing proinflammatory cytokines, chemokines and other inflammatory cells. Inflammation is a prominent feature of asthma. Dendritic cells present in airway catch allergens present in air and present them to naïve T-helper cells. After receiving stimulus from allergen, TH2 cells are brought into action to secrete IL-13 that increases the production of immunoglobulin-E (IgE) antibodies from B-cells. IL-5 secreted from TH2 cells sustains existence and maturation of eosinophils. The maturation of TH2 cells are controlled by Lymphokines released from T-cells. T-regulatory cells that are generally responsible for supressing immune response elicited by TH2 cells show disturbance in synthesis of IL-10 and tumour growth factor-β (TGF-β). All immune cells bring inflammation into play by secreting pro-inflammatory cytokines, chemokines, growth factor and Eicosanoids. TH2 cells secrete IL-4 and TGF-β that stimulate another variety of T-cells known as TH9 cells thus causing release of histamine, prostaglandins and elastase from mast cells. These mast cells are occupied by IGE antibodies produced from B-cells to secrete various mediators. TH1, TH17 and airway epithelia cells increase neutrophil numbers inside airways thus forcing the release of metalloproteinases and ROS from neutrophils [40].

Chemokines released from immune cells and airway epithelium attract other cell from location at diseased spot. Eicosanoids are a class of lipid mediators that cause many deleterious effects in epithelia of airway. These four principal mediators, namely lymphokine, pro-inflammatory cytokines, chemokines, growth factors and eicosanoids, generally drive pathogenesis of asthma [41]. All the constituents of lung including vascular endothelial cells, epithelial cells, fibroblast and smooth muscle cells are affected by them [40]. Asthma is generally caused as a part of response offered to allergens, infections in respiratory tract, presence of irritants in surroundings and non-steroidal anti-inflammatory drugs in some patients. Prolonged inflammation of airway is responsible for causing excessive secretion of mucus, and fibrosis in subepithelial portion, multiplication of blood vessels and invasion of inflammatory cells [42]. Respiratory epithelium is itself an important part of innate immune system that helps in mounting immune response and hence offers its share in inflammation by producing several mediators of inflammation [43]. Various immune cells comprising T-cells, macrophages, eosinophils, neutrophils, dendritic cells, mast cells and basophils are potential contributors of asthma. Both innate and adaptive immunity jointly play a decisive role in development of asthma. There is an imbalance in production of T-cells in asthma. Allergens excite and force naïve

T-cells to differentiate into TH2 cells. IL-4, 5, 9 and 17 secreted from T-cells increase the population of mast cells and eosinophils. Inflammation in airway is stimulated by TH2 cells that release IL-4, IL-13 and IL-15. Hyperreactivity of airway is negatively regulated by IFN-γ released from TH2 cells. The $CD4^{+}$T-cells of asthma patients are characterized by low levels of IFN-γ owing to decrease methylation. Cigarette smoke work as an allergen to promote progression of asthma. There is drop in methylation status of genes encoding IL-4, IL-5 and IL-13 that cause DNA methylation in whole lung and increase production of TH2 cytokines that help in the development of lung inflammation. Smoking by mothers increases methylation happening at AXL gene encoding a tyrosine receptor in newly born infants and ultimately induces symptoms of asthma. Increased inhalation of polycyclic aromatic hydrocarbons found in air pollutants alter methylation status of gene acyl-CoA Synthetase long-chain fatty family member-3 [44].

Peripheral blood mononuclear cells (PBMCs) of children suffering from asthma were exposed to CO, NO2 and particulate matter 2.5 (PM2.5) led to different patterns of methylation within FOXP3 and IL-10 DNA. All three pollutants cause hyperacetylation of H3K9 and H3K14 in IL-4 promoter of CD4+ T-cells [45]. When certain mice are repeatedly brought in contact with PM 2.5, then it enhances acetylation in IL-4 gene of $CD4^{+}$ T-cells [46]. Epithelial cells and smooth muscle cells present inside airways after getting exposed to allergen present in environment help to unleash IL-4, IL-5, IL-13 and thymic stromal lymphopoietin (TSLP). All these cytokines change naïve T-cells into THI and TH2 cells. Any deviation in normal levels of TH1 and TH2 cells invokes inflammation in lungs contains Airway cells including airway epithelial cells, goblet cells, fibroblast and smooth muscles constitute respiratory passage. All cells of respiratory passage are the sole and prime target of immune cells.

Several proinflammatory cytokines like IL-1β, TNF-α and IL-6 are produced from airway epithelia. Inflammatory genes are stimulated by these pro-inflammatory cytokines that create a vicious loop by enhancing secretion of cytokines and chemokines and thus amplify population of inflammatory cells [41]. Nonetheless, this airway also is an origin point for secretion of chemokines and hence attracting other immune cells at affected location thus causing accumulation of inflammatory cells in giant amount. CCL-2 or monocyte chemoattractant protein-1 is released from epithelial cells, macrophages and mast cells thus causing pro-inflammatory response. IL-1β and TNF-α released by communion of epithelial cells and macrophages elicit inflammatory response. TNF-α released from mast cells displays proinflammatory influence. G-protein-coupled receptors on surface of airway epithelia are occupied by leukotrienes released from T-cells. Secretion of mucus and enhancement in secretion of small blood vessels, contraction of smooth muscles and increased hire leukocytes in airway is caused by leukotrienes. Inflammatory cytokines, chemokines and proteases unleashing from mast cells induce inflammation in airway [47, 48]. Airway is characterised by high number of macrophages that excite the synthesis of cytokines and chemokines. People smoking regularly display a high count of neutrophils in airway [49, 50]. Principle lymphokines including IL-5, IL-4, IL-13, IL-9 and IL-17 are unleashed from T-cells and other innate immune cells. Production of IgE antibody is triggered by IL-4 and IL-13. High affinity receptors FcER1 receptors on mast cells and basophils are occupied by IgE antibodies released into circulation from B-cells [41]. Number of neutrophils is boosted by IL-17 releasing from T-cells. IL-17 also causes enhanced secretion of cytokines from airway epithelia. There is high amount of IL-17 found in blood and airway of asthma patient. Special T-cells called TH17 cells induce invasion of neutrophils into lungs post allergen attack [51].

Several immune cells of innate and adaptive immunity including mast cells, T-cells, basophils and eosinophils enhance the production of IgE antibodies and cause revamping of airway. IL-5 released from T-cells boosts the number of eosinophils [51, 52]. Growth factor- colony stimulated factor released from epithelial cells, macrophages and mast cells in turn enhances no. of neutrophils and eosinophils. TGF-β released from eosinophils, epithelial cells and macrophages increases deposition of fibrotic tissue in airway. SCF released from epithelial cells, smooth muscle cells, fibroblasts and eosinophils enhances number of mast cells. Vascular endothelial growth factors are released from epithelial cells that increase multiplication of blood vessels. Epithelial cell growth factors from epithelial cells help to stimulate remodelling of blood vessels and excessive secretion of mucus. Thymic stromal lymphopoietin (TSP) is a main cytokine unleashed from epithelium of airway that causes alterations in dendritic cells and allows them to present antigens efficiently to T-cells [53]. MHC-II on M1 macrophages are known for manufacturing cytokines generally secreted from TH1 and TH17 cells along with ROS, chemokines and monocyte attractant protein. Invasion of neutrophils inside lungs is encouraged by M1 pro-inflammatory macrophages. Another kind of macrophages named M2 macrophages show less amounts of MHC-II on their surface and show enhancement in levels of arginase, CD206 and CD163 that force intrusion of neutrophils into lungs [54, 55].

IL-4 production from T-cells brings about a rise in IgE antibodies and also enhances the number of TH2 cells. IL-17 released from T-cells enhances the quantity of neutrophils and positively impacts the production of cytokines from epithelium of airway. Several proinflammatory cytokines released from joint contribution of epithelial, macrophages and mast cells induce inflammation and also stimulate dendritic cells to enhance the number of TH2 cells [41]. T-cells bring inflammation at its peak by secreting cytokines and chemokines. These cytokines and chemokines enhance contraction of smooth muscles found within airway. Enhanced mucus secretion and hyperresponsiveness in airway abnormally excite T-cell multiplication in airways [56]. The inequality in TH1 and TH2 cells have a vital contribution in leading to progression of disease. Production of IFN-γ, IL-2 and lymphotoxin-α commences from TH1 cells that encourage type-I immunity. TH1 cells function as double-edged sword as they restrict inflammation caused by eosinophils but proceed the inflammation linked with neutrophils. Recruitment of TH1-cells towards inflammatory sites is promoted by CXC chemokine ligand 10 and CXC cognate receptor-3 hence providing shield against corticosteroid. Epithelium of airway secretes certain cytokines including IL-25, IL-33 and thymic stromal lymphopoietin that directly prompt TH2 cells. These stimulated TH2 cells further secrete more cytokines including IL-5, IL-4 and IL-13. B-cells tend to produce IgE antibodies. Excessive mucus is resulted after getting exposed to pollutants that stimulate aryl hydrocarbon receptor (AHR) under effect of IL-4 and IL-1. Inflammation resulting from eosinophils is sustained by IL-5 that maintains the existence and importance of eosinophils. Chemokine receptors and ligands present over TH2 cells play a central role in attracting eosinophils and AHR.

With the help of TLR and IFN-γ pathway, TH 17 cells produces IL-17F, IL-17A and IL-22 that exert proinflammatory effect on neutrophil stimulation. Generally, IL-17 has a curative role as it calls neutrophil at affected lung site to look after lungs, but this neutrophil recruitment worsens situation by causing abnormal immune response by linking it with asthma [57]. TH17 cells are increasingly called at injured site by controlling bond between CCR4 or CCR6 and CCL-220 [58]. Secretion of anti-inflammatory factors like IL-10 and TGF-β is accomplished from thymus natural T reg and peripheral induced T reg possessing CD25, CD4 and Foxp3. These T reg cells regulate immune cells from responding harshly [59].

4.6.2 COPD

A potential hazard factor for expansion of COPD is cigarette smoking. Smoking of cigarette causes release of CXCL9, CXCL10, CXCL11 and CXCL12 from epithelial cells, and macrophages attract TH1 cells and $CD8^+$ cells. There is increased production of IFN-γ from epithelial cells to destroy alveolar epithelial cells. Release of TGF-β and fibroblast growth factor (FGF) from epithelial cells increases fibrosis in lungs by promoting gathering of fibroblast inside lungs. Release of CXCL-8-chemokine from epithelial cells increases assembly of neutrophils inside airways. After getting exposed to smoke, epithelial cells also start to attract $CD8^+$ cells. Replacement of normal lung tissue by stiff fibrotic tissue is caused by macrophages that after receiving stimulus from smoke increase ROS and MMPs from neutrophils, macrophages and neutrophils [60]. Inflammatory reactions in airways and lungs parenchyma are triggered by cigarette smoke. PRR of innate immune system is activated under stimulation provided by smoke that in turn act as ligand for Toll-like receptors (TLRs). Sputum of COPD patients shows increased number of neutrophils [61]. Specifically, lower respiratory passage of COPD patients is inhabited by bacterial and viral pathogens thus causing provocation of inflammatory response [62]. Lymphoid tissue residing in bronchioles in lower respiratory tract exhibits an increase in quantity of B-cells as disease develops [62]. $CD8^+$ T-cells accumulation in airway showers destruction on alveolar epithelial cells. TNF-α, perforins, and granzymes coming from $CD8^+$ T-cells cause cell death in alveolar cells by stimulating Fas-Fas ligand pathway [63]. $CD8^+$ T-cells along with neutrophils and macrophages help in aggravating disease by causing spilling of various proteases named matrix metalloproteinases-9 and 12 (MMP-9, MMP-12) along with elastase which degrades the normal connective tissue constituting lung parenchyma [64]. Neutrophil don't show significant gathering in lungs as they make a rapid exit from airway and lung parenchyma. Serum proteinases comprising elastase, cathepsin-G and proteinase-3 along with MMP-8 and MMP-9 cause death of alveolar cells and provide stimulus for secretion of mucus. Increase in neutrophil count in blood vessels of lungs causes release of proteases and ROS that is detrimental for respiratory tract. Airway epithelial cells in patients with COPD show an increase in the number of E-selectin that allows neutrophil to bind with endothelial cells [65]. External factors like excessive smoking brings about a rise in count of neutrophils in blood. All neutrophils are attracted towards respiratory tract under stimulation of leukotriene B4, IL-8. CXC chemokines comprising growth-related oncogene-α and neutrophil attractant 78 also attracts neutrophils to respiratory tract [66]. Airways of COPD patients depict high levels of chemoattractant that acts as magnet to recruit inflammatory cells from various locations to respiratory tract. Smoking of cigarette uplifts level of neutrophils in blood of COPD patients and also confines neutrophils into capillaries of lungs [66]. The severity of COPD is decided by macrophage cells residing in airway. Lung parenchyma, airways and bronchoalveolar fluid of COPD patients bears about 5-10 times more macrophages than found in normal individuals [67]. Generally, COPD is divided into emphysema, chronic bronchitis and small airway disease. Inflammation in epithelium belonging to central airways and mucus-secreting glands are major features associated with chronic bronchitis. Inflammation of airway is accompanied by enhanced secretion of mucus, decrease in cilia count present in airway and enhanced permeability in blood vessels of epithelial barrier. However, secretion of excessive mucus as a cause of development in COPD is still suspected. Emphysema is another subtype of COPD that is caused by loss of elasticity in lungs and hence produce detrimental effects on expiratory air flow. There is expansion or destruction of bronchioles in centrilobular or centriacinar form of emphysema, while panlobular and panacinar form of emphysema version of emphysema is even more severe in causing destruction and expansion of the whole acinus and is linked with decrease in levels of α1 anti-trypsin [68]. Fibroblast and myofibroblast are central cells that cause pathogenic lung remodelling by causing the replacement of normal lung tissue by connective tissue deposition. When alveolar epithelium of lungs is subjected to recurring attacks of cigarette smoke along with toxicants present in environment, then it showers destruction an alveolar cell residing in epithelium of lungs. Gastro-oesophageal reflex also makes a person susceptible to develop COPD by destroying epithelial cells housed inside alveoli. Smaller air conduction are adversely affected in small airway disease of COPD. There is increase in number of $CD8^+$ T-cells and macrophages in bronchial tissues of COPD patients. CD8+ T-cells are found in elevated counts in lungs and blood of COPD patients. Peripheral airway also exhibits increased presence of $CD8^+$ T-cells that bear CXCR3 on their surface [62].

Cigarette smoking unleashes IL-8, TNF-α and other CXC chemokines comprising monocyte chemotactic peptide-1, and ROS from clutches of macrophages. Secretion of proteases like MMP-2, 9 and 11 cathepsin K, L and M is elicited from macrophages. Macrophages of COPD patients bear vigorous proteolytic activity which is further amplified by inhalation of cigarette smoke [69]. Airways and alveolar wall of smokers depict an enhanced population of dendritic cells. Inflammation is sustained in COPD patients by ensuring elevation of NF-κβ pathway in macrophages of alveoli. Peripheral airways in COPD patients show accumulation of fibrotic tissues in small airways. Peripheral airways and parenchyma of lungs depict high level of $CD4^+$ and $CD8^+$ lymphocytes [70, 71]. There is substantial destruction of gas exchange surface areas lying within lungs in COPD. MiR-15b, miR-223 and miR-1274A are confined within fibrotic and emphysematic spots of lungs. Severity of emphysema is defined by miRNA34c. Inflammatory reactions in lungs is exaggerated by $CD8^+$ cells whose existence is maintained by $CD4^+$ T-cells. IgG antibodies are present in blood of about 70% patients of COPD that are generally created against epithelial cells by B-cells under stimulation provided by CD4+ T-cells [72].

4.7 Idiopathic Pulmonary Fibrosis

Increased deposition of fibrotic tissues in lungs ruins the functioning of lungs and ultimately results in catastrophic events. There is a characteristic pattern of honeycomb-like structures that are found in lungs of these patients. Insult caused in alveolar epithelial cells helps them to release growth factors, cytokines and coagulant. Identity of mesenchymal cells is maintained through release of all these factors that pool lungs with myofibroblast and fibroblast causing accumulation of excessive extracellular matrix (ECM) in lungs. Any insult happening in lung tissue causes releases lysophosphatidic acid. Lysophosphatidic acid (LPA) occupies the receptor of LPA called LPAR-1 and attracts fibroblast at the site of action in lungs [73]. There is increase in levels of $\alpha_5\beta_6$ integrin after damage of lung tissue that stimulates the dormant TGF-β [74]. Fibrocytes are present inside blood are attracted towards lung after getting alarming signal from damage of lung tissue. There is enhancement in expression of *MUC5B* and *MMP-7* found inside genes of cilia residing in the lungs. Inflammatory and mesenchymal cells invading in lungs are cleared by apoptosis and phagocytosis. Idiopathic pulmonary fibrosis is characterised by injury to alveolar epithelial cells. Long-term inflammation found in lung fibrosis is associated with increased deposition of connective tissue containing fibrous cells. Tissue repair and fibrosis are majorly regulated by macrophages. Macrophages play both detrimental and protective role as they help in tissue healing and also contribute their share in injury. Normally INF-γ and TNF-α lead to stimulation of M1 cells. Exposure of Il-4, IL-13, IL-10 and TGF-β1 leads to stimulation of M2 macrophages. Patients afflicted with idiopathic pulmonary fibrosis show difficulty in breathing as gas exchanging region of lung is replaced by some other components of extracellular matrix. Experimental models of pulmonary fibrosis prove the possibility of gathering of fibroblast in lung tissue by causing epithelial to mesenchymal transition. Major features of lung fibrosis include deposition of extracellular matrix apparatuses, loss of basement membrane and rapid multiplication of mesenchyme that increases its amount in this disease. Failure in regeneration of epithelial cells belonging to lungs is responsible for unusual curative response. There is increased accumulation of leukocytes and enhanced blood vessel growth that is responsible for causing disintegration of integrity of endothelial cells and epithelial cells that in turn causes oedema [75].

4.8 Epigenetic Regulation of Inflammatory Lung Diseases

4.8.1 Asthma

There is a substantial involvement of diet taken by mother in development of diseases like asthma. Mothers ingesting diet rich in methylation-promoting supplements exhibit a strong increase in inflammation of airways by rising recruitment of eosinophils in airway along with spike in levels of IL-4 and IL-13. The mother receiving methylation supplement containing diets shows high levels of IgE in serum and extreme hypersensitivity in airways. After bringing person in contact with house dust mites, there is hypermethylation in gene encoding IL-4 and IFN-γ inside $CD4^+$ T-cells [76, 77].

Patients of asthma are marked by enhanced acetylation of histone H4 that in turn brings inflammation into picture by positively influencing inflammatory gene expression. There is drop-in activity of HDACs along with enhanced expression of HATs in asthma. By changing acetylation and deacetylation level of histones, glucocorticoids provide relief from inflammation happening in asthma [78]. Latest studies provide evidence that when mothers are exposed to folate which is a methylating agent, then it brings about a rise in susceptibility to develop asthma. Alteration in pattern of methylation is found in genes belonging to major histocompatibility complex-I, eotaxins, ILs, cytokines, eosinophil major granule protein and IgE receptors on mast cells following exposure to black carbon sulphate [79] Intake of methylation-promoting diet causes methylation of 82 gene-associated CpG islands and exaggerated hypermethylation of promoter of *RUNX-3* that in turn exerts its control over gene expression. Asthma disease shows an upregulation in 13 sites by utilising differential methylation [80]. Lung tissue of patients suffering from asthma exhibits rise in expression of HDAC-4 that removes acetyl group from Kruppel-like factor-4 KLF-E5 to bring about a surge in expression of slug and CXCL-12 thus revamping lung tissue by replacing it with fibrous tissue [81]. The different types of epigenetic mechanisms and their effects on expression of genes underlying various inflammatory lung diseases is outlined in Table 4.1. Addition of phosphate on Smad-3 under effect of TGF-β. Addition of phosphate on Smad-3 is often correlated with exaggerated expression of Sirtuin-6. There is a drop in transcriptional activity of C-Jun promoter by downregulating TGF-β expression [82]. There is a central involvement of methylation in development of T-regulatory cells that in turn channelise its effects by influencing progression of childhood airway disease [83]. Generally, asthma patients show high expression of H3K18 Ac and H3K9Me3 in patients suffering from asthma. Start sites of genes EGFR, STAT-6 and TP63 are controlled by H3K18Ac and H3K9Me. Sputum of asthmatic patients shows increase in expression of miR-223-3p, miR-629-3p and miR-142-3p. Blood eosinophil count (BEC) of asthmatic patients exhibits increased

Table 4.1 Epigenetic mechanism and their effects on gene expression with response caused in inflammatory lung disease

Disease	Mode of epigenetic regulation and their effects on genes	Response caused in disease	References
Asthma	Decreased methylation in genes of histidine decarboxylase that in turn brings a rise in secretion of histamine brings out inflammation	Rise in secretion of histamine	[88]
	Production of IL-13 and IL-15 exerts its influence on mRNA of TGF-β	Controls TGF-β secretion	[90]
	Rise in expression of HDAC-4 that removes acetyl group from Kruppel-like factor-4 KLF-E5 to bring about a surge in expression of slug and CXCL-12	Replacement of normal lung tissue by fibrous tissue	[81]
	Alteration in pattern of methylation is found in genes belonging to major histocompatibility complex-I, eotaxins, ILs, cytokines, eosinophil major granule protein and IgE receptors on mast cells following exposure to black carbon sulphate	Increases degranulation of mast cells	[79]
	H3K27Me3 decreases production of IL-4 in TH1 cells while in TH2 cells it supresses IFN-γ	Decreases IL-4 secretion in TH1 cells and repress secretion of IFN-γ in TH2 cells	[90]
	Genes*IL4, CCDC80, DAPK3, LOXL1, PROC, FUCA2, SP100, ITCH* characterised by rise in methylation	Increased presentation of antigen	[88]
	Decrease in methylation of *NDFIP2* and *RIPK-2*	Increases activity of TH cells	[87]
	Hypomethylation of *CCL-6, EPX* and *IL-13* in eosinophils	Increases activity of eosinophils	[91]
	Basophils exhibit hypermethylation in genes *PRG-2* and *PRG-3*	Increased activity of basophils	[87]
	Cytotoxic T-cells depict hypermethylation in *PRF-1* and hypomethylation in *CTSC*	Increase in activity of cytotoxic T-cells	[87]
	Decreased level of miR-449 brings a spike in mucus secretion	Increase in mucus secretion	[92]
	MiR-221 forces uses sirtuin signalling	Injury in alveolar cells	[93]
	Excessive expression of miR-221-3p	Block anti-inflammatory response	[94]
COPD	Alteration in methylation of HSH2D, SNX10CLIP-4 and TYKZ	Effects ventilation in upper and lower lobes lying within lungs	[102]
	Increase in methylation of *SERPINA1*	Excessive mucus secretion	[96]
			[98]

(continued)

Table 4.1 (continued)

Disease	Mode of epigenetic regulation and their effects on genes	Response caused in disease	References
	11 CpG islands inside foxaA2 show decrease in methylation	Malfunctioning at level of differentiation in goblet cells	
	Loss of 2 CpG islands in IGF	Negative influence on lung development	[101]
	Altered expression of S1 gene in smokers owing to abnormal methylation of S1P	Hypersecretion of proinflammatory cytokines from macrophages	[99]
	Decrease in miR-146a	Increases inflammatory cytokines by affecting toll-like receptor	[104]
	Decreased expression of mi-R186a	Increases HIF-1α and inhibits cell death of fibroblast	[106]
	Decrease in expression of miR-27-3p	Excessive secretion of pro-inflammatory cytokines	[107]
	Decreased expression of miR-503.	Stimulates release of VEGF from fibroblast located inside lung	[116]
	Decrease in levels of miR-483-5p. miR-1236, miR-206, miR-34a, miR-199a-5p, miR-424-5p show increase in their levels inside lungs of COPD patients	There is abnormal rise in levels of fibronectin and α-SMA in fibroblast	[110]
IPF	Diametric increase in both let-7d and miR-21 that exaggerate TGF-β1 that is considered to be an epicentre of fibrosis	Increases TGF-β to cause fibrosis	[111]
	Increase in miR155 and miR-21	Increase in fibrosis	[111, 112]
	miR-199, miR-377 and miR-299 show increase in their expression inside lungs of diseased patients and inhibits VEGF	Increases VEGF	[111]
	There is increase in fibrosis as shown by miR-21	Increase in fibrosis	[117]
	Let-7d, miR-200, miR-26a, miR-375	Changes epithelial cells into mesenchymal cells	[113]
	MiR-29 controls expression of many genes named *ELN, FBN-1, COL-1A1, COL1A-2, COL3A-1*	Control synthesis of ECM components	[113]
	lnRNAs uc.77 and 2700086A05Rik	Causes change of epithelial cell	[114]

expression of miR-629-3p. Neutrophils present inside sputum of asthmatic patients exhibit rise in expression of three miRNAs.

Different expression of 24 enzymes inside alveolar epithelial cells (AECs) and fibroblast present in bronchial lining are reported in asthmatic patients in comparison to healthy individuals. AECs of asthmatic patients show alteration in expression of six modifiers of histones in asthma patients [84]. Certain viral infections like human rhinovirus and influenza virus infection makes host prone to development of asthma by controlling gene expression by manipulating epigenetics. There is a malfunctioning in expression of miR-22 after getting infected from influenza virus infection. Nasal secretions of persons suffering from human rhinovirus infection causes emergence of mi-R55 in nasal secretions. These viral expressions are marked by changes happening at the level of epigenetics inside alveolar epithelial cells. Transition of mesenchymal cells into epithelial cell is an important event in development of pulmonary fibrosis as mesenchymal cells such as myofibroblasts are known for redesigning or replacing gaseous exchange alveolar surface of lungs by fibroblast and connective tissue accumulation. This transition is also strengthened by epigenetic alterations. The transition of epithelial cells into mesenchymal cells also regulates asthma development [85, 86]. Genes that control phagocytosis in blood involves *SERPINC1*. There are other genes found in granulocytes comprising *COL15A1*, *RB1*, *FOXP1* and *CCDC19* that control remodelling of airways. Genes like *ACOT7* and *PPT2* controls secretion of surfactant; meanwhile, *IL-5RA* and *DICER-1* control manufacture and signalling emerging from cytokines. All the genes show their engagement in processes of phagocytosis of blood, remodelling of airway, nitric oxide manufacture, and secretion of surfactant, and cytokines show drop in methylation [87]. Basophils and mast cells are characterised by decreased methylation in genes of histidine decarboxylase. Decreased methylation in turn brings a rise in the secretion of histamine bring out inflammation. There is demethylation of gene promoter inside TH2 cells that brings about re-secretion of IFN-γ from these cells [88]. There is enhancement in methylation of *FOXP3* present inside T-reg cells of blood following their exposure to pollutants found in environmental air. The genes that are linked with enhancement in the presentation of antigens include genes of *IL-4, CCDC80, DAPK3*, *LOXL1*, *PROC*, *FUCA2*, *SP100*, *ITCH* characterised by rise in methylation [88]. B-cells elicit production of IgE antibodies in asthmatic patients. A gene called *CYP26A1* offers its share in clearing retinoic acid thus causing hypermethylation of B-cells in patients suffering from asthma [89]. Severe asthma generally involves HDAC-1 that revamps and repairs the airway cells. There is spike in expression of *EGFR* and *STAT-6* inside lung epithelium of patients owing to increased acetylation at histone H3 [89]. Transcription of both IFN-γ and IL-4 is enhanced after trimethylation of H3K4. Differentiation of $CD4^+$ T-cells is positively stimulated by methylation of histones. Depending on the location of methylation of histone 3 H3K27Me3, it brings about different response. H3K27Me3 decreases production of IL-4 in TH1 cells while in TH2 cells it supresses IFN-γ [90]. Smooth muscles present inside airways of asthmatic patients show drop in methylation in H3K27Me3 by employing JMJD2D promoter demethylase at promoter of VEGF gene. Around genes of dendritic cells, the place at which H3K4me3 and H3K27me3 are found decides their change into APCs. Reconstruction of airways is caused owing to rise in expression of TGF-β2. Elevation in expression of TGF-β2 is caused by miR-19 elevation. Patients of asthma show rise in levels of miR-21 and miR-26. The downregulation of miR-21, miR-625-5p and miR-513a-5p in peripheral blood cells of asthma patients leads to alterations in production and multiplication of eosinophils. An abnormal rise in expression of VEGFA has been reported in sputum and serum of asthmatic patients. VEGFA increases growth of blood vessels that as a result brings more immune cells into lungs utilising blood as vehicle thus casting an epic catastrophe on lungs that ultimately leads to inflammation. Genes like *CBL*, *PPARGC1B* and *ESR-1* belonging to PI3K-AKt and NF-κβ signalling pathways are decreased by effects of these miRNAs that decrease their expression which in turn lower the expression of IFN-γ, IL-12, TNF-α and IL-10 in plasma. miR-221 are responsible for increasing inflammation of airways and also causes mast cell degranulation. Expression of miR-21 in bronchial epithelium is found its association with rise in expression of IL-13. miR-19 present inside T-cells of airway and lung tissue increases the manufacture of IL-13 and IL-15 and also produces its influence on mRNA of TGF-β [90]. Neutrophils of asthmatic patients show addition of methyl residues in genes. There is decrease in methylation of genes *PSD-4* and *SLC25A33* and decrease in methylation of *LAG-3* and *ID-1* in dendritic cells. There is hypomethylation of *CCL-6, EPX* and *IL-13* in eosinophils [91]. Basophils exhibit hypermethylation in genes *PRG-2* and *PRG-3*. Cytotoxic T-cells depict hypermethylation in *PRF-1* and hypomethylation in *CTSC*. TH cells exhibit decrease in methylation of *NDFIP2* and *RIPK-2*. B-cells show decrease in methylation in genes named *ARID3A, BANK-1, GPI, IL-13,LFNG,MMP-14* and *PRKCH* [87]. Control of ciliated cells lying in respiratory epithelium falls under miR-449. Decreased level of miR-449 brings a spike in mucus secretion [92]. miR-221 forces injury in alveolar cells by utilising sirtuin signalling [93]. Exaggerated expression of miR-221-3p negatively influences anti-inflammatory response and worsens inflammation caused by eosinophils [94].

4.8.2 COPD

The functioning of the lungs is ruined owing to the presence of defective histone methylation happening at the promoter of genes *p16* or *GATA-4* as found in sputum. Health level of persons suffering from COPD is assessed by methylation happening in genes of *GATA-4* belonging to airway. Increased inhalation of smoke enhances methylation occurring at *SULF-2* thus causing long-term secretion of mucus. DNMT-1, DNMT-3 and DNMT-3b are responsible for performing defective methylation in COPD. Inspiration of air rich in cigarette smoke causes decrease in expression of DNMT-1 and heightens DNMT-3b in normal airway epithelial cells and human bronchial epithelial cells. Hypomethylation happening in genes *D4Z4, NBL-2* genes and LINE-1 repetitive DNA sequences is forced by cigarette smoking in time-reliant manner. Expression of let-7c and miR-125b decreases in COPD patients in comparison to healthy persons. There is increase in hypermethylation of mitochondrial transcription factor-A (mtTFA) promoter following inspiration of cigarette smoke. The presence of particulate matter in air having size less than <10 μm is responsible for causing alteration in methylation of about 12 probes and 27 CpGs. These alterations in methylation of CpG are reported to occur in certain genes including *NEGR1, ARID5A, FOX12, WDR46, AKNA and SYTL-2* [90]. There are about 45 differentially methylated probes and 57 differentially methylated regions in CpG of genes of fibroblast present in airways comprising ERI3, *RPL5, CPLX1* and *STON1* [95]. About 652 differentially methylated regions are present inside genes of TMEM44, *RPH3AL*, *WNT3A*, *HLA-DP1* and *HLA-DRB5*. Patients of COPD exhibit increase in methylation of three CpG sites found in *LX* (H2.0 like Homeobox) genes in concordance with decrease in methylation of *NXN* (nucleoprotein) gene. Smoking causes methylation of *SERPINA* gene at two CpG sites. Exaggerated multiplication of goblet cells combined with secretion of mucus in abundance together causes COPD [96]. Genes like *FOXA2* and transcription factors SPDEF play an important role in controlling differentiation of goblet cells. About 11 CpG islands inside *FOXA2* show decrease in methylation. Six CpG islands in SPDEF also show drop in methylation. Differentiation of goblet cells in lungs is negatively influenced by foxA2 while manufacture of goblet cells and their differentiation falls under control of SPDEF [97, 98]. Macrophages isolated from COPD patients show altered expression of S1 gene in smokers owing to abnormal methylation of S1P and disturb functioning of macrophages to elicit release of pro-inflammatory cytokine [99]. Persons suffering from COPD and doing cigarette smoke show rise in methylation of *NOS1AP*, *TNFAIP2*, *GABRB1* and *BID* [100]. Hypomethylation is observed in genes belonging to *AHRR* and *SERPINA-1*. Hypermethylation of *SERPINA1* causes spike in secretion of mucus and manufacture of goblet cells [101, 102]. Ninety-seven per cent of DNA methylation probes show rise in methylation of small airway cells genes including three cholinergic receptors *CHRND*, *CHRNB2* and *CHRNB1* [96]. Smoking of cigarette causes acetylation of H3 in macrophages and lungs of patients. There is rise in expression of HDAC-1 and -2 in mice breathing air rich in cigarette smoke. Smoke inhalation causes loss of methylation of two CpG sites lying in *IGFR-2* promoters in COPD and negatively influences development of lungs [101]. Arginine methyl transferases (PRMTs) are responsible for causing methylation of arginine residues in histone and non-histone proteins. Antagonistic to PRMTs, CARM1 removes methyl residues from both histone and non-histone resides. Insult of epithelial cells as seen in COPD patients causes attenuation in levels of CARM [102, 103]. Sirtuins are important in controlling inflammation and translate their effect at epigenetic level by means of utilising their HDAC activity. There is increased partnership between foxo-1 and sirtuins in COPD patients which together exerts their effect by negatively influencing NF-κβ pathway. Cigarette smoking increases the level of ROS and increases the resistance to glucocorticoid by uplifting levels of HDAC-2. There is increase in the accumulation of collagen in bronchial epithelium following rise in expression of IL-17A and HDAC-2. miRNAs also offer their share in development of COPD. Certain long non-coding RNAs show rise in their expression in COPD. Expression of miR-34a and miR-199a-5p exhibits a rise in their expression inside lungs of COPD patients. In a nutshell, there is a decrease in expression of miR-146a, miR-186, miR-181a-2-3p, miR-197 and miR-27-3p. Inflammatory cytokines are unleashed by TLR as miR-146a controls it. In adenocarcinoma alveolar basal epithelial cell lines, downregulation of miR-146a results in upregulation of *IRAK-*

6, TNF and TRAF-6. miR-146a excessively stimulates toll-like receptor to release pro-inflammatory cytokines [104]. miR-181a-2-3p forces an increase in stimulation of inflammasome to elicit inflammatory response [105]. Fibroblasts escape from cell death by utilising decrease in expression of miR-186. miR-186 also increases the expression of HIF-1α in endothelial cells and smooth muscle cells constituting respiratory passage [106]. Decrease in expression of miR-27-3p is associated with excessive secretion of pro-inflammatory cytokines [107]. Decreased expression of miR-503 positively stimulates release of VEGF from fibroblast located inside lung. There is boost in levels of fibronectin and α-SMA following attenuation of miR-483-5p. mi-R1236, miR-206, miR-34a, miR-199a-5p and miR-424-5p show increase in their levels inside lungs of COPD patients. Both miR-34a and miR-199a-5p aim for Notch-1 gene in endothelial cells by binding with 3' UTR region of this gene. By utilising TLR, IRAK-4, IRAK-6 and TRAF-6 can aggravate pathology of COPD. Rise in expression of miR-34a, miR-26 and miR-199a blocks the activity of Notch-intracellular domain (NICD) thus worsening pathology of COPD [103]. There is prominent participation of NF-κβ signalling pathway in development of COPD and asthma. Seven lysine residues of NF-κβ undergo acetylation under instructions received from CBP/p300 transferases. Activity of NF-κβ is positively affected by HDAC-3 that removes acetyl residues from lysine 122, 23, 314 and 315. NF-κβ also interacts with HDAC-1 and HDAC-2.

4.8.3 Idiopathic Pulmonary Fibrosis

Genetic factors, environmental factors and deterioration of healthy cells with increased changes force alterations at epigenetic levels. IPF is caused by alterations in genes encoding TERT, PARN and RTEL, desmoplakin, dipeptidyltransferase-9 and *MUC-5B*. These alterations are associated with emergence of IPF. Air rich in Cd and carbon black causes citrullation of vimentin protein inside lungs that increases invasion of fibroblast inside lungs. Concerted effect of environmental toxicants, increased age and alteration in gene manifest its effect by increasing harm imposed on lung epithelial cell. Wide variety of growth factors named PDGF, CTGF and TGF-β increase the dysfunction happening in lung. Smoking of cigarette is the main environmental risk factor that worsens the progression of diseases by manipulating epigenetics to exert its effects at various genes. Smoke inhalation causes addition of methyl residues in genes such as Wnt-7a [108]. Myofibroblasts are the main culprit cells responsible for accumulation of fibrotic tissues inside lungs. They differ from normal healthy fibroblast owing to presence of different kind of α-SMA that is non-contractile in its function. Increased methylation of *Thy-1 gene* along with the genes encoding α-SMA also positively influences the differentiation of myofibroblast and enhances accumulation of collagen inside lung tissue. Replacement of normal gaseous exchanging lung tissue by fibrotic tissue decreases the efficiency of respiratory exchange inside lungs. This replacement involves accumulation of components of ECM including collagen, laminin and integrin which are secreted at diseased site by myofibroblast and fibroblast. The histone modifications such as H3K9Me3 and H3K27Me3 in the fibroblasts further leads to structural changes in IPF. IPF is also a kind of pulmonary fibrosis whose causes are generally undefined. Human fibroblast present already inside lung tissue shows increased methylation at *Thy-1* gene promoter and gene encoding *α-SMA* to switch identity of fibroblast residing inside lungs towards myofibroblast [109]. Fibroblast, myofibroblast and AEC-II exhibits alteration in three CpG sites lying within gene promoters of α-SMA that coincides with gene expression activity of α-SMA [109]. Myofibroblasts are forced to synthesise components of ECM under influence of increased HDAC-4 activity. Inhalation of smoke worsens the pathology of IPF by increasing change of epithelial cells into mesenchymal cells as it downregulates expression of E-cadherin on epithelial cells. TGF-β1 is the main cytokine that is responsible for causing fibrosis and inflammation. DNMT-1 adds methyl residues on gene encoding miR-17–92 cluster to pool lungs with excessive amounts of collagen. H3K9 methylation and acetylation of histone H3 cause increase in cell death of fibroblast by targeting genes encoding *BAK, Bcl-xL* and *Fas*. Fibrosis is marked by change in identity of epithelial cell in which they lose their epithelial characteristics to change into mesenchymal cells. These mesenchymal cells differ from normal lung epithelial cells from the fact that they actively participate in fibrosis by increasing accumulation of cells found in ECM. Genes encoding *TLR-9* found on surface of lung fibroblast show decrease in methylation of CpG islands that in turn helps epithelial cells to change into mesenchymal cells which synthesise more fibrotic tissues. TLR-9 are also important constituent of innate immunity. A gene called Tollip show its participation in innate immunity and inflammation. Hypermethylation of two intronic DNA-methylated regions decreases expression of *Tollip* by about 11%. There is disturbance in key processes including autophagy and apoptosis owing to alteration in methylation of genes like *NOS1AP* and *BID* thus boosting their expression. Sputum of fibrosis patients exhibits increase in methylation of gene promoter of *SULF-2* [110]. Acrolein is an essential constituent of cigarette that brings about decrease in acetylation of H3K9 and H3 K14 of histone-3 residing at several genes.

Lungs of IPF patients differ from normal healthy individuals by showing difference in expression of about 10% of miRNAs. Let-7d is an miRNA that is required for maintaining normal vital capacity of lungs. Promoter of Let-7d contains an SMAD binding element (SBE) in its promoter. There is decrease in expression of Let-7d in lungs of COPD patients that changes structure of epithelial cells into mesenchymal cells. Lungs of IPF patients show diametric increase in both let-7d and miR-21 that exaggerates TGF-β1 that is considered to be an epicentre of fibrosis. miR-199, miR-377 and miR-299 show increase in their expression inside lungs of diseased patients and inhibit VEGF. miR-21 also shows an increase in expression of lungs of IPF patients that negatively influences SMAD-7. miR-495 also show spike in their expression and stop *FZD8* and *THBS-2* from expressing themselves. miR-181, miR-203, miR-17/20, miR-29, miR-30, miR-338, miR-224 and let-7 show downfall in their expression. miR-29 inhibits genes including *VANGL-1* and *IGF-1*. Let-7 blocks the activity of HMGA2, *STEAP-3* and *RPF-2*. MiR-181 negatively influences expression of GLI-2 human glioma-associated homologue-2, VANGL-1 and

SMAD-7. miR-203 stops activity of SFRP-1 and PXN. miR-17/20 inhibits expression of SMAD-7, RRM-2 and HGMA-2. miR-224 negatively regulates genes called SFRP-2 and SUFU [111]. There is increase in fibrosis as shown by increase in miR155 and miR-21 [111, 112]. Transformation of epithelial identity into mesenchymal is caused by certain miRNAs including let-7d, miR-26a, miR-200, miR-375. These micro-RNAs controls progression of IPF. Family of miR-30 show decrease in their expression inside lungs of patients suffering from IPF. MiR-29 controls expression of many genes named *ELN, FBN-1, COL-1A1, COL1A-2, COL3A-1* that in turn control ECM components [113]. There is decrease in synthesis of ECM induced by TGF-β1 as performed by miR29. DNMT-1 (DNA methyltransferase-1) adds methyl residues on genes named *CTGF, COL1A1 and COL13A1* to decrease their levels following exposure of lung fibroblast with miR-17–92. Transcription of most of miRNAs comprising miR-154 and let-7d is controlled by Smad-3. Several lnRNAs uc.77 and 2700086A05Rik causes change of epithelial cells into mesoderm by controlling expression of genes named *Zeb-2* and *Hoxa-3*[114]. There about 358 ln-RNAs that show drop in their expression while about 210 lnRNAs are increased in patients with pulmonary fibrosis [115]. A long non-coding -RNA named MRAK088388 controls N4bp2. whereas another lncRNA named MRAK081523 binds with let-7i-5p and controls the expression of plxna4 [6]. Binding sites of several Mi-RNAs comprising miR-21, miR-199 miR-101, miR-31, miR-29 and let-7d are found inside 34 lnRNAs [6].

4.9 Targeting Epigenetics: Novel Epigenetic Therapy for Inflammatory Lung Disease

The importance of epigenetics comes into picture as our respiratory tract is constantly exposed to different types of environmental constituents with each inhalation and exhalation. Respiratory tract is composed of buccal, mucosal, nasal and bronchial epithelium. When DNA methylation and acetylation are targeted, they can give rise to side effects by extending their effects over immune cells function and other cell types. MiRNA targeting is preferrable as it minimises the number of genes and is much more specific that DNA methylation and acetylation targeting. These T-regulatory guide macrophages to differentiate along anti-inflammatory lane and also influence functioning of neutrophils. T-regulatory cells exert their effect on T-cells by downregulating expression of pro-inflammatory cytokines like TNF-α and IFN-γ. Lung injury is alleviated by using DNA methylase inhibitor named 5-aza-29 deoxycytidine that exerts its protective effects by regulating effects of Treg cells. Number of T-reg cells is increased following treatment with DAC [118]. There is substantial amount of evidence that indicates γδ T-cells, Tc-17 cells and ILC that counteract the tissue damage evoked by inflammation achieve their resolving property by utilising reprogramming of chromatin at epigenetic level. HDAC-2 inhibitors are chosen for treating asthma and hence provide relief. Certain drugs known for targeting HDAC-2 are used in treatment of COPD. These drug-targeting activity of HDAC-2 includes trichostatin, CARM-1, sirtuins, LL-23, theophylline, statin and curcumin. Inhibitors of HDAC-3 are preferred for treating COPD and asthma due to their ability to attenuate inflammation. Expression of anti-inflammatory cytokine IL-10 is attributed to erinostat that inhibits HDAC-1,2 and 3 thus causing spike in acetylation of NF-κβ gene. PCI-34051 is an HDAC-8-inhibitor that reduces hyperresponsiveness, remodelling of lung tissue and inflammation. Generally, HDAC-6 and HDAC-8 are targeted to improve health of COPD and lung fibrosis patients. Specifically, HDAC-6 is downregulated by tubastatin-A and HDAC-8 is downregulated by PCI-34051. Targeting of HDAC-1,2 and 3 is accomplished by o-aminoanilide zinc-binding group (ZBG) [1]. Ivermectin is also a kind of HDAC inhibitor that decreases production of inflammatory cytokines and hence provides relief against asthma. Acetylation of non-histone proteins like heat shock protein-20 (HSP20) and cortactin is promoted by certain HDAC inhibitor that causes dilation of blood vessels inside bronchioles. COPD patients show decrease in mRNA encoding HDAC-2, 5 and 8 in macrophages. Macrophages of COPD patients have lesser levels of HDAC-2. Increased H3 and H4 acetylation is found in patients of COPD. Smooth muscles of airways found in VEGF promoter are associated with drop in expression of VEGF in COPD. There are certain DNMT inhibitors that are used in treatment of asthma and COPD. Drugs like 5-azacytidine and decitabine are some of known DNMT inhibitor that are used in treatment of both COPD and asthma [119, 120]. SAHA, OSU-HDAC-44 and LBH589 are the drugs used in treatment of COPD and asthma. Quercetin and theophylline are plant constituents that brings about a rise in level of HDACs in diseases like COPD [121, 122]. Histone methyl transferase inhibitors called BIX-1294 and 3-Deazaneplanocin inhibits activity of this enzyme in fibroblast present inside lungs [123]. JQ-1 is an inhibitor of bromodomain containing proteins that show its treatment effects by targeting fibroblast present inside lungs [124]. Generally systemic delivery of MiRNA is preferred in comparison to enzymatic inhibitors as HDAC and DNMT inhibitor causes methylation and acetylation changes outside genome of diseased cells also thus eliciting some undesirable effect. mi-RNAs are generally chosen for treatment as they target certain genes only and hence their effects are very specific and limited to diseased location only. Antagomirs are certain single-stranded nucleic acids that bind with members of mi-RNAs thus causing decrease in expression of that specific miRNAs. Delivery of mi-RNAs inside body of individual is very difficult in comparison to blocking their activity [125]. Hence, in a nutshell, miRNA targeted therapy is preferred over HDAC inhibitors and DNMT inhibitors to treat lung inflammatory disease.

4.10 Conclusion

Lung inflammatory disease includes immune cells as their key regulators in showering disease. Earlier only alteration in gene expression along with aggressive response caused by immune cells was thought to be responsible for causing COPD. The aggressive response offered by immune cells creates an environment rich in cytokines, chemokines and other inflammatory mediators that manifests its effect by comprising normal healthy cells of lungs along with diseased cells of lungs. Generally immune cells are protective but exaggerated, and overamplified response of immune cell sacrifices healthy cells along with malfunctioning cells in process. Epigenetics generally sits at back sit of genetic portion to control its activity by utilising its weapons including maternal diet and quality of air inhaled by a person. Epigenetics generally includes non-coding portions of genome that engender its effect on coding portion of DNA. Development and differentiation of immune cells also lies in hand of epigenetic alterations happening in genes. Growth of blood vessel caused by presence of VEGF is an epic event that brings all the immune cells from circulation to their site of action. This blood brings with itself a whole population of innate and adaptive immune cells that aggravates the progression of most lung disease. Besides, many MiRNAs are known to control several genes and signalling pathway and hence cause complete stoppage in expression of genes or degrade them. Some enzymes including histone methylases and demethylases along with histone acetyltransferase and deacetyl transferase control conversion of transcriptionally inactive heterochromatin into euchromatin and vice versa. Hence, role of epigenetics in addition to genetics is increasingly recognised as it controls the activity of several genes involved in lung diseases. Several inhibitors of these epigenetics controlling enzymes have shown their potential contribution in progression of several disease. miRNAs are generally a favourable option for treating lung inflammatory disease as they target genes specifically and control their activity without causing an undesirable side effect.

References

1. Zwinderman MRH, de Weerd S, Dekker FJ. Targeting HDAC complexes in asthma and COPD. Epigenomes. 2019;3(3):19.
2. Dharmage SC, Perret JL, Custovic A. Epidemiology of asthma in children and adults. Front Pediatr. 2019;7:246.
3. Sauleda J, et al. Idiopathic pulmonary fibrosis: epidemiology, natural history, phenotypes. Med Sci (Basel). 2018;6(4):110.
4. Navaratnam V, et al. The rising incidence of idiopathic pulmonary fibrosis in the U.K. Thorax. 2011;66(6):462–7.
5. Lederer DJ, Martinez FJ. Idiopathic pulmonary fibrosis. N Engl J Med. 2018;378(19): 1811–23.
6. Huang C, Yang Y, Liu L. Interaction of long noncoding RNAs and microRNAs in the pathogenesis of idiopathic pulmonary fibrosis. Physiol Genomics. 2015;47(10):463–9.
7. Van den Bossche J, et al. Macrophage polarization: the epigenetic point of view. Curr Opin Lipidol. 2014;25(5):367–73.
8. Kumaki Y, et al. Analysis and synthesis of high-amplitude cis-elements in the mammalian circadian clock. Proc Natl Acad Sci U S A. 2008;105(39):14946–51.
9. Heinz S, et al. Simple combinations of lineage-determining transcription factors prime cis-regulatory elements required for macrophage and B cell identities. Mol Cell. 2010;38(4): 576–89.
10. Patel DJ. A structural perspective on readout of epigenetic histone and DNA methylation Marks. Cold Spring Harb Perspect Biol. 2016;8(3):a018754.

11. Lavin Y, et al. Tissue-resident macrophage enhancer landscapes are shaped by the local microenvironment. Cell. 2014;159(6):1312–26.
12. Li R, et al. MEG3-4 is a miRNA decoy that regulates IL-1beta abundance to initiate and then limit inflammation to prevent sepsis during lung infection. Sci Signal. 2018;11(536): eaao2387.
13. Akira S, Uematsu S, Takeuchi O. Pathogen recognition and innate immunity. Cell. 2006;124 (4):783–801.
14. Yan Q, et al. Nuclear factor-kappaB binding motifs specify toll-like receptor-induced gene repression through an inducible repressosome. Proc Natl Acad Sci U S A. 2012;109(35): 14140–5.
15. Chen X, et al. The NF-kappaB factor RelB and histone H3 lysine methyltransferase G9a directly interact to generate epigenetic silencing in endotoxin tolerance. J Biol Chem. 2009;284(41):27857–65.
16. Seeley JJ, et al. Induction of innate immune memory via microRNA targeting of chromatin remodelling factors. Nature. 2018;559(7712):114–9.
17. Bullwinkel J, et al. Epigenotype switching at the CD14 and CD209 genes during differentiation of human monocytes to dendritic cells. Epigenetics. 2011;6(1):45–51.
18. Quintin J, et al. Candida albicans infection affords protection against reinfection via functional reprogramming of monocytes. Cell Host Microbe. 2012;12(2):223–32.
19. Medoff BD, Thomas SY, Luster AD. T cell trafficking in allergic asthma: the ins and outs. Annu Rev Immunol. 2008;26:205–32.
20. Sidler C, et al. Immunosenescence is associated with altered gene expression and epigenetic regulation in primary and secondary immune organs. Front Genet. 2013;4:211.
21. Degerman S, et al. Immortalization of T-cells is accompanied by gradual changes in CpG methylation resulting in a profile resembling a subset of T-cell leukemias. Neoplasia. 2014;16 (7):606–15.
22. Murayama A, et al. A specific CpG site demethylation in the human interleukin 2 gene promoter is an epigenetic memory. EMBO J. 2006;25(5):1081–92.
23. Northrop JK, et al. Epigenetic remodeling of the IL-2 and IFN-gamma loci in memory CD8 T cells is influenced by CD4 T cells. J Immunol. 2006;177(2):1062–9.
24. Wright KL, Ting JP. Epigenetic regulation of MHC-II and CIITA genes. Trends Immunol. 2006;27(9):405–12.
25. Schoenborn JR, et al. Comprehensive epigenetic profiling identifies multiple distal regulatory elements directing transcription of the gene encoding interferon-gamma. Nat Immunol. 2007;8 (7):732–42.
26. Cohen CJ, et al. Human Th1 and Th17 cells exhibit epigenetic stability at signature cytokine and transcription factor loci. J Immunol. 2011;187(11):5615–26.
27. Akimzhanov AM, Yang XO, Dong C. Chromatin remodeling of interleukin-17 (IL-17)-IL-17F cytokine gene locus during inflammatory helper T cell differentiation. J Biol Chem. 2007;282(9):5969–72.
28. Baron U, et al. DNA demethylation in the human FOXP3 locus discriminates regulatory T cells from activated FOXP3(+) conventional T cells. Eur J Immunol. 2007;37(9):2378–89.
29. van Loosdregt J, et al. Regulation of Treg functionality by acetylation-mediated Foxp3 protein stabilization. Blood. 2010;115(5):965–74.
30. Maier H, et al. Early B cell factor cooperates with Runx1 and mediates epigenetic changes associated with mb-1 transcription. Nat Immunol. 2004;5(10):1069–77.
31. Crouch EE, et al. Regulation of AID expression in the immune response. J Exp Med. 2007;204 (5):1145–56.
32. Caganova M, et al. Germinal center dysregulation by histone methyltransferase EZH2 promotes lymphomagenesis. J Clin Invest. 2013;123(12):5009–22.
33. Good-Jacobson KL. Regulation of germinal center, B-cell memory, and plasma cell formation by histone modifiers. Front Immunol. 2014;5:596.

34. Luckey CJ, et al. Memory T and memory B cells share a transcriptional program of self-renewal with long-term hematopoietic stem cells. Proc Natl Acad Sci U S A. 2006;103(9): 3304–9.
35. Weng NP, Araki Y, Subedi K. The molecular basis of the memory T cell response: differential gene expression and its epigenetic regulation. Nat Rev Immunol. 2012;12(4):306–15.
36. Steinfelder S, et al. Epigenetic modification of the human CCR6 gene is associated with stable CCR6 expression in T cells. Blood. 2011;117(10):2839–46.
37. Araki Y, et al. Genome-wide analysis of histone methylation reveals chromatin state-based regulation of gene transcription and function of memory CD8+ T cells. Immunity. 2009;30(6): 912–25.
38. Kuraoka M, et al. Activation-induced cytidine deaminase mediates central tolerance in B cells. Proc Natl Acad Sci U S A. 2011;108(28):11560–5.
39. Ogbomo H, et al. Histone deacetylase inhibitors suppress natural killer cell cytolytic activity. FEBS Lett. 2007;581(7):1317–22.
40. Pelaia G, et al. Cellular mechanisms underlying eosinophilic and neutrophilic airway inflammation in asthma. Mediat Inflamm. 2015;2015:879783.
41. Barnes PJ. The cytokine network in asthma and chronic obstructive pulmonary disease. J Clin Invest. 2008;118(11):3546–56.
42. Castillo JR, Peters SP, Busse WW. Asthma exacerbations: pathogenesis, prevention, and treatment. J Allergy Clin Immunol Pract. 2017;5(4):918–27.
43. Schleimer RP, et al. Epithelium: at the interface of innate and adaptive immune responses. J Allergy Clin Immunol. 2007;120(6):1279–84.
44. Perera F, et al. Relation of DNA methylation of 5'-CpG island of ACSL3 to transplacental exposure to airborne polycyclic aromatic hydrocarbons and childhood asthma. PLoS One. 2009;4(2):e4488.
45. Prunicki M, et al. Exposure to NO(2), CO, and PM(2.5) is linked to regional DNA methylation differences in asthma. Clin Epigenetics. 2018;10:2.
46. Zhou J, et al. PM(2.5) exposure and cold stress exacerbates asthma in mice by increasing histone acetylation in IL-4 gene promoter in CD4(+) T cells. Toxicol Lett. 2019;316:147–53.
47. Boyce JA. Mast cells: beyond IgE. J Allergy Clin Immunol. 2003;111(1):24–32. quiz 33
48. Galli SJ, et al. Mast cells as "tunable" effector and immunoregulatory cells: recent advances. Annu Rev Immunol. 2005;23:749–86.
49. Peters-Golden M. The alveolar macrophage: the forgotten cell in asthma. Am J Respir Cell Mol Biol. 2004;31(1):3–7.
50. Wenzel SE, et al. Bronchoscopic evaluation of severe asthma. Persistent inflammation associated with high dose glucocorticoids. Am J Respir Crit Care Med. 1997;156(3 Pt 1): 737–43.
51. Oh CK, et al. Biology of the interleukin-9 pathway and its therapeutic potential for the treatment of asthma. Inflamm Allergy Drug Targets. 2011;10(3):180–6.
52. Stirling RG, et al. Interleukin-5 induces CD34(+) eosinophil progenitor mobilization and eosinophil CCR3 expression in asthma. Am J Respir Crit Care Med. 2001;164(8 Pt 1):1403–9.
53. Ishmael FT. The inflammatory response in the pathogenesis of asthma. J Am Osteopath Assoc. 2011;111(11 Suppl 7):S11–7.
54. Karta MR, et al. LPS modulates rhinovirus-induced chemokine secretion in monocytes and macrophages. Am J Respir Cell Mol Biol. 2014;51(1):125–34.
55. Girodet PO, et al. Alternative macrophage activation is increased in asthma. Am J Respir Cell Mol Biol. 2016;55(4):467–75.
56. Ayakannu R, et al. Relationship between various cytokines implicated in asthma. Hum Immunol. 2019;80(9):755–63.
57. Chesne J, et al. IL-17 in severe asthma. Where do we stand? Am J Respir Crit Care Med. 2014;190(10):1094–101.
58. Louten J, Boniface K, de Waal Malefyt R. Development and function of TH17 cells in health and disease. J Allergy Clin Immunol. 2009;123(5):1004–11.

59. Strickland DH, Holt PG. T regulatory cells in childhood asthma. Trends Immunol. 2011;32(9): 420–7.
60. Rodrigues SO, et al. Mechanisms, pathophysiology and currently proposed treatments of chronic obstructive pulmonary disease. Pharmaceuticals (Basel). 2021;14(10):979.
61. Keatings VM, et al. Differences in interleukin-8 and tumor necrosis factor-alpha in induced sputum from patients with chronic obstructive pulmonary disease or asthma. Am J Respir Crit Care Med. 1996;153(2):530–4.
62. Kim WD, et al. Abnormal peripheral blood T-lymphocyte subsets in a subgroup of patients with COPD. Chest. 2002;122(2):437–44.
63. Majo J, Ghezzo H, Cosio MG. Lymphocyte population and apoptosis in the lungs of smokers and their relation to emphysema. Eur Respir J. 2001;17(5):946–53.
64. Shapiro SD, et al. Neutrophil elastase contributes to cigarette smoke-induced emphysema in mice. Am J Pathol. 2003;163(6):2329–35.
65. Di Stefano A, et al. Upregulation of adhesion molecules in the bronchial mucosa of subjects with chronic obstructive bronchitis. Am J Respir Crit Care Med. 1994;149(3 Pt 1):803–10.
66. Traves SL, et al. Increased levels of the chemokines GROalpha and MCP-1 in sputum samples from patients with COPD. Thorax. 2002;57(7):590–5.
67. Russell RE, et al. Release and activity of matrix metalloproteinase-9 and tissue inhibitor of metalloproteinase-1 by alveolar macrophages from patients with chronic obstructive pulmonary disease. Am J Respir Cell Mol Biol. 2002;26(5):602–9.
68. Kim WD, et al. Centrilobular and panlobular emphysema in smokers. Two distinct morphologic and functional entities. Am Rev Respir Dis. 1991;144(6):1385–90.
69. Lim S, et al. Balance of matrix metalloprotease-9 and tissue inhibitor of metalloprotease-1 from alveolar macrophages in cigarette smokers. Regulation by interleukin-10. Am J Respir Crit Care Med. 2000;162(4 Pt 1):1355–60.
70. Finkelstein R, et al. Alveolar inflammation and its relation to emphysema in smokers. Am J Respir Crit Care Med. 1995;152(5 Pt 1):1666–72.
71. Saetta M, et al. CD8+ve cells in the lungs of smokers with chronic obstructive pulmonary disease. Am J Respir Crit Care Med. 1999;160(2):711–7.
72. Nurwidya F, Damayanti T, Yunus F. The role of innate and adaptive immune cells in the Immunopathogenesis of chronic obstructive pulmonary disease. Tuberc Respir Dis (Seoul). 2016;79(1):5–13.
73. Tager AM, et al. The lysophosphatidic acid receptor LPA1 links pulmonary fibrosis to lung injury by mediating fibroblast recruitment and vascular leak. Nat Med. 2008;14(1):45–54.
74. Munger JS, et al. The integrin alpha v beta 6 binds and activates latent TGF beta 1: a mechanism for regulating pulmonary inflammation and fibrosis. Cell. 1999;96(3):319–28.
75. Thannickal VJ, et al. Mechanisms of pulmonary fibrosis. Annu Rev Med. 2004;55:395–417.
76. White GP, et al. CpG methylation patterns in the IFNgamma promoter in naive T cells: variations during Th1 and Th2 differentiation and between atopics and non-atopics. Pediatr Allergy Immunol. 2006;17(8):557–64.
77. Shang Y, et al. Epigenetic alterations by DNA methylation in house dust mite-induced airway hyperresponsiveness. Am J Respir Cell Mol Biol. 2013;49(2):279–87.
78. Barnes PJ. Targeting the epigenome in the treatment of asthma and chronic obstructive pulmonary disease. Proc Am Thorac Soc. 2009;6(8):693–6.
79. Sofer T, et al. Exposure to airborne particulate matter is associated with methylation pattern in the asthma pathway. Epigenomics. 2013;5(2):147–54.
80. Ren Y, et al. Identification of histone acetylation in a murine model of allergic asthma by proteomic analysis. Exp Biol Med (Maywood). 2021;246(8):929–39.
81. Wei W, Chen W, He N. HDAC4 induces the development of asthma by increasing slug-upregulated CXCL12 expression through KLF5 deacetylation. J Transl Med. 2021;19(1):258.
82. Liu F, Shang YX. Sirtuin 6 attenuates epithelial-mesenchymal transition by suppressing the TGF-beta1/Smad3 pathway and c-Jun in asthma models. Int Immunopharmacol. 2020;82: 106333.

83. Haberg SE, et al. Folic acid supplements in pregnancy and early childhood respiratory health. Arch Dis Child. 2009;94(3):180–4.
84. Stefanowicz D, et al. Epigenetic modifying enzyme expression in asthmatic airway epithelial cells and fibroblasts. BMC Pulm Med. 2017;17(1):24.
85. Yang ZC, et al. MiR-448-5p inhibits TGF-beta1-induced epithelial-mesenchymal transition and pulmonary fibrosis by targeting Six1 in asthma. J Cell Physiol. 2019;234(6):8804–14.
86. Alashkar Alhamwe B, et al. Epigenetic regulation of airway epithelium immune functions in asthma. Front Immunol. 2020;11:1747.
87. Hudon Thibeault AA, Laprise C. Cell-specific DNA methylation signatures in asthma. Genes (Basel). 2019;10(11):932.
88. Kuramasu A, et al. Mast cell−/basophil-specific transcriptional regulation of human L-histidine decarboxylase gene by CpG methylation in the promoter region. J Biol Chem. 1998;273(47):31607–14.
89. Moheimani F, et al. The genetic and epigenetic landscapes of the epithelium in asthma. Respir Res. 2016;17(1):119.
90. Wei G, et al. Global mapping of H3K4me3 and H3K27me3 reveals specificity and plasticity in lineage fate determination of differentiating CD4+ T cells. Immunity. 2009;30(1):155–67.
91. Wen T, Rothenberg ME. The regulatory function of eosinophils. Microbiol Spectr. 2016;4(5): 4–5.
92. Marcet B, et al. Control of vertebrate multiciliogenesis by miR-449 through direct repression of the Delta/notch pathway. Nat Cell Biol. 2011;13(6):693–9.
93. Zhang H, et al. miR-221 participates in the airway epithelial cells injury in asthma via targeting SIRT1. Exp Lung Res. 2018;44(6):272–9.
94. Zhang K, et al. Decreased epithelial and sputum miR-221-3p associates with airway eosinophilic inflammation and CXCL17 expression in asthma. Am J Physiol Lung Cell Mol Physiol. 2018;315(2):L253–64.
95. Clifford RL, et al. Altered DNA methylation is associated with aberrant gene expression in parenchymal but not airway fibroblasts isolated from individuals with COPD. Clin Epigenetics. 2018;10:32.
96. Hazari YM, et al. Alpha-1-antitrypsin deficiency: genetic variations, clinical manifestations and therapeutic interventions. Mutat Res Rev Mutat Res. 2017;773:14–25.
97. Pfaff M, et al. Activation of the SPHK/S1P signalling pathway is coupled to muscarinic receptor-dependent regulation of peripheral airways. Respir Res. 2005;6(1):48.
98. Song J, et al. Aberrant DNA methylation and expression of SPDEF and FOXA2 in airway epithelium of patients with COPD. Clin Epigenetics. 2017;9:42.
99. Barnawi J, et al. Potential link between the Sphingosine-1-phosphate (S1P) system and defective alveolar macrophage phagocytic function in chronic obstructive pulmonary disease (COPD). PLoS One. 2015;10(10):e0122771.
100. Epaud R, et al. Knockout of insulin-like growth factor-1 receptor impairs distal lung morphogenesis. PLoS One. 2012;7(11):e48071.
101. Liu JP, et al. Mice carrying null mutations of the genes encoding insulin-like growth factor I (Igf-1) and type 1 IGF receptor (Igf1r). Cell. 1993;75(1):59–72.
102. Nakahira K, Hisata S, Choi AM. The roles of mitochondrial damage-associated molecular patterns in diseases. Antioxid Redox Signal. 2015;23(17):1329–50.
103. Zhang L, et al. Epigenetic modifications and therapy in chronic obstructive pulmonary disease (COPD): an update review. COPD. 2020;17(3):333–42.
104. Qi X, et al. LncRNAs NR-026690 and ENST00000447867 are upregulated in CD4(+) T cells in patients with acute exacerbation of COPD. Int J Chron Obstruct Pulmon Dis. 2019;14:699–711.
105. Long YJ, et al. miR-34a is involved in CSE-induced apoptosis of human pulmonary microvascular endothelial cells by targeting Notch-1 receptor protein. Respir Res. 2018;19(1):21.
106. Jalali S, et al. Mir-206 regulates pulmonary artery smooth muscle cell proliferation and differentiation. PLoS One. 2012;7(10):e46808.

107. Karch A, et al. The German COPD cohort COSYCONET: aims, methods and descriptive analysis of the study population at baseline. Respir Med. 2016;114:27–37.
108. Tennis MA, et al. Methylation of Wnt7a is modulated by DNMT1 and cigarette smoke condensate in non-small cell lung cancer. PLoS One. 2012;7(3):e32921.
109. Sanders YY, et al. Thy-1 promoter hypermethylation: a novel epigenetic pathogenic mechanism in pulmonary fibrosis. Am J Respir Cell Mol Biol. 2008;39(5):610–8.
110. Helling BA, Yang IV. Epigenetics in lung fibrosis: from pathobiology to treatment perspective. Curr Opin Pulm Med. 2015;21(5):454–62.
111. Li H, et al. MicroRNAs in idiopathic pulmonary fibrosis: involvement in pathogenesis and potential use in diagnosis and therapeutics. Acta Pharm Sin B. 2016;6(6):531–9.
112. Pottier N, et al. Identification of keratinocyte growth factor as a target of microRNA-155 in lung fibroblasts: implication in epithelial-mesenchymal interactions. PLoS One. 2009;4(8): e6718.
113. Maurer B, et al. MicroRNA-29, a key regulator of collagen expression in systemic sclerosis. Arthritis Rheum. 2010;62(6):1733–43.
114. Song X, et al. Analysing the relationship between lncRNA and protein-coding gene and the role of lncRNA as ceRNA in pulmonary fibrosis. J Cell Mol Med. 2014;18(6):991–1003.
115. Cao G, et al. Differential expression of long non-coding RNAs in bleomycin-induced lung fibrosis. Int J Mol Med. 2013;32(2):355–64.
116. Taganov KD, et al. NF-kappaB-dependent induction of microRNA miR-146, an inhibitor targeted to signaling proteins of innate immune responses. Proc Natl Acad Sci U S A. 2006;103(33):12481–6.
117. Liu G, et al. miR-21 mediates fibrogenic activation of pulmonary fibroblasts and lung fibrosis. J Exp Med. 2010;207(8):1589–97.
118. Singer BD, et al. Regulatory T cell DNA methyltransferase inhibition accelerates resolution of lung inflammation. Am J Respir Cell Mol Biol. 2015;52(5):641–52.
119. Wu Y, et al. Therapeutic delivery of MicroRNA-29b by cationic Lipoplexes for lung cancer. Mol Ther Nucleic Acids. 2013;2(4):e84.
120. Dakhlallah D, et al. Epigenetic regulation of miR-17~92 contributes to the pathogenesis of pulmonary fibrosis. Am J Respir Crit Care Med. 2013;187(4):397–405.
121. Ganesan S, et al. Quercetin prevents progression of disease in elastase/LPS-exposed mice by negatively regulating MMP expression. Respir Res. 2010;11(1):131.
122. Cosio BG, et al. Theophylline restores histone deacetylase activity and steroid responses in COPD macrophages. J Exp Med. 2004;200(5):689–95.
123. Coward WR, et al. Repression of IP-10 by interactions between histone deacetylation and hypermethylation in idiopathic pulmonary fibrosis. Mol Cell Biol. 2010;30(12):2874–86.
124. Filippakopoulos P, et al. Selective inhibition of BET bromodomains. Nature. 2010;468(7327): 1067–73.
125. Pandit KV, Milosevic J, Kaminski N. MicroRNAs in idiopathic pulmonary fibrosis. Transl Res. 2011;157(4):191–9.

Epigenetics in Asthma

5

Waleed Hassan Almalki

Abstract

Asthma is one of the highest incidences and disbursed respiratory diseases the world over. There is incomprehensible knowledge of pathophysiological epigenetic pathways and causal interaction in asthma. Hereby, with a focus on DNA methylation, we examine human investigations on the epigenetic processes of asthma. On analysis, it was found that epigenetic research on childhood asthma has uncovered distinct methylation profiles linked to allergic inflammation in immune cells and the airways, demonstrating that methylation plays a regulatory function in the pathogenesis of asthma. Considering these ground-breaking findings, additional research is needed to understand how epigenetic pathways contribute to the endotypes of asthma. Studies on histone alterations in asthma are very few. Future investigations into the epigenetic causes of asthma will be aided by the inclusion of data from groups with accurate phenotyping.

5.1 Introduction

Waddington coined the term "epigenetics," which is formed by the Greek words "epi" and "genome." Epigenetics means phenotype changes without a corresponding change in genotype [1]. Although the mechanisms behind were not known when the term was first used, advancement in genetics and genomics has revealed significant epigenetic mechanisms involved in cell function, including both controlling the homeostasis of healthy cells and in the pathophysiology of the disease [2]. Numerous studies on asthma have demonstrated that epigenetic pathways affect the disease's

W. H. Almalki (✉)
Department of Pharmacology, College of Pharmacy, Umm Al-Qura University, Makkah, Saudi Arabia

G. Gupta et al. (eds.), *Targeting Epigenetics in Inflammatory Lung Diseases*,
https://doi.org/10.1007/978-981-99-4780-5_5

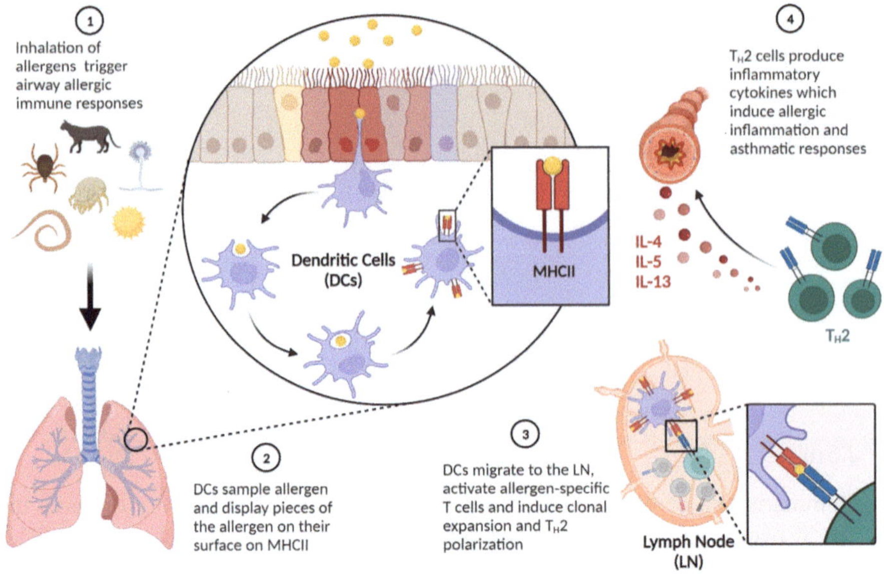

Fig. 5.1 Allergic airway sensitization

numerous manifestations, including asthma in children and asthma in adults. Epigenetic regulatory mechanisms, whether directly affected by inflammation and the modulation of respiratory function or indirectly influenced by pharmaceutical therapies or environmental factors, can provide insights into the substantial heterogeneity observed in asthma [3, 4]. This chapter highlights recent advancements in the perception of the epigenetics of asthma in individual research by framing the topic with well-known instances of epigenetic mechanisms in asthma (Fig. 5.1).

5.2 Overview of Epigenetic Pathways

5.2.1 DNA Methylation

DNA methylation involves the addition of methyl group on the cytosine C5 site to form 5-methylcytosine (5mC). The existing 5mC in specific genome sites causes alteration in organic phenomenon by the binding transcription factor [5]. Near the gene transcription start site (TSS), abundant CG sequences are found at the DNA regions which is called CpG island (CGI). As a result, both increased and decreased gene expression are correlated with methylation alterations in CGIs [6]. Hence, gene suppression and CGI methylation are frequently linked; moreover, the role of DNA methylation is complex and includes methylation in the gene and CGIs [7]. Heritable modifications to the DNA that affect gene function are known as epigenetic alterations. DNA methylation is a sophisticated gene regulatory system that is passed along cellular division through generations. It also plays a main role in the transcriptional elongation formation of the permutable elements, and the creation and slicing of mutation elements [8, 9].

5.2.2 Histone Alteration

Most of the proteins in the DNA-protein complex chromatin are histones. It is crucial for packing, ensuring genomic integrity, and controlling gene expression so that proteins are closely associated with DNA [10]. A crucial epigenetic mechanism is the regulation of histone function through post-translational changes [11]. Histone acetyltransferases (HATs) and histone deacetylases (HDACS) are the two major categories among the numerous enzymes that are involved in histone modifications. HATs and HDACs are differentiated according to where they are found and how particular they are and can be distinguished accordingly [12]. The contrasting effects on lysine acetylation make it simple to comprehend the functions of HATs and HDACs in histone modifications [13]. HDACs function as transcriptional suppressors by maintaining the chromatin structure once lysine acetylation is reversed. Additionally, gene expression is regulated by events like histone acetylation, methylation, and phosphorylation: by DNA effectiveness histone methylation and nucleosome unfolding [14]. DNA histone interaction efficiency is altered by nucleosome unwrapping. Histone methylation causes modification of histone protein either by adding or deleting a methyl group which is catalyzed by a group of demethylase and methyltransferase. It is important to remember that histone methylation affects DNA methylation in a similar manner [11, 15, 16].

5.2.3 Silencing of Transcriptional Genes and Non-coding RNA

Non-coding RNA (ncRNAs) are transcription that regulates the biological process. Based on their nucleotide size, ncRNAs are divided into two classes, i.e., small nucleotide has a size of less than 200 and long nucleotide has a size of more than 200 [17]. According to research, ncRNAs are known to contribute to cell differentiation and organogenesis. Despite being categorized as ncRNAs, small and big ncRNAs are different and display distinctive characteristics that define their function [18]. Non-coding RNA includes transfer RNA (tRNAs), ribosomal RNA (rRNAs), and small RNAs (sRNAs) such as microRNA, small nucleolar RNA (snRNAs), small interfering RNA (siRNA), small RNA derived from transfer RNA (tsRNA), etc. These small RNAs play numerous roles in regulating cellular processes [19].

Long non-coding RNAs (lncRNAs) are frequently used to control nearby protein-coding genes. Further complicating their role in the epigenetic regulation of cell activity is the fact that lncRNAs target histone methyltransferases and demethylases [20]. This chapter deals with asthma in adults, children, and its vulnerability. Since DNA methylation has been the subject of the bulk of recent high-quality human research on asthma, this study focuses on epigenetic mechanisms and provides an overview of microRNAs and histone alterations [18, 21] (Fig. 5.2).

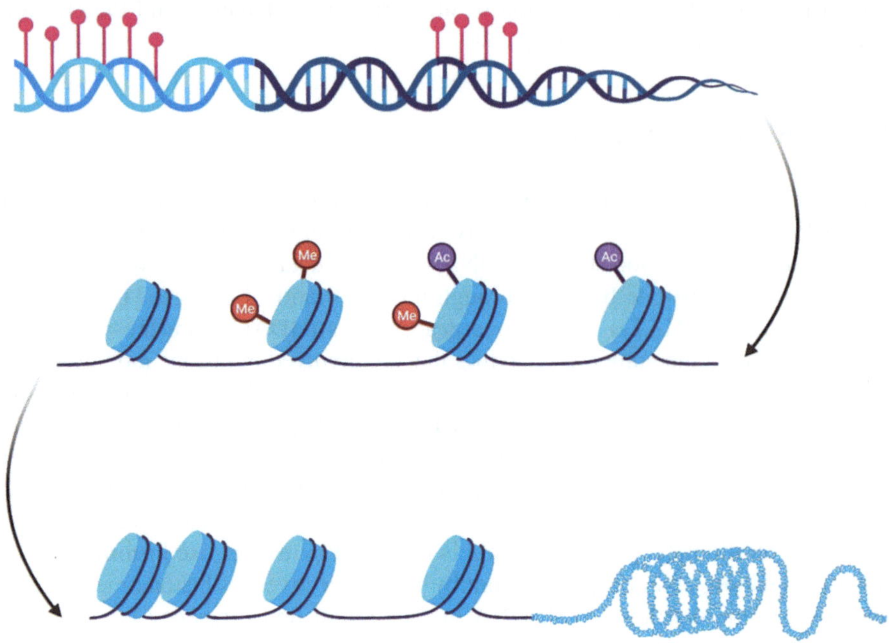

Fig. 5.2 Epigenetic level

5.2.4 Asthma in Adults

There are few investigations into DNA methylation in adult asthma. In an asthmatic study involving smokers, it was demonstrated that Protocadherin-20 (PCDH20) exhibited higher levels of methylation in sputum cells [22]. Even after taking environmental factors into account, this connection remained strong. A connection between CGIs and particular asthma endotypes was found after methylation alterations in the airway epithelium were assessed [23]. Interleukins induced alterations in DNA methylation have been linked to eosinophils. The inhaled corticosteroids block IL-13 signaling through the IL-13Ra1 receptor and result in the reduction of FENO [24]. Separate research on the methylation of blood found methylation variations in gene networks linked to asthma subtypes, with purine metabolism and calcium signaling genes being enriched in eosinophilic asthma and neutrophil asthma had significant accumulation for SUMOylation, basal cell carcinoma signaling, and Wnt/−catenin pathways [25, 26].

5.2.5 Asthma in Children

DNA methylation serves as a biomarker in all asthma investigations during human studies, involving the sampling of various biological compartments, such as epithelial and immunological, and utilizing individual genes and genomes [27, 28]. These studies showed a relationship between methylation patterns in the blood and various processes and features of pediatric asthma [29]. Studies on candidate genes have demonstrated that nitrogen dioxide (NO2) and particulate matter can affect the methylation of the adrenergic β2 receptor agonist (ADRB2) in the gene associated with asthma [30]. Moreover, the methylation of ARDB2 was connected to lessened dyspnea in young asthmatics, according to a related investigation. The idea that these variances are caused by cohort-specific traits must be further investigated [31]. Hypomethylation at CpG (cytosine guanine) dinucleotide site of arachidonate12-lipoxygenase (ALOX12) gene associated with persistent wheezing. Genome-wide methylation investigations have demonstrated that the presence of CGIs in blood samples is associated with an elevated risk of chronic wheezing. Asthma is associated with hyper-responsiveness mediated by hypomethylation of important cytokine IL-13, transcription factor RUNX3 modulates Th1/Th2 balance, microRNA, and TIGIT in blood [32–34]. Like this, asthma in children was connected to hypomethylation of CGIs linked to IL5RA. Cell-specific and epigenetic methylation in pediatric asthma is interrelated; the DNA of 14 CpG sites in eosinophils has been found to have methylated cytosine and adenosine residues [35]. PCSK6 was hypomethylated in children with atopic and atopy asthma. Assessments of blood spots and umbilical blood have been carried out to evaluate links between allergic disease and asthma [36]. There was a reduced incidence of asthma in toddlers by higher methylation of the GATA3 CpGs transcription factor. In mid-childhood, DNA methylation of cord blood is linked to IgE in genes involved in cell signaling, growth, and development. After accounting for the difference in DNA methylation between birth and mid-childhood, two methylation sites, i.e., C7orf50 and ZAR1, remained steady [37, 38]. Two methylation sites (C7orf50 and ZAR1) remained stable after adjusting for the change in DNA methylation from birth to mid-childhood. Methylation of DNA sequence on AXL receptor at birth was linked to an increased incidence of wheezing, demonstrating that methylation has an impact on gender [39].

Therefore, the risk that infants would develop asthma, wheeze, and high IgE levels are correlated with the methylation patterns at birth seen in cord blood and blood spots [40]. According to studies on the methylation of buccal DNA, increased methylation of arginase 2 (ARG2) has been associated with a decrease in the fraction of exhaled nitric oxide (FeNO) in asthmatic children [41]. Similarly, to this, reducing allergen exposure resulted in less methylation of the FOXP3 promoter. Together, these findings point to a function for methylation in the regulation of immune signaling, and airway obstruction CGIs in the STAT5A transcription factor were differently methylated in asthmatics' airway epithelial cells (AECs), which caused STAT5A expression to be downregulated [42]. Following stimulation by certain ligands like Interleukin-2,3,7 and GM-CSF, this pathway is important in the regulation of downstream signaling. Asthma in nasal epithelial is associated with hypomethylation of ALOX15 and POSTN genes. Th2 cytokines induce the upregulation of the POSTN gene in asthmatic airway epithelium, which can be exploited as a biomarker for long-term prediction in the treatment of severe asthma [34, 43, 44].

5.2.6 Methylated Differentially in the Nasal Epithelial of African American Newborns

This study also demonstrated that in children of African American, European, and Hispanic heritage, a nasal epithelial classifier based on methylation may be able to differentiate between atopy and allergic asthma. This methylation pattern is applied in the epigenetic regulation of FeNO (fractional exhaled nitric oxide) and Immunoglobulin E [45, 46].

5.2.7 Asthma Risk and Vulnerability

Environmental factors have a big impact on asthma and they also affect DNA methylation. Acyl-CoA synthetase (ACSL3) long-chain family member methylation has been linked to asthma symptoms in children under the age of five, according to research on prenatal exposure to polycyclic organic matter(POM) from traffic [47]. An association between elevated FOXP3 methylation and defective regulatory T cells was found in distinct research on asthmatic children exposed to higher levels of ambient pollution [48]. Multiple studies have demonstrated a link between air pollution, variable DLG2 methylation, and corresponding changes in blood expression [49]. Like how prenatal tobacco exposure is linked to significant changes in children's blood DNA methylation, tobacco smoke affects DNA methylation, as shown by differential methylation of placental and fetal lung tissue [50]. These differentially methylated sites show the importance of environmental factors in methylation patterns, although this is unclear how they are connected to asthma. SMAD3 methylation at birth was linked to a higher risk of childhood asthma in the offspring of women with asthma, who also had higher levels of it [51]. In another study, the blood methylome of children showed a comparable impact on maternal asthma. In children, maternal blood eosinophils, FeNO, and total IgE all had a negative correlation with MAPK8IP3 methylation. According to this research, exposure to an asthmatic mother while the child is still in the uterus may influence DNA methylation, which may have an impact on how asthma develops [52, 53].

5.3 Role of MicroRNAs in Asthma

MiRNAs play a significant role in the control of both healthy and unhealthy cellular responses. Let-7 family miRNAs have been linked to regulating Th2 inflammation in human, animal, and in vitro research. MiR-21 controls the cytokine IL-12, which is involved in the polarization of Th1 cells [54]. MiR-146a has been identified as a potential asthmatic molecule in studies on the effects of genetic variation on miRNA function and their relationship to asthma; variation in HLA-G in children influences asthma risk through interaction with miR-152. Several cellular elements of the allergic response, including eosinophils, macrophages, and mast cells in both asthma and allergic rhinitis, have been linked to Th2 responses by two miRNAs, miR-155 and miR-221 [55, 56]. Several miRNAs, including miR-26, 133a, 140, 206, and 221, have been linked to a role in smooth muscle cell proliferation and function. Few studies have examined the impact of miRNAs on severe asthma despite their crucial role [57]. These studies on patients with severe asthma showed that miR-221 governs the proliferation of airway smooth muscle cells, miR-28-5p and miR-146a/b led to the activation of circulating CD8+ T cells in severe asthma, and miR-223-3p, miR-142-3p, and miR-629-3p were linked to serious neutrophilic asthma [58]. In the Childhood Asthma Management Program (CAMP) cohort, eight serum microRNAs, including miR-296-5p, were connected to PC20 [59]. Inhaled corticosteroid budesonide, used to treat asthma, had a negligible impact on the miRNA profile in steroid-naive asthmatics' bronchial epithelium after therapy. These findings on the activity of several miRNAs in asthma show their impact on the regulation of inflammatory events, Th1/Th2 polarization, cell function, disease severity, and treatment response [55, 60, 61].

5.4 Role of Histone Modifications in Asthma

A poorly understood mechanism for asthma is an epigenetic modification of the histones [62]. The amount of cellular acetylation activity was related to the severity of bronchial hyperresponsiveness in allergic asthmatic children, and the ratio of HDAC/HAT activity was biased toward greater histone acetylation in those cases [63]. Adults with asthma who had severe asthma had lower nuclear HDAC and HAT activity in their mononuclear cells than those who had milder cases of the condition [64]. An analysis of the genome-wide histone changes in T cell subsets from asthma patients and healthy controls revealed variations in cell enhancers involved in T cell development that were specific to asthma. Furthermore, compared to cells from healthy people, human bronchial epithelial cells (HBECs) from adult asthmatics showed impaired tight junction integrity [65, 66]. In HBECs from asthmatic patients, there was greater expression of Sirtuins 6 and 7, HDACs 1 and 9, and HDACs. Tight junction molecules increased to levels like those reported in healthy controls when HDAC was inhibited. Together, these findings point to an overlap that is particular to cells and asthma and is caused by various histone changes [67].

5.5 Aspects of the Future

By combining omics data from microRNAs, genome-wide variation, methylome, and transcriptome cohorts with known traits, it may be feasible to better comprehend these associations. The importance of histone changes in asthma, environmental data, and the intensity of relationships with certain asthma endotypes all lay considerable limitations on our understanding of asthma. In light of these findings, it is necessary to create better prediction models, biomarkers, and medicines that target dysregulated pathways.

References

1. Abdel-Aziz MI, et al. Omics for the future in asthma. Semin Immunopathol. 2020;42(1): 111–26.
2. Agache I. Severe asthma phenotypes and endotypes. Semin Immunol. 2019;46:101301.
3. Agache I, et al. Advances and highlights in asthma in 2021. Allergy. 2021;76(11):3390–407.
4. Alashkar Alhamwe B, et al. The role of epigenetics in allergy and asthma development. Curr Opin Allergy Clin Immunol. 2020;20(1):48–55.
5. Alashkar Alhamwe B, et al. Extracellular vesicles and asthma-more than just a co-existence. Int J Mol Sci. 2021;22:9.
6. Alizadeh Z, et al. Role of epigenetics in the pathogenesis of asthma. Iran J Allergy Asthma Immunol. 2017;16(2):82–91.
7. Bélanger É, Laprise C. Could the epigenetics of eosinophils in asthma and allergy solve parts of the puzzle? Int J Mol Sci. 2021;22:16.
8. Benincasa G, et al. Epigenetics and pulmonary diseases in the horizon of precision medicine: a review. Eur Respir J. 2021;57(6):2003406.
9. Benjamin S, et al. Phthalates impact human health: epidemiological evidences and plausible mechanism of action. J Hazard Mater. 2017;340:360–83.
10. Boulet LP. Airway remodeling in asthma: update on mechanisms and therapeutic approaches. Curr Opin Pulm Med. 2018;24(1):56–62.
11. Brook PO, et al. Epigenome-modifying tools in asthma. Epigenomics. 2015;7(6):1017–32.
12. Camoretti-Mercado B, Lockey RF. Airway smooth muscle pathophysiology in asthma. J Allergy Clin Immunol. 2021;147(6):1983–95.
13. Castro-Rodríguez JA, et al. Epigenetics in allergic diseases and asthma. Rev Chil Pediatr. 2016;87(2):88–95.
14. Cevhertas L, et al. Advances and recent developments in asthma in 2020. Allergy. 2020;75(12): 3124–46.
15. Choi BY, et al. Genetics and epigenetics in allergic rhinitis. Genes (Basel). 2021;12(12):2004.
16. Chowdhury NU, et al. Sex and gender in asthma. Eur Respir Rev. 2021;30(162):210067.
17. Clapp PW, Jaspers I. Electronic cigarettes: their constituents and potential links to asthma. Curr Allergy Asthma Rep. 2017;17(11):79.
18. Conrad LA, Cabana MD, Rastogi D. Defining pediatric asthma: phenotypes to endotypes and beyond. Pediatr Res. 2021;90(1):45–51.
19. Davidson EJ, Yang IV. Role of epigenetics in the development of childhood asthma. Curr Opin Allergy Clin Immunol. 2018;18(2):132–8.
20. Devries A, Vercelli D. Epigenetics of human asthma and allergy: promises to keep. Asian Pac J Allergy Immunol. 2013;31(3):183–9.
21. DeVries A, Vercelli D. Early predictors of asthma and allergy in children: the role of epigenetics. Curr Opin Allergy Clin Immunol. 2015;15(5):435–9.
22. DeVries A, Vercelli D. Epigenetics in allergic diseases. Curr Opin Pediatr. 2015;27(6):719–23.
23. DeVries A, Vercelli D. Epigenetic mechanisms in asthma. Ann Am Thorac Soc. 2016;13(Suppl 1(Suppl 1)):S48–50.
24. Durham A, et al. Epigenetics in asthma and other inflammatory lung diseases. Epigenomics. 2010;2(4):523–37.
25. Durham AL, Wiegman C, Adcock IM. Epigenetics of asthma. Biochim Biophys Acta. 2011;1810(11):1103–9.

26. Fang L, Sun Q, Roth M. Immunologic and non-immunologic mechanisms leading to airway remodeling in asthma. Int J Mol Sci. 2020;21(3):757.
27. Godfrey KM, et al. Influence of maternal obesity on the long-term health of offspring. Lancet Diabetes Endocrinol. 2017;5(1):53–64.
28. Gomez JL. Epigenetics in asthma. Curr Allergy Asthma Rep. 2019;19(12):56.
29. Gruzieva O, et al. An update on the epigenetics of asthma. Curr Opin Allergy Clin Immunol. 2021;21(2):175–81.
30. Guo C, et al. Serum sphingolipid profile in asthma. J Leukoc Biol. 2021;110(1):53–9.
31. Han X, et al. LncRNA PTPRE-AS1 modulates M2 macrophage activation and inflammatory diseases by epigenetic promotion of PTPRE. Sci Adv. 2019;5(12):eaax9230.
32. Harb H, et al. Recent developments in epigenetics of pediatric asthma. Curr Opin Pediatr. 2016;28(6):754–63.
33. Heijink IH, et al. Epithelial cell dysfunction, a major driver of asthma development. Allergy. 2020;75(8):1902–17.
34. Hellings PW, Steelant B. Epithelial barriers in allergy and asthma. J Allergy Clin Immunol. 2020;145(6):1499–509.
35. Hernandez-Pacheco N, Kere M, Melén E. Gene-environment interactions in childhood asthma revisited; expanding the interaction concept. Pediatr Allergy Immunol. 2022;33(5):e13780.
36. Ho SM. Environmental epigenetics of asthma: an update. J Allergy Clin Immunol. 2010;126(3): 453–65.
37. Holgate ST, et al. Asthma. Nat Rev Dis Primers. 2015;1(1):15025.
38. Hudon Thibeault AA, Laprise C. Cell-specific DNA methylation signatures in asthma. Genes (Basel). 2019;10(11):932.
39. Isidoro-García M, et al. Pharmacogenetics and the treatment of asthma. Pharmacogenomics. 2017;18(13):1271–80.
40. Kabesch M. Epigenetics in asthma and allergy. Curr Opin Allergy Clin Immunol. 2014;14(1): 62–8.
41. Kabesch M, Adcock IM. Epigenetics in asthma and COPD. Biochimie. 2012;94(11):2231–41.
42. Kabesch M, Tost J. Recent findings in the genetics and epigenetics of asthma and allergy. Semin Immunopathol. 2020;42(1):43–60.
43. Koefoed HJL, Vonk JM, Koppelman GH. Predicting the course of asthma from childhood until early adulthood. Curr Opin Allergy Clin Immunol. 2022;22(2):115–22.
44. Kong LD, Wu QP. Effect of ketogenic diet on obesity asthma. Zhonghua Jie He He Hu Xi Za Zhi. 2022;45(2):222–6.
45. Koppelman GH, Nawijn MC. Recent advances in the epigenetics and genomics of asthma. Curr Opin Allergy Clin Immunol. 2011;11(5):414–9.
46. Lebold KM, Jacoby DB, Drake MG. Inflammatory mechanisms linking maternal and childhood asthma. J Leukoc Biol. 2020;108(1):113–21.
47. Li CY, Guo XJ, Gan LX. The epigenetics in asthma. Zhonghua Jie He He Hu Xi Za Zhi. 2009;32(10):759–61.
48. Lira G, et al. Psychological stress in asthma: repercussions on epigenetics-genetics, immune responses, and pulmonary function in the pediatric population. Allergol Immunopathol (Madr). 2022;50(2):78–88.
49. Long A, et al. Epigenetics and the environment in airway disease: asthma and allergic rhinitis. Adv Exp Med Biol. 2020;1253:153–81.
50. Lovinsky-Desir S, Miller RL. Epigenetics, asthma, and allergic diseases: a review of the latest advancements. Curr Allergy Asthma Rep. 2012;12(3):211–20.
51. Lu X, Li R, Yan X. Airway hyperresponsiveness development and the toxicity of PM2.5. Environ Sci Pollut Res Int. 2021;28(6):6374–91.
52. Lu Y, et al. Eosinophil extracellular traps drive asthma progression through neuro-immune signals. Nat Cell Biol. 2021;23(10):1060–72.

53. Lynch SV, Vercelli D. Microbiota, epigenetics, and trained immunity. Convergent drivers and mediators of the asthma trajectory from pregnancy to childhood. Am J Respir Crit Care Med. 2021;203(7):802–8.
54. Maneechotesuwan K. Role of microRNA in severe asthma. Respir Investig. 2019;57(1):9–19.
55. Martino D, Prescott S. Epigenetics and prenatal influences on asthma and allergic airways disease. Chest. 2011;139(3):640–7.
56. McKenzie C, et al. The nutrition-gut microbiome-physiology axis and allergic diseases. Immunol Rev. 2017;278(1):277–95.
57. Mekov E, et al. Update on asthma-COPD overlap (ACO): a narrative review. Int J Chron Obstruct Pulmon Dis. 2021;16:1783–99.
58. Miller RL, Ho SM. Environmental epigenetics and asthma: current concepts and call for studies. Am J Respir Crit Care Med. 2008;177(6):567–73.
59. Mims JW. Asthma: definitions and pathophysiology. Int Forum Allergy Rhinol. 2015;5(Suppl 1):S2–6.
60. Murphy SK, Hollingsworth JW. Stress: a possible link between genetics, epigenetics, and childhood asthma. Am J Respir Crit Care Med. 2013;187(6):563–4.
61. Noutsios GT, Floros J. Childhood asthma: causes, risks, and protective factors; a role of innate immunity. Swiss Med Wkly. 2014;144:w14036.
62. Ntontsi P, et al. Genetics and epigenetics in asthma. Int J Mol Sci. 2021;22(5):2412.
63. Ober C. Asthma genetics in the post-GWAS era. Ann Am Thorac Soc. 2016;13(Suppl 1(Suppl 1)):S85-90.
64. Potaczek DP, et al. Epigenetics and allergy: from basic mechanisms to clinical applications. Epigenomics. 2017;9(4):539–71.
65. Qi C, Xu CJ, Koppelman GH. The role of epigenetics in the development of childhood asthma. Expert Rev Clin Immunol. 2019;15(12):1287–302.
66. Rico-Rosillo G, et al. Epigenetics, environment and asthma. Rev Alerg Mex. 2014;61(2): 99–109.
67. Rosa MJ, Lee AG, Wright RJ. Evidence establishing a link between prenatal and early-life stress and asthma development. Curr Opin Allergy Clin Immunol. 2018;18(2):148–58.

Epigenetic Optimization in Chronic Obstructive Pulmonary Disease (COPD)

6

Khalid Saad Alharbi, Samiyah Mohammed Alshehri, and Sattam Khulaif Alenezi

Abstract

As of 2020, chronic obstructive pulmonary disease (COPD), remains the third common principal cause of morbidity and mortality in adults. The various pathological processes governing COPD are chronic lung inflammation caused by increasing levels of environmental insults (particle matter, cigarette smoke, chemical fumes, ROS, etc.), inflammatory mediators, and protease. Alpha-1 antitrypsin deficiency causes lung tissue damage (apoptosis). The epigenetic mechanism includes post-translational methylation and acetylation of histone proteins and DNA, as well as modulation of miRNA production. In this chapter, we address all the recent research on the up- or down-regulation of methylation in various genes linked to COPD. A significant part of preventing and slowing the progression of COPD is played by inhibiting histone deacetylase activity, which is brought up by several variables and miRNAs. Additionally, some COPD treatment plans focus miRNAs and HDAC2 for therapeutic effects.

K. S. Alharbi (✉)
Department of Pharmacology and Toxicology, Unaizah College of Pharmacy, Qassim University, Qassim, Saudi Arabia
e-mail: khalid.alharbi9@qu.edu.sa

S. M. Alshehri
Department of Pharmacology and Toxicology, College of Pharmacy, King Saud University, Riyadh, Saudi Arabia

S. K. Alenezi
Department of Pharmacology and Toxicology, Unaizah College of Pharmacy, Qassim University, Qassim, Saudi Arabia

G. Gupta et al. (eds.), *Targeting Epigenetics in Inflammatory Lung Diseases*,
https://doi.org/10.1007/978-981-99-4780-5_6

6.1 Introduction

COPD is characterized by mucociliary dysfunction, pneumonia, chronic bronchitis, airway fibrosis, and alveolar disintegration. Because of the rising smoking epidemic, it is a prevalent chronic adult disorder, and will become the third biggest cause of mortality worldwide [1]. The pathophysiology of COPD is influenced by several variables, including chronic inflammation, the elastase/anti-elastase concept, proteinase-3 and cathepsin G-induced emphysema, apoptosis, oxidant-antioxidant imbalancing, and infectious repair [2]. With more than 7357 chemical components, 400 plus toxins, and 1014 reactive species, smoking of cigarette is the primary source of COPD and inflammatory processes via stimulating inflammatory cells declines lung function [3]. In COPD, the activation of inflammatory mediators such as IL1B, TNF-alpha, and TLR leads to the activation of NF-kB and other redox-sensitive epigenetic regulators. This, in turn, results in an upregulation of cytokines, cell adhesion molecules, and pro-inflammatory chemokines [4, 5]. An enzymatic modifications made to DNA proteins such as histone and others post-translationally followed by protein biosynthesis without altering the primary set of genes is called epigenetic modification [6]. A gene's expression may be influenced by DNA and histone methylation. DNA methylation is one significant epigenetic technique that alters the chemical structure of the DNA and controls transcription by interacting with microRNA and histone modification [7]. The post-translational modifications that affect the histone proteins include phosphorylation, methylation, ubiquitination, sulfonation, and acetylation. By adding methyl groups to the cytosine residues in loci, DNA methylation, the primary method of epigenetic control, is carried out by enzymes of the DNA ethyl transferase (DNMT) class [8, 9]. In CpG island gene promoter regions, DNA hypomethylation stimulates transcription, whereas DNA hypermethylation typically results in gene silencing [10]. Hypermethylation of DNA in CpG island gene promoter regions frequently cause gene silencing, whereas hypomethylation causes transcriptional activation of DNA. One of the essential elements of epigenetic mechanisms is histone modifications, which produce hereditary changes in gene expression without altering the DNA sequence [11] (Fig. 6.1).

The important functional unit of chromatin, is nucleosome, is composed of a core of two molecules of H2A, H2B, H3, and H4 each histone, which is surrounded by approximately 150 base pairs of DNAs [12]. The histone modifications H3K4me3, H3K9me3, and H3K27me3 are crucial for gene regulation. H3K4me3 and the initiation of gene expression are related, whereas H3K27me3 and H3K9me3 are related to organic phenomenon. The expression of protein-coding genes is strictly regulated by ncRNAs, which do not act as templates for protein synthesis [13]. Non-coding RNAs, often known as microRNAs, are another type of post-transcriptional regulator of gene expression by degradation of mRNA and disruption. RNA polymerases can create miRNAs, which are 21–23 nucleotides long and can be produced by transcription of parent RNA [14]. Non-coding RNA (ncRNA) controls the expression of protein-coding genes in a significant way. Non-coding RNA are made of small ncRNAs like microRNA and long ncRNAs [15]. Studies show that miRNAs regulate maximum of the protein-coding genes. To boost

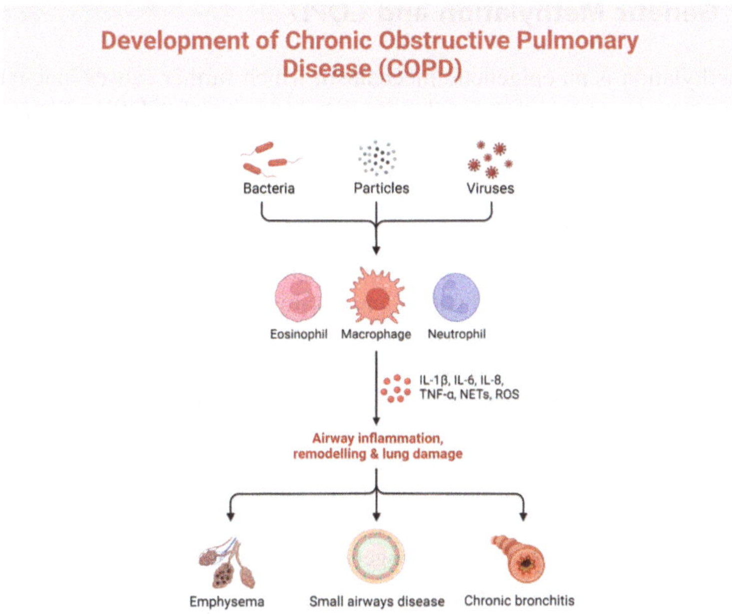

Fig. 6.1 Development of COPD

fragmentation or reduce translation, respectively, they do this by binding to the 3'-UTR or 5'-UTR of the nucleotide. While lncRNA are connected to genetic variations linked to disease, microRNA act as a negative regulator of gene expression [16].

6.2 Inheritance and COPD

Gene expression can alter in a heritable way through the modification of histones or DNA sequences. It has been shown that epigenetic alterations in COPD include abnormal inflammatory gene activity, aberrant DNA methylation, aberrant histone acetylation and deacetylation, and dysregulated miRNAs [17, 18]. In COPD patients' and smokers' lungs, epithelial cells and macrophages may react differently to cigarette smoke and oxidants due to epigenetic events. Asthma and COPD are chronic lung disorders that may be exacerbated or prevented by changes in DNA methylation, histone acetylation, and miRNA activity [19, 20]. Cigarette smoke extract (CSE) increased COPD by aggravating genetic modifications, particularly those in DNA of mitochondria. Investigations reveal that the CpG region of COPD patients (CpG) site displays a range of methylation signals in response to air pollution [21].

6.3 Genetic Methylation and COPD

DNA methylation is an epigenetic mechanism which further causes inactivation of pro-inflammatory genes and results in development of COPD. Pleural space macrophages and epithelium from COPD patients have been found to have DNA methylation of the genes of proinflammatory cytokines [22]. The existence and progression of COPD are discovered to be strongly influenced by DNA methylation; cigarette smoking can also change DNA methylation by inducing inflammatory responses and resulting diseases like COPD. The division of innate and adaptive immunity, along with the identification and abolition of pathogens, are all important functions of lung macrophages, which are innate immune cells [23]. Studies have shown physiological differences between the upper and lower parts of the lung in ventilation and respiration, while COPD typically appears as upper lobe predominance [24]. Several alveolar macrophage genes, like HSH2D, have been shown to have 95 CpG sites with significant methylation dysregulation via ventilation and oxygen variation between the upper and lower lobes. Similar mechanism is seen in SNX10/CLIP4 (Sorting Nexin 10) and TYKZ genes [25]. Mitochondrial transcription factor A is a key player in the pathogenesis of disorders like immune responses, necrosis, and inflammation. The mtTFA expression in the skeletal muscles is markedly lower in COPD patients. To regulate mtDNA nucleotide sequence and mitochondrial transcription initiation, mtTFA binds to the promoter regions of the HSP1 and LSP of mtDNA [26, 27].

Smoking increased the mtTFA promoter's hypermethylation, which led to the development of COPD. There is a link between air pollution and various lung diseases, including COPD [28]. Promoters of air pollution, such as toxic particles or gases, can cause epigenetic alterations, including altered DNA methylation [29]. Air pollution promoters can cause epigenetic changes, including altered DNA methylation. Air pollutant of size less than 10um are called particle matter 10 (PM10). Air pollutants such as PM10 and NO2 lead to the dysregulation of methylation at target genes like ARID5A (AT-Rich Interaction Domain 5 A), NEGR1 (Neuronal growth regulator 1), RPL5 (Ribosomal Protein L5), FOXl2 (Forkhead Box 12), CPLX1 (Complexin 1), STON1, and others [30, 31]. PM10 is closely associated with 12 differentially methylated probes (DMP) and 27 differentially methylated probes (DMR) while NO2 is associated at 57 DMRs and 45 DMPs in CpG region [32]. Fibroblasts can be found in a variety of endothelium tissues, including the adventitia of the vascular system, the alveolar ducts, and the elderly respiratory muscles. Lung fibroblasts are important for ECM homeostatic mechanisms, lung recovery, and cell of stem repairs [33]. According to TGFß response, rate of proliferation, and ECM, airway and parenchymal fibroblasts behave differently in COPD patients (ECM). A total of six hundred fifty-two regions with differential methylation, some of which are within gene regions [34]. In reply to TGFB and physiological extracellular matrix, the targeted genes HLA-DP1, RPH3AL, TMEM44, WNT3A, and HLA-DRB5 experience dysregulation of methylation by airway and parenchymal fibroblasts. HLX genes, which are hypomethylated in NXN (Nucleoredoxin) genes but hypermethylated in COPD, have at least three CpG sites, according to research that showed 44 DNA differentially methylated areas [35, 36]. Variants in the expression of the SERPINA1 gene may raise the risk of COPD and its related lung function abnormalities. SERPINA1 expression of gene alternatives might increase COPD hazard and linked function of lung phenotypes [37]. In smoking adults with COPD, there is altered methylation at two CpG sites within the SERPINA1 gene. This abnormal methylation may lead to excessive mucus secretion and goblet cell metaplasia, contributing to the development of COPD [38].

Human airways' tracheal epithelium consists of ciliated, basal, neuroendocrine, Clara cells, and basal cells. FoxA2 and transcription factors are both forkhead box proteins. Genes encoding SAM-pointed domains that include ETS-like factors (SPDEF) are two essential controllers of goblet cell development [39]. SPDEF is accountable for both mucus production and the development of goblet cells. On the other hand, dysregulated goblet cell differentiation in the lungs involves targeted genes regulated by FoxA2 [40]. The foxA2 promoter has 11 CpGs that are hypomethylated, while the SPDEF promoter has 6 CpGs that are hypomethylated. Sphingosine-1 phosphate (S1P) promotes maturation of macrophages and is essential for their phagocytosis [41]. In COPD, alveolar macrophages are associated with dysregulated S1P gene expression. S1P methylation in smokers is lower than in non- or ex-smokers. In individuals with COPD and smokers, targeted genes such as GABRB1, NOS1AP, TNFAIP2, and BID show hypermethylation compared to the healthy group [42]. The genes SERPINA1 and AHRR (Aryl-Hydrocarbon Receptor Repressor) have considerably lower levels of methylation in people with COPD and smokers [43]. The IGF system, particularly IGF1 and IGF1R, is essential for lung development. Smoke causes a CpG site-specific hypomethylation in the IGF1R promoter leading to suppression of lungs development [44]. Cigarette smoking can disturb DNA methylation, potentially initiating COPD. Several genes, including GPR126, three cholinergic receptors (CHRND, CHRNB1, and CHRNB2), EPHX1, and three glutathione S-transferases, exhibited hypermethylation. Conversely, only 3% of the genes, including KSR1, showed hypomethylation in response to cigarette smoke [45, 46].

6.4 HDAC2 Deregulation and COPD Induce Histone Modification

HDACs and HAT coenzymes jointly regulate activation (by histone acetylation) and silencing (by deacetylation), which has a significant effect on the emergence of inflammation in COPD [47]. Lysine residues are used to acetylate histones H3 and H4. Studies have shown that in individuals with COPD and the resulting inflammation, the balance between acetylation and deacetylation can change in favor of acetylation [48]. Smoke from cigarettes can cause acetylation of H3 in macrophages and human lung tissue. Histone deacetylases (HDACs) control the amount of histone acetylation to control protein function and gene transcription. HDACs control numerous major macromolecular complexes and epigenetic regulation during biological processes [49]. Previous studies have demonstrated that inhibiting HDAC1 and HDAC2 during the perinatal period in muscle tissue causes muscle fiber degeneration and mitochondrial abnormalities, which lead to the death of some mouse pups. Mice exposed to tobacco smoke have increased levels of HDAC1/2 (CS) [50]. HDAC blocker can be used to treat COPD patients by decreasing HDAC1/2, which ultimately prevents muscle fiber degeneration and histomorphological changes [51]. Trichostaina (TSA) comes under the category of HDACs blocker. Protein arginine methyl transferases (PRMTs), found in both histone and non-histone proteins, add a methyl group to arginine residues as the second significant histone alteration [52]. Coactivator-associated arginine methyl transferase 1 (CARM1) affects the regulation of gene expression by demethylating arginine residues in several non-histone proteins, including histone H3 and many others. CARM1 expression is increased in healthy epithelial cells but downregulated following epithelial cell damage in COPD [53, 54]. Therefore, via controlling cellular senescence, the regeneration and repairing of airway epithelial cells depend on CARM1. Lowering HDAC activities allows PM, such as fine and semi ultra-fine particles (UFP), to enhance the HAT/HDAC ratio [55]. The COPD-affected human bronchial epithelial (DHBE) group and the group stated above both had significant levels of H3K9 histone acetylation. Two key factors to limit inflammation in the airways and lung parenchyma in COPD are oxidative stress and cigarette smoking [56]. Oxidative stress can trigger the activation of nuclear factor kB (NFkB), which raises levels of pro-inflammatory factors and causes COPD. Sirt1 (Silent Information Regulator 1) belongs to the family of class 3 histone/protein deacetylases and is associated with inflammation, cell aging, senescence, as well as COPD/emphysema [57]. Research has revealed that SIRT1 controls NFkB and lessens inflammatory reactions. In COPD, erythromycin increased SIRT1 expression, which in turn reduced NFkB acetylation and pro-inflammatory cytokines. FoxO3 is a member of the Fox family, and it has been shown to decrease NFkB activity in COPD patients when SIRT1 and foxO interact [58]. CSE interferes with the function of SIRT1/FoxO3, which in turn dysregulates the NFkB activity and increases inflammatory responses. Glucocorticoid-dependent anti-inflammatory effect depends on reducing HDAC2, yet when HDAC2 is lowered, oxidative stress and inflammation increase. Smoking elevated oxidative stress and promoted COPD glucocorticoid resistance, both of which were associated with higher acetylation (HDAC2). A peptide called LL-37/hCAP18 has the power to inhibit the C-Jan N-terminal kinases (JNK), the AKt phosphorylation, and the activity of pro-inflammatory cytokines in macrophages [59, 60].

According to research, LL-37 boosted HDAC2 activity and expression in COPD patients via blocking the PI3K/AKt pathway. A phosphodiesterase isoenzyme inhibitor called theophylline reduces activity by preventing pro-inflammatory transcription factor expression and its entry into the nucleus. In rodent skeletal muscle cells, tobacco smoke can increase NFkBp65 polypeptide, TNF-a, and IL-8 levels in COPD [61]. By increasing HDAC2 expression and decreasing pro-inflammatory transcription factor expression, theophylline decreases inflammation in COPD patients (NFkBp65). Inhibitor kappa B (IkB) is phosphorylated and degraded to activate

NFkB, which causes transcription of NFkB-dependent genes [52, 62]. The pulmonary vasculature of COPD patients has been reported to have higher levels of the cytokine thymic stromal lymphopoietin (TSLP), which influences T cell survival, activation, and expression [62]. The pairing between IKKa and acetyl-histone H3 (Lys14) proteins is promoted by IL-17A, and IKKa protein silencing can drastically lower the expression of TSLP. A cytokine that helps combat bacterial infection is called IL-17A. HDAC2 may be involved in the differentiation of T cells that produce IL-17 [63]. In COPD patients' lung tissue samples, it was found that the expression of lymphocyte-associated protein and histone deacetylase was associated with the accumulation of collagen and thickening of the bronchial walls [64]. This study shows that in COPD patients, activating deacetylase enzyme can decrease lymphocyte-associated protein production and stop airway remodeling. Statins decrease the formation of 3-hydroxy-3-methylglutaryl coenzyme A (HMG-CoA), which is involved in the production of cholesterol, through the post-translational modification of the small G-proteins Ras and Rho [53, 65]. In patients with COPD, statins reduce the risk of death rates, respiratory infections, and hospitalization. The statins restore the expression and operation of weakened HDAC2 [66]. Type II alveolar epithelial cells (AECII), which also release inflammatory chemokines like interleukin-8, monocyte chemoattractant protein-1 (MCP-1), and macrophage inflammatory protein-2a, are necessary for lung remodeling and development (MIP-2 a) [67]. *Curcuma longa* plant produces curcumin a pale-yellow pigment with anti-inflammatory properties. In COPD, curcumin can restore corticosteroids and reduce inflammatory chemokines by modulating the expression of HDAC2 [68].

6.5 mRNAs as a COPD Disease Risk

Long non-coding RNAs (lncRNAs) undergo continuous processes of splicing, capping, and polyadenylation. Some lncRNAs may act as indicators for the evaluation and prognosis of COPD, according to studies [69]. Peripheral blood mononuclear cells from COPD patients have lower levels of the lncRNA ENST00000502883. Enst00000447867 and NR-026690, two lncRNAs that may serve as diagnostic indicator (biomarkers) for COPD, were both elevated in the disease. Prolonged artificial breathing can cause ventilator-associated pneumonia (VAP) in critical care unit's patients [70]. According to studies, TLR4 and the risk of VAP are correlated, and COPD is linked to increased rates of VAP and ICU mortality. In COPD patients, miR-1236 can enhance the prevalence of VAP by binding to the 3'-UTR of the TLR4 mRNA [71]. Patients with COPD had miR-206 that was elevated in their skeletal muscle and plasma, whereas it was dysregulated in cases of gastric, lung, and colorectal cancer. It has been shown that Notch signaling is highly expressed in the airway epithelium and is involved in controlling cell fate, including apoptosis [72]. Reduced Notch signaling, particularly Notch 3, has been observed in smokers with COPD, according to studies. The transcription of miR-206 was elevated in the lung parenchyma of the COPD group, which suppressed the transcription of the mRNAs for Notch 3 and VEGFA. In COPD lung tissues, miR-34a and miR199a-5p expression was noticeably higher [73]. Alveolar epithelial cells and pulmonary endothelial cells were more numerous in COPD-affected lungs than in healthy lungs, and CSE caused human umbilical vein endothelial cells to undergo apoptosis in a time- and dose-dependent manner [74]. The binding of the miR34a to the 3/UTR of the CSE-related Notch-1 gene can target the Notch-1 gene in endothelial cells. Growth cancer, atrophy, and hypertrophy can all be caused by impaired insulin-like growth factor (IGF) signaling, which also affects the quantity of ribosomes and protein synthesis [75]. As building blocks for other tissues, amino acids, energy, and carbon are all provided by protein turnover. In addition to controlling protein synthesis pathways, miRNAs also regulate ribosome function or the synthesis of ribosomal proteins. MiR-424-5p expression in COPD patients prevents protein synthesis by regulating polymerase I, which lowers muscle mass [76].

6.6 miRNAs as a COPD Treatment Strategy

miRNAs (micro RNAs) are short non-coding RNA which are not translated into proteins; they have a certain impact on the development of COPD. Numerous studies have demonstrated the impact of miRNAs on lung disorders which shows a correlation between specific miRNA profiles and both the prevention and acceleration of the development of COPD [77]. The control of cellular pathways has a significant impact on lung cancer development in COPD patients by miR-320b and miR-150-5p, two miRNAs that have been formerly proved to be anti-cancer. The expression of miR-320 and miR-150-5p has two effects on the onset of COPD: it protects the cancers linked to COPD, such as lung cancer, and it lowers inflammation and tissue damage. Inflammatory cytokine production is suppressed by miR146a [78, 79]. TNFa and proinflammatory mediators (IRAK1) are responsible for the negative regulation of the Toll-like receptor (TLR) and Interleukin-1 (IL-1) signaling components IL-8, IL-6, and IL-1b. In human adenocarcinoma alveolar basal epithelial cell line, miR146a with nanoparticles (NPs) activity lowered IRAK1 and TRAF6, aiding breathing in the management and treatment of COPD. One of the main traits of COPD is persistent hypoxia, and hypoxia-inducible factor-1 (HIF-1) regulates how the body reacts to chronic hypoxia, where HIF-1a is significant in COPD [80]. Previous research has identified miR-186 as one of the most important factors influencing cell proliferation in many malignancies. When lung fibroblast cell lines are transfected with miR-186, HIF-1a is impacted and its expression is reduced, which causes inflammatory fibroblasts to apoptosis. Cadmium (Cd), one of the harmful substances in cigarette smoke, causes lung impairment and inflammation in persons with COPD. Cd is linked to the development of COPD as well [53, 61, 70]. After exposure to Cd, human bronchial epithelial cells' transcription of MiR-181a-2-3p was downregulated, although proinflammatory activity and inflammatory reactions were both elevated. Consequently, miR-181a-2-3p may be used as a treatment for COPD. Intimal proliferation of dedifferentiated vascular smooth muscle cells in COPD is the key cellular factor contributing to pulmonary artery remodeling (SMCs) [67]. During vascular remodeling, miRNAs regulate the fate of both endothelial cells (ECs) and smooth muscle cells (SMCs). MiR-197 transcription must be negatively regulated for COPD to establish contractile activity. There is a correlation between SMC contractile markers and their respective phenotypes. Immune cells called alveolar macrophages (AMs) influence both acute and chronic inflammatory responses [70].

6.7 Conclusion

Cytokines that are involved in the pathophysiology of COPD can be released when AMs are activated. Lung inflammation can be brought on by the nuclear hormone receptor subfamily member peroxisome proliferator-activated receptor gamma (PPARc). By concentrating on the 3′-UTR regions of PPARc and inhibiting PPARc activation, miR-27-3p expression can restrict the production of pro-inflammatory cytokines and regulated TLR2/4 signaling. It demonstrated that miR-27-3p can be used as a COPD treatment strategy. The functionality of lung fibroblasts is altered in a variety of ways by the production of lung fibroblasts, including growth factors, fibronectin, and inflammatory mediators. COPD lung fibroblasts expressed miR-503 less highly. The loss of vasculature in COPD is aided by vascular endothelial growth factors (VEGF). Reduced expression of miR-503 in COPD patients increases the release of VEGF from lung fibroblasts, suggesting that it may be used as a COPD treatment. Expression of miR-483-5p prevents the suppression of cell development brought on by a-Smooth muscle actin (a-SMA), fibronectin, and transforming growth factor-b (TGF-b). Some miRNAs may serve as crucial indicator for COPD therapy targets. COPD severity is linked to inflammation indicators, including dysregulation of miRNAs. MiR-183-5p and miR-3177-3p, which are important biomarkers for the diagnosis of COPD, were downregulated during the onset and disease progression. It has been shown that miR-218-5p is suppressed in smokers or people with COPD insufficiency, and there is a reported negative association between miR-218-5p and the severity of COPD.

References

1. Agustí A, et al. Pathogenesis of chronic obstructive pulmonary disease: understanding the contributions of gene-environment interactions across the lifespan. Lancet Respir Med. 2022;10 (5):512–24.
2. Ahmad S, et al. Epigenetic underpinnings of inflammation: connecting the dots between pulmonary diseases, lung cancer and COVID-19. Semin Cancer Biol. 2022;83:384–98.
3. Alfahad AJ, et al. Current views in chronic obstructive pulmonary disease pathogenesis and management. Saudi Pharm J. 2021;29(12):1361–73.
4. Avci E, et al. Epigenetic mechanisms in parenchymal lung diseases: bystanders or therapeutic targets? Int J Mol Sci. 2022;23:1.
5. Balasubramanian S, et al. MicroRNAs and xenobiotic toxicity: an overview. Toxicol Rep. 2020;7:583–95.

6. Balestro E, et al. Lung tumors, COPD and immune response: is epigenetics the bottom line? Minerva Med. 2016;107(6 Suppl 1):1–8.
7. Barnes PJ. Senescence in COPD and its comorbidities. Annu Rev Physiol. 2017;79:517–39.
8. Barnes PJ. Pulmonary diseases and ageing. Subcell Biochem. 2019;91:45–74.
9. Barreiro E, Gea J. Epigenetics and muscle dysfunction in chronic obstructive pulmonary disease. Transl Res. 2015;165(1):61–73.
10. Baßler K, et al. Alveolar macrophages in early stage COPD show functional deviations with properties of impaired immune activation. Front Immunol. 2022;13:917232.
11. Benincasa G, et al. Epigenetics and pulmonary diseases in the horizon of precision medicine: a review. Eur Respir J. 2021;57:6.
12. Caramori G, et al. Molecular links between COPD and lung cancer: new targets for drug discovery? Expert Opin Ther Targets. 2019;23(6):539–53.
13. Chen X, et al. DNA methylation in chronic obstructive pulmonary disease. Adv Exp Med Biol. 2020;1255:83–98.
14. Clapp PW, Jaspers I. Electronic cigarettes: their constituents and potential links to asthma. Curr Allergy Asthma Rep. 2017;17(11):79.
15. Conlon TM, et al. Inhibition of LTβR signalling activates WNT-induced regeneration in lung. Nature. 2020;588(7836):151–6.
16. Corlăţeanu A, et al. From smoking to COPD—current approaches. Pneumologia. 2016;65(1): 20–3.
17. Dunham-Snary KJ, et al. Hypoxic pulmonary vasoconstriction: from molecular mechanisms to medicine. Chest. 2017;151(1):181–92.
18. Easter M, et al. Targeting aging pathways in chronic obstructive pulmonary disease. Int J Mol Sci. 2020;21(18):6924.
19. Forder A, et al. Mechanisms contributing to the comorbidity of COPD and lung cancer. Int J Mol Sci. 2023;24(3):2859.
20. Goh F, et al. Personalizing and targeting therapy for COPD: the role of molecular and clinical biomarkers. Expert Rev Respir Med. 2013;7(6):593–605.
21. Gruzieva O, et al. An update on the epigenetics of asthma. Curr Opin Allergy Clin Immunol. 2021;21(2):175–81.
22. Gruzieva O, Merid SK, Melén E. An update on epigenetics and childhood respiratory diseases. Paediatr Respir Rev. 2014;15(4):348–54.
23. Günes Günsel G, et al. The arginine methyltransferase PRMT7 promotes extravasation of monocytes resulting in tissue injury in COPD. Nat Commun. 2022;13(1):1303.
24. Herrington CS, Poulsom R, Coates PJ. Recent advances in pathology: the 2020 annual review issue of the journal of pathology. J Pathol. 2020;250(5):475–9.
25. Hey J, et al. Epigenetic reprogramming of airway macrophages promotes polarization and inflammation in muco-obstructive lung disease. Nat Commun. 2021;12(1):6520.
26. Hikichi M, et al. Pathogenesis of chronic obstructive pulmonary disease (COPD) induced by cigarette smoke. J Thorac Dis. 2019;11(Suppl 17):S2129–s2140.
27. Hoang TT, et al. Epigenome-wide DNA methylation and pesticide use in the agricultural lung health study. Environ Health Perspect. 2021;129(9):97008.
28. Huertas A, Palange P. COPD: a multifactorial systemic disease. Ther Adv Respir Dis. 2011;5 (3):217–24.
29. Huo X, et al. DNA methylation in chronic obstructive pulmonary disease. Epigenomics. 2021;13(14):1145–55.
30. Kabesch M, Adcock IM. Epigenetics in asthma and COPD. Biochimie. 2012;94(11):2231–41.
31. Kapellos TS, et al. Dysregulated functions of lung macrophage populations in COPD. J Immunol Res. 2018;2018:2349045.
32. Kapellos TS, et al. Human monocyte subsets and phenotypes in major chronic inflammatory diseases. Front Immunol. 2019;10:2035.
33. Kemp P, Natanek A. Epigenetics and susceptibility to muscle wasting in COPD. Arch Bronconeumol. 2017;53(7):364–5.

34. Lagoumtzi SM, Chondrogianni N. Senolytics and senomorphics: natural and synthetic therapeutics in the treatment of aging and chronic diseases. Free Radic Biol Med. 2021;171:169–90.
35. Leader BA, et al. Epigenetics of obstructive sleep apnea syndrome: a systematic review. J Clin Sleep Med. 2021;17(12):2533–41.
36. Lee HW, Jose CC, Cuddapah S. Epithelial-mesenchymal transition: insights into nickel-induced lung diseases. Semin Cancer Biol. 2021;76:99–109.
37. Lee M, et al. Pulmonary function and blood DNA methylation: a multiancestry Epigenome-wide association meta-analysis. Am J Respir Crit Care Med. 2022;206(3):321–36.
38. Li R, Zhou R, Zhang J. Function of PM2.5 in the pathogenesis of lung cancer and chronic airway inflammatory diseases. Oncol Lett. 2018;15(5):7506–14.
39. Ma J, Rubin BK, Voynow JA. Mucins, mucus, and goblet cells. Chest. 2018;154(1):169–76.
40. Mahrooz A, Mackness M. Epigenetics of paraoxonases. Curr Opin Lipidol. 2020;31(4):200–5.
41. Malhotra R, Olsson H. Immunology, genetics and microbiota in the COPD pathophysiology: potential scope for patient stratification. Expert Rev Respir Med. 2015;9(2):153–9.
42. Mao Y, et al. Genome-wide methylation and expression analyses reveal the epigenetic landscape of immune-related diseases for tobacco smoking. Clin Epigenetics. 2021;13(1):215.
43. Mekov E, et al. Update on asthma-COPD overlap (ACO): a narrative review. Int J Chron Obstruct Pulmon Dis. 2021;16:1783–99.
44. Melén E, et al. Allergies to food and airborne allergens in children and adolescents: role of epigenetics in a changing environment. Lancet Child Adolesc Health. 2022;6(11):810–9.
45. Mostafalou S, Abdollahi M. Pesticides and human chronic diseases: evidences, mechanisms, and perspectives. Toxicol Appl Pharmacol. 2013;268(2):157–77.
46. Nana-Sinkam SP, Choi AM. Epigenetics and the unfolded protein response in the lung: emerging role for microRNAs. Am J Respir Crit Care Med. 2014;189(3):239–40.
47. Nedeljkovic I, et al. Understanding the role of the chromosome 15q25.1 in COPD through epigenetics and transcriptomics. Eur J Hum Genet. 2018;26(5):709–22.
48. Ortiz-Quintero B, Martínez-Espinosa I, Pérez-Padilla R. Mechanisms of lung damage and development of COPD due to household biomass-smoke exposure: inflammation, oxidative stress, MicroRNAs, and gene polymorphisms. Cells. 2022;12:1.
49. Pantazopoulos I, et al. Incorporating biomarkers in COPD management: the research keeps going. J Pers Med. 2022;12(3):379.
50. Parris BA, et al. Chronic obstructive pulmonary disease (COPD) and lung cancer: common pathways for pathogenesis. J Thorac Dis. 2019;11(Suppl 17):S2155–s2172.
51. Qi C, Sun SW, Xiong XZ. From COPD to lung cancer: mechanisms linking, diagnosis, treatment, and prognosis. Int J Chron Obstruct Pulmon Dis. 2022;17:2603–21.
52. Rajendrasozhan S, et al. Deacetylases and NF-kappaB in redox regulation of cigarette smoke-induced lung inflammation: epigenetics in pathogenesis of COPD. Antioxid Redox Signal. 2008;10(4):799–811.
53. Rao W, et al. Regenerative metaplastic clones in COPD lung drive inflammation and fibrosis. Cell. 2020;181(4):848–64.e18
54. Regan EA, et al. Omics and the search for blood biomarkers in chronic obstructive pulmonary disease. Insights from COPDGene. Am J Respir Cell Mol Biol. 2019;61(2):143–9.
55. Rider CF, Carlsten C. Air pollution and DNA methylation: effects of exposure in humans. Clin Epigenetics. 2019;11(1):131.
56. Rosário Filho NA, et al. Air pollution and indoor settings. World Allergy Organ J. 2021;14(1): 100499.
57. Rosenwasser Y, Berger I, Loewy ZG. Therapeutic approaches for chronic obstructive pulmonary disease (COPD) exacerbations. Pathogens. 2022;11(12):1513.
58. Saco TV, et al. Epigenetics of mucus hypersecretion in chronic respiratory diseases. Am J Respir Cell Mol Biol. 2018;58(3):299–309.
59. Sakao S, Tatsumi K. The importance of epigenetics in the development of chronic obstructive pulmonary disease. Respirology. 2011;16(7):1056–63.

60. Schamberger AC, et al. Epigenetic mechanisms in COPD: implications for pathogenesis and drug discovery. Expert Opin Drug Discov. 2014;9(6):609–28.
61. Schwartz DA. Epigenetics and environmental lung disease. Proc Am Thorac Soc. 2010;7(2): 123–5.
62. Sczepanik FSC, et al. Periodontitis is an inflammatory disease of oxidative stress: we should treat it that way. Periodontol. 2020;84(1):45–68.
63. Shanmugam G, Rakshit S, Sarkar K. HDAC inhibitors: targets for tumor therapy, immune modulation and lung diseases. Transl Oncol. 2022;16:101312.
64. Shanmugam MK, Sethi G. Role of epigenetics in inflammation-associated diseases. Subcell Biochem. 2013;61:627–57.
65. Silverman EK. Applying functional genomics to chronic obstructive pulmonary disease. Ann Am Thorac Soc. 2018;15(Suppl 4):S239–s242.
66. Song Q, Chen P, Liu XM. The role of cigarette smoke-induced pulmonary vascular endothelial cell apoptosis in COPD. Respir Res. 2021;22(1):39.
67. Wielscher M, et al. DNA methylation signature of chronic low-grade inflammation and its role in cardio-respiratory diseases. Nat Commun. 2022;13(1):2408.
68. Wu DD, et al. The potential for targeted rewriting of epigenetic marks in COPD as a new therapeutic approach. Pharmacol Ther. 2018;182:1–14.
69. Wu H, et al. Regulation of lung epithelial cell senescence in smoking-induced COPD/emphysema by microR-125a-5p via Sp1 mediation of SIRT1/HIF-1a. Int J Biol Sci. 2022;18(2): 661–74.
70. Xie Z, et al. Perspectives on epigenetics alterations associated with smoking and vaping. Function (Oxf). 2021;2(3):zqab022.
71. Xu PW, Jin YT. Lung cancer and its epigenetics association with chronic obstructive pulmonary disease. Zhonghua Yi Xue Yi Chuan Xue Za Zhi. 2013;30(1):70–3.
72. Xu X, et al. Arachidonic acid 15-lipoxygenase: effects of its expression, metabolites, and genetic and epigenetic variations on airway inflammation. Allergy Asthma Immunol Res. 2021;13(5):684–96.
73. Yang D, et al. Ambient air pollution and biomarkers of health effect. Adv Exp Med Biol. 2017;1017:59–102.
74. Yao H, Rahman I. Current concepts on the role of inflammation in COPD and lung cancer. Curr Opin Pharmacol. 2009;9(4):375–83.
75. Yao H, Rahman I. Role of histone deacetylase 2 in epigenetics and cellular senescence: implications in lung inflammaging and COPD. Am J Physiol Lung Cell Mol Physiol. 2012;303(7):L557–66.
76. Yuan C, et al. Genetic polymorphism and chronic obstructive pulmonary disease. Int J Chron Obstruct Pulmon Dis. 2017;12:1385–93.
77. Zhai T, et al. Potential micronutrients and phytochemicals against the pathogenesis of chronic obstructive pulmonary disease and lung cancer. Nutrients. 2018;10(7):813.
78. Zhang L, et al. Epigenetic modifications and therapy in chronic obstructive pulmonary disease (COPD): an update review. COPD. 2020;17(3):333–42.
79. Zhang Z, et al. Hypermethylation of the Nrf2 promoter induces ferroptosis by inhibiting the Nrf2-GPX4 Axis in COPD. Int J Chron Obstruct Pulmon Dis. 2021;16:3347–62.
80. Zhong S, et al. Identification and validation of aging-related genes in COPD based on bioinformatics analysis. Aging (Albany NY). 2022;14(10):4336–56.

Epigenetics of Lung Cancer

7

Vibhav Varshney, Ahsas Goyal, and Neetu Agrawal

Abstract

Lung cancer is the biggest cause of cancer-related death, with a 5-year survival rate of just 18%. Thanks to recent developments in targeted pharmacological medications and immunotherapies, the survival rates of some patients have increased considerably. However, patients are usually only provided standard chemotherapy, which has a history of providing mediocre outcomes in the clinic. Accordingly, innovative methods of therapy are urgently required. Recent advances in epigenetic assessments and their use in research of cancer have shown the significance in the regulation of epigenetic lung cancer's progression, development, initiation, and treatment. There are several epigenetic modifications that occur in lung cancer at different phases of progression, and some of them are essential for tumour growth. Improved considerate of the natural science behind lung cancer growth and the facilitation of the creation of novel treatment options are both dependent on the continued development of state-of-the-art technologies like single-cell epigenomics. As a consequence of this strategy, treatments that use either a single medication or a combination of medicines to target epigenetic modifiers have been created and are now being explored in clinical trials. In this chapter, we discuss the function of epigenetics in lung cancer at different stages of the disease and how this knowledge is being used in clinical practice.

Keywords

Lung cancer · Oxidative stress · Epigenetics · Inflammation · Therapy

V. Varshney · A. Goyal (✉) · N. Agrawal
Institute of Pharmaceutical Research, GLA University, Mathura, UP, India
e-mail: ahsas.goyal@gla.ac.in

G. Gupta et al. (eds.), *Targeting Epigenetics in Inflammatory Lung Diseases*,
https://doi.org/10.1007/978-981-99-4780-5_7

7.1 Introduction

Lung cancer is the foremost cause of death due to cancer worldwide, with only around an 18% 5-year survival rate. In the majority of instances (80%), lung cancer is fatal if not caught early [1]. There have been only slight improvements in survival rates as a consequence of the theory. About 220,000 instances of lung and bronchus cancer are diagnosed each year in the United States, and use of tobacco is the major cause. The two most frequent types of lung cancer are small cell lung cancer (SCLC) and non-small cell lung cancer (non-SCLC) (NSCLC). NSCLC accounts for around 85% of all lung cancer diagnoses; its subtypes include squamous cell carcinoma (40%), adenocarcinoma (40%), and giant cell carcinoma (10%). SCLC accounts for around 15% of lung cancer diagnoses [2, 3]. The histologic classification of LC has been revised in light of the development of novel targeted therapies and chemotherapy regimens. Variations in kinases like HER2, BRAF, EGFR, ALK, RET, ROS1, and NTRK have been associated with therapeutically relevant adenocarcinomas. When it comes to treating SCLC, PI3K, FGFR1, and DDR2 inhibitors have shown very modest efficacies in clinical trials [4]. Immune frontier drugs have newly also shown groundbreaking results in the treatment of lung cancer, with some patients seeing durable responses. However, the wide variety of the condition means that even with these advancements, a single, effective treatment method cannot be applied to everyone. For this reason, specialised analyses of molecules are essential to decipher the complex network of relevant clinical phenotypes that regulate lung cancer growth, to define effectual methods of diagnostic for initial lung cancer recognition, and to advance novel approaches to increase clinically beneficial efficiencies for lung cancer [5]. Developments in high-throughput and high-resolution molecular technologies, our considerate mechanisms of molecular study of causal lung cancer progression, especially the complexity of its driving variables, have increased. Epigenetic alterations, which regulate gene expression and preserve genomic integrity, have been linked to lung cancer and may play a significant role in the disease's progress. In this chapter, we concentrate on lung cancer and address the potential therapeutic and diagnostic applications of epigenetic (de)regulation in the clinic [6, 7] (Fig. 7.1).

7.2 Cancer Epigenetics

Without changing the DNA sequence itself, epigenetic mechanisms allow for the guideline of gene expression and genome steadiness. Examples of alterations that may work together include modification of histone, nucleic acid methylation, chromatin remodelling, and changes in the production of non-coding RNAs. Many illnesses and disorders have been connected to epigenetic changes that are passed down via families [8, 9]. In a typical environment, DNA methylation is a strictly regulated process. This process, which occurs at cytosines and especially at CpG nucleotides and is often seen in gene regulatory elements, is mediated by the enzymes DNA methyltransferases (DNMTs) and demethylases. However, a

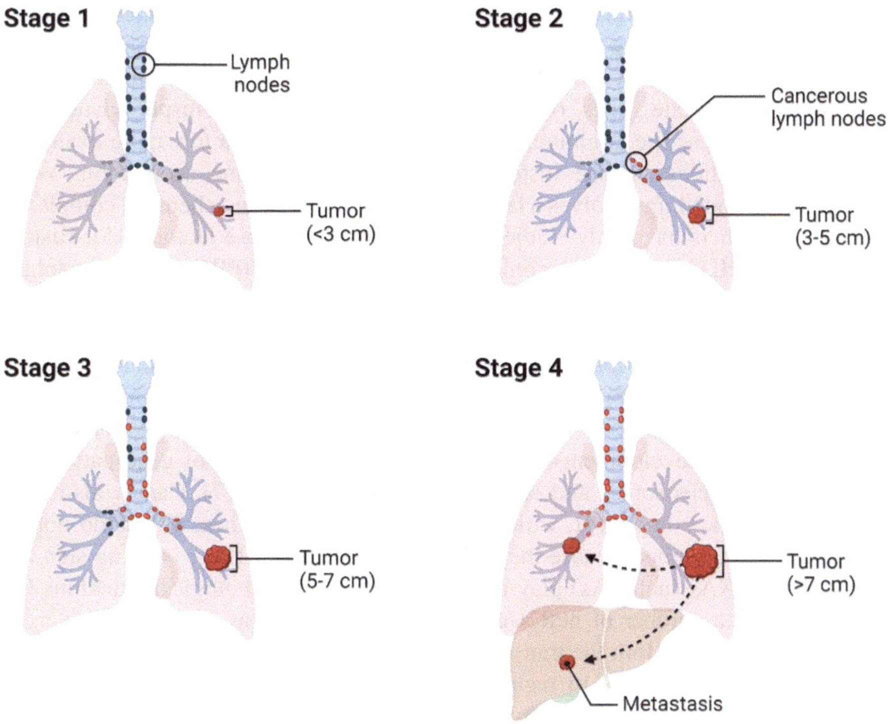

Fig. 7.1 Lung cancer stages

disturbance in DNA methylation equilibrium in tumour tissue may lead to modification of epigenetic key regulatory sections linked with transforming gene or tumour suppressor genes [10, 11]. Malignancies may have a more prominent hypo- to hypermethylation spectrum than noncancerous tissues. Hypomethylation has been related to a wide variety of pathologies, including chromosomal instability and aneuploidy, imprinting loss, transposon reactivation, and most importantly, oncogene activation. In contrast, DNA hypermethylation, the opposite process, may have dual roles depending on its position in the human genome [12]. Numerous tumour suppressor genes have been shown to be silenced by hypermethylation in CpG-rich regions inside gene promoters. Indirectly or directly, DNA hypermethylation may bring in complexes of protein affinity which is high for DNA methylation, which can decrease transcription factor binding. This might lead to the recruitment of chromatin remodellers like histone deacetylases and methylases thatmay result in the genes silencing like tumour suppressor genes or RNAs non-coding that have been linked to malignant transformation [13]. Increased expression of a gene could be correlated with high levels of methylation not only in the promoter region but also in the gene body. The fascinating potential is raised by our findings that hypermethylation of gene bodies may increase oncogene production [14]. Potentially, the state of chromatin is vigorously changed by a system of proteins with a wide variety of modes of action. Eukaryotic chromatin is composed mostly of nucleosomes, which communicate with histone binding proteins. Histones may be covalently changed to adjust their interface with DNA and so recruit histone modifiers to alter chromatin structure, mute or activate genes, or do both. The glycosylation, phosphorylation, and ubiquitination of serine, arginine, and threonine remains are a few more examples. Other modifications include histone methylation and lysine acetylation. While chromatin remodelling involves a large number of proteins and chemical complexes, cancer researchers have only studied a small portion of those factors [15]. Histone deacetylases (HDACs) and polycomb repressive complexes (PRCs) may both inhibit gene expression in ways unrelated to their enzymatic activity by altering histones or by directly compacting chromatin. Two other examples of chromatin remodelling with a greater relevance for cancer are the variants of histone, which can replace the canonical core of histones in nucleosomes and confer diverse functional and structural possessions that may distress chromatin compaction, and Switch/Sucrose non-fermentable developments, which can modify expression of gene patterns by repositioning nucleosomes and chromatin remodelling [16, 17]. Cancer also has a strong epigenetic component, which is noncoding RNAs. These RNAs do not get translated into proteins because they serve other functions. There is rising sign that non-coding RNAs (ncRNAs) play an important role in pathological situations like cancer, where a number of ncRNAs have been identified as oncogenic drivers or tumour suppressors. Both the transcriptional and post-transcriptional stages of gene expression are susceptible to regulation by ncRNAs. MicroRNAs (miRNAs) are small RNAs that target mRNAs for destruction, hence lowering gene target expression, while non-coding long RNAs interrelate with modifying chromatin enzymes and remodelling of chromatin factors, altering their activity [18]. In addition, scientists' focus on modifications of epigenetic RNA molecules is growing. Changes like cytosine or adenine methylation are only two of the numerous known variations that may influence the stability, translation, localization, splicing, and/or targeting of RNA. These changes may play a role in carcinogenesis by amplifying or dampening tumour characteristics like invasion and proliferation [19, 20] (Fig. 7.2).

7.3 Epigenetics of Lung Cancer

Genome-wide investigations of huge number of patient associates have shown the relevance of epigenetic changes in many cancer types, most notably lung cancer. Lung cancers are among those most often affected by the high mutation frequency in SWI/SNF and HDAC gene components (about 20%). Tobacco use is the primary cause of lung cancer deaths, and these changes might be related to this behaviour [21]. Further evidence suggests that smoking causes far-reaching impacts on DNA methylation, which seem to be long-lasting even after stopping. Changes of methylation in disorder-related genes has been seen in complete DNA blood samples from lung cancer patients. Several genes, including the aryl hydrocarbon repressor receptor and F2R like trypsin receptor 3 or thrombin, have had their methylation levels reduced in people who smoke and are at a higher risk of growing lung cancer, as

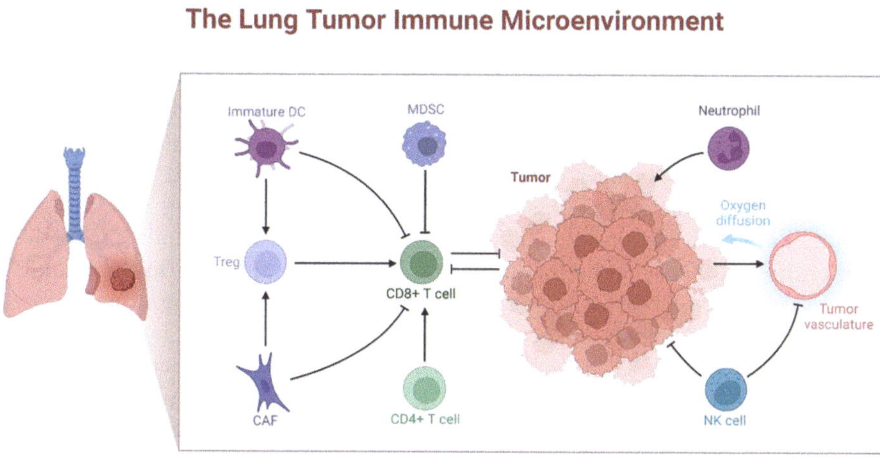

Fig. 7.2 The lung tumour immune microenvironment

shown by blood testing (F2RL3) [22]. Bossé et al. found a significant association between alterations in the expression of AHRR in non-tumour tissues of lung and cigarette smoking. Research by others has shown that BECs derived from lung cancer patients exhibit hypermethylation of the p16 and, to a lower extent, the death-associated protein (DAP) kinase promoters. This abnormal methylation occurred in BECs from present and previous smokers who did not develop lung cancer, suggesting a connection between tobacco use and cancer [23]. Certain expression of miRNA patterns in BECs have also been accompanying with cigarette smoking, perhaps because to the presence of intrinsic toxins in cigarettes. DNA methylation at individual CpG sites may have a dose- and cell-type-specific association with smoking; therefore, these results should be interpreted cautiously [24].

7.4 Epigenetics of NSCLC

Many studies have looked at how epigenetic events and interactions with tumour microenvironment (TME) affect the development and progression of NSCLC [25].

7.5 Epigenetics of NSCLC Tumour Initiation

The downstream fate-specifying of cell and other factors of transcription may contribute to tumour initiation, as shown by multi-omics investigations. The most common cause of lower gene expression is hypermethylation. The homeobox genes HOXA4, HOXA2, and NKX2-1, along with other factor of transcription genes, including GATA2 and ZNF132, were demonstrated to have a role in the discovery of possible tumour gene suppressor for SCLC [26]. The hypermethylation of genes including ZEB2, KCNIP4, and FOXF1 was found in lung cancer. Also, transcription factor genes hypermethylation like HOXA5, TAL1, FOXJ1, FOXA2, and HLF has been seen in both types of lung tumours COPD is a risk factor for lung cancer, and the hypermethylation of certain of these genes has been associated to COPD as well. Multiple additional epigenetic mechanisms besides hypermethylation have also been associated with lung cancer [27]. Overexpression of DNMT3A, DNMT1, and DNMT3B, for instance, has been associated to the onset of lung cancer. These enzymes have been demonstrated to contribute to NSCLC risk by inducing methylation errors in DNA and chromatin remodelling at an early stage in the disease's progression. When DNMT1 is upregulated during the onset of lung squamous cell carcinoma, tumour suppressor genes including RASSF1A, which is intricate in RAS signalling, and CDKN2A, a cyclin-dependent kinase blocker intricate in arrest of cell cycle at the G1/S phase, are silenced by hypermethylation (LUSC) [28, 29]. Also, it has been found that in lesions of preneoplastic squamous histology, aberrant promoter methylation affects the expression of multiple genes, including FHIT, whose degradation is associated with invasiveness and proliferation, and miR47b, a tumour suppressor whose downregulation is linked with tumour growth via the stemness-related Wnt pathway. The involvement of chromatin-remodelling complexes in LUSC development has also been identified. Aberrant chromatin remodellers reduce levels of H4K20me3 and elevate EZH2 expression leading to Silencing of tomour suppressor miRNAs [30]. Lung cancer has been connected to the downregulation of genes involved in apoptosis (DAPK), immortalization (hTERT), insulin metabolism (PTPRN2), and DNA repair (MGMT). The preneoplastic lesions of this histologic category have been reported to express CDKN2A, p741, RASSF1A, and many Wnt pathway antagonists [31, 32].

7.6 Epigenetic Determinants of NSCLC Progression and Metastasis

Aggressive NSCLC has been associated to changes in DNA methylation throughout the whole genome. Teschendorff et al. discovered a novel methylation of DNA pattern in cells of buccal that correlates with smoking and has the potential to be used to differentiate between advanced and lung in situ cancers. In lung cancer, hypermethylation was shown to downregulate 164 genes involved in differentiation, EMT, and cell cycle progression, whereas hypomethylation upregulated 57 genes. An epigenomic signature indicative of invasive lesions was identified by Teixeira et al. among 12,064 differentially methylation sites associated with 2695 genes in squamous cell carcinoma. In pro-metastatic phenotypic samples, many members of the homeobox family, including NKX2-1, were shown to exhibit hypermethylation and, as a result, to have reduced expression among the identified impacted genes [33, 34]. Others have recognized potential targets of key methylation associated with the acquisition of an invasive phenotype in NSCLC, including PTGDR, MAGE family members, FBP1, CDO1, and AJAP1, and alterations in the methylation pattern of PGC-related coactivator (PRC) members of family involved in regulation of cell cycle, invasion, and proliferation [35].

The epigenetic regulation of miRNA expression has been implicated as a cause of NSCLC. Highly aggressive NSCLC has been related to oncogenic miRNAs like miR-135b, which target components of the Hippo pathway and de-methylate their promoters. For instance, the miR-200 family controls EMT in NSCLC by targeting various effectors of this process such as miR-132, GATA3, ZEB1, and miR-149, which control FOXM1 and ZEB2. Thus, these miRNAs suppress invasion, migration, and metastasis in NSCLC by inhibiting the mesenchymal conversion process [36, 37].

NSCLC invasion and metastasis have been linked to the overexpression of certain epigenetic "writers" and "erasers" involved in chromatin remodelling. Proteins like H3K36 demethylase activating mitogen-activated protein kinase (KDM2A), H3K9 methyltransferase inducing Wnt (SETDB1), and H3.3 histone variant encoding (H3F3A) are all examples [38]. Mutations in BAF and PBAF, which are also components of SWI/SNF chromatin-remodelling complex of the human, have also been associated with NSCLC. There is evidence that mutations in the SMARCA4 gene contribute to tumour growth through altering the expression of other genes in the body [39].

Numerous lncRNAs have been identified to have tissue-specific expression patterns; several of them are downregulated in lung tumours. Genes like RCC2 and LCAL1 and KPNA2 have been connected to the growth and metastasis of lung tumours, while ENST00000439577 and LOC146880 have been correlated with the expression of these genes [40]. Other lncRNAs have been discovered to have a role in metastasis of lung cancer and/or EMT, in addition to HOTAIR and MALAT1, which produce a pro-metastatic gene expression profile. EZH2 downregulates a number of genes that play a role in modulating EMT, including BANCR and, most significantly, SPRY4-ITI, which does so via activating E-cadherin and repressing vimentin. Metastasis of lung cancer and EGFR-based therapy resistance has been related to the BC087858, lncRNAs, UCA1, and GAS5, perhaps via the activation of Akt signalling. Both EMT and treatment resistance may be influenced by these lncRNAs [41, 42]. Several epigenetic mechanisms have been associated with treatment resistance in NSCLC. It has been demonstrated, for example, that a drug-tolerant subset of NSCLC cells dependent on anti-EGFR treatment requires a KDM5A demethylase-induced altered chromatin state for survival of histone [43].

7.7 Epigenetics of Interactions Between NSCLC Cells and the TME

Numerous researches have been conducted to determine what role epigenetics plays in the interaction between cancer cells and the tumour microenvironment (TME). Researchers have discovered that connections between tumours and TMEs, which may be produced epigenetically, contribute to the development of lung tumours [44]. For instance, in NSCLC, miRs targeting TIMP3, a protein involved in the regulation of cytokines and growth hormones, are considered to contribute to tumour development by downregulating their expression. It is possible that TME, such as prostaglandin, will be found to have a role in the body. The overexpression of c-Myc in stromal cells is controlled by prostaglandin, which in turn affects the synthesis of the miR-17-92 cluster, which inhibits apoptosis in NSCLC tumour cells by targeting the tumour suppressor PTEN [45]. Other connections between tumours and TMEs have been identified to promote angiogenesis as well. There is an uptick in the expression of let-7b and miR-126 in cancer cells and their surrounding tissue. Similar to this, cancers may have the capacity to change the TME in ways that promote spread [46]. The interaction between a tumour and its TME may have a tumour-suppressing impact in certain situations. Exosomes released from lung tumours inhibited invasion and metastasis by decreasing ICAM, TME IL-8, and CXCL1 production, according to one research [47].

Multiple lines of evidence suggest that epigenetics plays a role in the control of antitumour immune responses in lung cancer. It has been shown that downregulating the DAP12 expression, a crucial transduction of signal receptor in NK cells, is one way in which TGF induces the miR-183 release from lung cancer cells, hence reducing the antitumour cytotoxic activity of NK cells. A second miRNA, miR-9, is upregulated in lung cancer and inhibits the immune system from recognising downregulating tumour cells by major histocompatibility complex (MHC) class I gene [48, 49].

7.8 Epigenetics of SCLC

Epigenetic changes have been hypothesised to have a role in the growth of SCLC. Neuroendocrine markers like NEUROD1 have been shown to be present in 75% of instances of SCLCs, which distinguishes SCLCs from other kinds of lung cancer. Samples of SCLC with very similar histological and genetic features have been shown to cluster with varied DNA methylation and gene expression, indicating that these biomarkers may be helpful for discriminating among SCLC subtypes [50]. The tumour-suppressor genes methylation is common in small cell lung cancer. Hypermethylation of the RASSF1A promoter was seen in almost all SCLC tumours, suggesting a function for this mechanism in tumour development. The DAPK tumour suppressor gene is methylated in 30% of patients with SCLC. RASSF1A promoter and DAPK tumour suppressor are only two examples of genes that play a role in signalling pathways linked with death receptor-mediated apoptosis [51].

Besides methylation-level anomalies, changes in chromatin remodelling enzymes have been detected in SCLC, signifying that they may play a key role in the tumour's progression. Lung neuroendocrine tumours like SCLCs have been shown to lose H4 methylation, which has been associated to increased proliferation. It has also been shown that several mutations exist in the genes that code for remodelling enzymes [52]. The MLL2 gene has the most frequent somatic mutations (8% of SCLC tumours). These modifications are associated with dysfunctional enhancers, which in turn cause genes to become dormant. EP300, CREBBP, and KAT6B are histone acetyltransferases; ARID1A, PBRM1, and ARID1B are chromatin remodelling factors; and inactivating mutations in these genes have been observed less often. Researchers have also shown that chromatin remodellers are expressed abnormally in these malignancies, in addition to the mutations already known to be present [53, 54]. In a subset of SCLC tumours, overexpression of the histone methyltransferase EZH2 accelerates E2F-driven carcinogenesis, and PCR2-related protein is increased. ASXL3, has been found in main SCLC tumours and is linked with enhanced cell lines growth of SCLC [55].

In the latter phases of small cell lung cancer growth, epigenetic changes may potentially play a role. Nfib, discovered in a recent research to have a role in increasing chromatin accessibility in several intergenic areas and in inducing neuronal gene expression programmes that drive the metastatic capacity of SCLC cells, is a transcription factor. Overexpression of MYCL, a member of the EMT-involved MYC family, is common in SCLC, and has been shown to be regulated by great enhancers and dependent on acetylation of histone [56, 57].

Although chemotherapy is the backbone of care for treating advanced SCLC tumours, drug resistance may develop quickly in certain patients. It has been suggested that one epigenetic modification causing treatment resistance in this collection of cells is the overexpression of EZH2, a DNA damage repair protein that is connected to the silencing of SLFN11 chemoresistance [58].

7.9 Translation of Epigenetic Knowledge to Clinical Practice

The development and course of lung cancer may be influenced significantly by epigenetic alterations, which might potentially be vital in the progress of successful treatments. It has been hypothesised that the epigenetic inactivation of specific tumour suppressor genes lies at the heart of chemoresistance, and many biomarkers of epigenetic have been found as possible predictors of chemo-resistance [59]. Through methylation, GSTP1 and RAR2 expression is silenced in SCLC and adenocarcinoma cells. Not only that, but the IGFB3 promoter in cisplatin-resistant lung cancer cell lines is hypermethylated. One such cell is the cancer stem cell (CSC), which, if it survives therapy, may repopulate the tumour and cause resistance. Among stem-related genes, unmethylated expression is inversely correlated with treatment resistance in NSCLC cells in vitro for OCT4 and SLUG [60, 61].

It is possible to classify therapeutic strategies for epigenetic regulation as either focusing on writers, focusing on erasers, or focusing on readers. To reactivate silenced tumour suppressor genes or revive production of tumour suppressor proteins, researchers have created DNMTi and histone-modifying enzyme inhibitors for the first category [62]. Tumour cells are more effectively arrested in their growth cycle, induced to undergo apoptosis, and differentiated when treated with either kind of inhibitor. Two types of DNMTis are used in research that are nucleoside analogues and non-nucleoside analogues. AZA and Decitabine, two cytidine nucleoside analogues, belong to the first group and work by inducing DNA hypomethylation and inhibiting DNMTs [63]. Both the Food and Drug

Administration and the European Medicines Agency have approved these drugs for the treatment of myelodysplastic syndrome (MDS). AZA was the first treatment to prolong life expectancy in people with MDS. The effectiveness of decitabine in treating acute myeloid leukaemia is also acknowledged by the European Union [64]. Downregulating the epigenetic miR-200/Z1EB axis with decitabine prevented TGF-1 from inducing epithelial-mesenchymal transition and mortality in lung cancer preclinical trials, as described by Zhang et al. In NSCLC, however, both decitabine and AZA have been demonstrated to be somewhat ineffective when administered alone. Promising results have been shown with the use of zebularine and other DNMTis since they provide a less risky option to conventional therapy. Highly selective for cancer cells yet harmless to healthy tissue, zebularine is a potent cytidine deaminase inhibitor [65]. In studies employing adenocarcinoma-derived cell lines, zebularine was shown to induce cell death by lowering intracellular reactive oxygen species (ROS) and increasing glutathione levels, and to halt cell development by causing a cell cycle arrest. Several current clinical trials are evaluating the finished chemical for its effectiveness against various cancers [66]. NCT01696032 is for patients with advanced SCLC, while NCT02131597 is for patients with leukaemia (NCT03085849). There have been five studies (NCT01534598, NCT01479348, NCT00359606, NCT01041443, and NCT00978250) looking at the effects of 5-fluoro-2′-deoxycytidine (FdCyd) on patients with solid tumours, acute myeloid leukaemia, and myelodysplastic syndromes [67].

These drugs do not deliver the best therapeutic outcomes when taken alone, despite their targeted nature and lower toxicity. That's why researchers are now interested in gauging the cumulative benefits of several therapies. Inhibitors of poly (ADP-ribose) polymerase (DNMTis) and PARP1 have been demonstrated to enhance chromatin PARP1 binding in myeloid leukaemia and breast cancer triple-negative. Adding the Bcl-2 inhibitor venetoclax to decitabine or AZA has been shown to increase patients response with relapsed/refractory and formerly untreated severe myeloid leukaemias [68]. The combination of azacitidine with entinostat, on the other hand, has not been found to enhance the efficacy of chemotherapy in any clinical trials. In the beginning, irinotecan showed promising results in xenograft models taken from patients with lung cancer. NSCLC patients in stages IIIb and IV have a new therapeutic option according to the publication of research data on the combination of decitabine and genistein (NCT01628471). It has been postulated that genistein's anticancer activity may be attributable, in part, to its ability to decrease DNMT expression. The combination of AZA and erlotinib showed encouraging effectiveness in treating erlotinib-resistant lung cancer [69, 70].

HDAC are a family of drugs that have been shown to be effective against both solid tumours and blood malignancies. There are several different kinds of HDACis, including benzamides, cyclic peptides, hydroxamic acids, fatty acids, and electrophilic ketones. In contrast, trials with single HDACis have shown either poor replies or transient effects, and these drugs may be rather dangerous. In an in vitro model of lung tumour growth, the cyclin tetrapeptide romidepsin was able to restore wild-type gene expression [71]. It has been postulated that romidepsin acts by inhibiting the overexpressing p21Waf1/Cip1, PI3K/Akt pathway, and lowering Rb phosphorylation. The findings of animal research on romidepsin for SCLC were encouraging, but the drug has not been effective in aiding humans (SCLC). It has been shown by Peifer et al. that mutations of histone are amongst the most prevalent changes in SCLC. The possibility of using them as therapeutic targets has stoked renewed interest in their investigation and advancement [72]. Together, cisplatin and a hydroxamate-based HDACi (vorinostat) had a stronger anticancer effect in vitro. The anti-cancer effects of valproic acid on NSCLC cells have been shown as well. Recent research suggests that panobinostat (a pan-HDACi) may greatly decrease NSCLC and render tumours more susceptible to carboplatin. Meanwhile, TAZ is inhibited by panobinostat. The compensatory of EGFR mechanism is reduced, making NSCLC more sensitive to gefitinib. These results demonstrate that the combination of gefitinib and panobinostat inhibits the growth of tumours with both KRAS mutations and EGFR wild-type sensitivity [73]. It has been demonstrated that the benzamide-based HDACi entinostat may slow the development of cancers that express the stem cell factor SALL4. Patients with high E-cadherin expression had a long median overall survival when linked to those treated with erlotinib plus placebo, despite the fact that a phase II study of entinostat in combination with erlotinib for progressive NSCLC showed no improvement in clinical outcome. The progress and division of SCLC cells have also been inhibited in vitro by the thrichostatin valpromide, HDAC, and valproic acid, with thrichostatin [74]. Researchers have shown that A thrichostatin, like valproic acid, may boost Notch signalling. Dibenzazepine, an inhibitor of Notch signalling, combined with either HDACi, improved efficacy above the use of either drug alone. For this reason, numerous different histone deacetylase inhibitors (HDACis) may be useful in treating this disease. From this knowledge, studies involving HDAC inhibitors in lung cancer, such as NCT02728492, NCT01935947, NCT00738751, and NCT01059552, have greatly benefited [75]. Among the latest epigenetic medications in development are those that target the enzymes enhancer of disruptor of telomeric silencing1-like (DOT1L), zeste homolog 2 (EZH2), protein arginine nine N-methyltransferase, lysine-specific demethylase 1 (LSD1), and bromodomain and extraterminal motif. Evanno et al. discovered that epithelial-to-mesenchymal transition in NSCLC cells may be partially restored using a combination deacetylase inhibitor of histone and an inhibitor of BET. Studies using the LSD1 Inhibitor T-3775440 have showed encouraging results in slowing the development and spreading of SCLC cells [76, 77].

The microenvironment inside tumours has been hypothesised to have a significant role in tumour development. Besides tumour cells, the tumour microenvironment consists of vascular endothelial cells, adipocytes, pericytes, and immune system cells. Tumour cells may primarily generate immunological checkpoint dysregulation by dampening the body's natural defences against them. Learning more about these procedure has re-energized the lung cancer immunotherapeutic treatment [78]. The potential for immune-based therapies has so expanded. Furthermore, epigenetics has been proven to play a role in modulating immunological signals. There is evidence that combining epigenetic therapy with immunological checkpoint therapy may reawaken the immune system and effectively combat several forms of cancer, like lung cancer. Topper et al. discovered that mocetinostat, givinostat, and entinostat all had a potent effect of anti-tumour, and that givinostat displayed a considerable antiproliferative action in NSCLC when paired with AZA [79]. AZA was hypothesised to make tumour cells more vulnerable to HDACis, to cause a downregulation of MYC expression, and to alter the phenotype of T cells such that they become memory and effector cells. Based on these findings, researchers designed a clinical study in which patients with advanced NSCLC received escalating dosages of mocetinostat, guadecitabine, and an anti-PD1 antibody (NCT03220477; pembrolizumab). Numerous recent medical trials are looking at the effectiveness of combining epigenetic medicine with immune inhibitors checkpoint for the therapeutic effect on lung cancer [80]. The effects of decitabine in combination with immunological check point inhibitors and tetrahydrouridine are being studied in two-stage I/II clinical trials (NCT03233724 and NCT02664181, respectively) including patients with NSCLC. Moreover, two clinical trials are investigating the efficacy and safety of entinostat coupled with pembrolizumab in patients with advanced NSCLC (NCT02909452and NCT02437136). The common goal of all of these studies is to find a way to increase immune activation via epigenetic change and so improve the efficiency of treatment [81, 82].

7.10 Conclusions and Future Perspectives

Epigenetic study has been a huge help in understanding the molecular processes underlying cancer genesis, which has huge therapeutic ramifications. Epigenetic regulation is highly organised in lung cancer and plays a role at every step of the disease's course, from the earliest stages of formation through treatment resistance in the later stages. Evidence suggests that epigenetic therapies may be useful for treating cancer. To combat haematological cancers, researchers first studied these medicines and then altered them. Learning more about the proteins that play a role as epigenetic regulators has led to the development of novel drugs and treatment methods. In addition, a patient's distinctive characteristics may be employed to enhance the beneficial effects of a certain therapy. Tyrosine kinase inhibitors, chemotherapy, and/or immunotherapy in combination with epigenetic drugs have demonstrated beneficial effects in the prevention of lung cancer. The potential for reviving latent immune responses via the use of epigenetic medications in combination with immunotherapy has far-reaching consequences for the current state of immunotherapy. To identify successful strategies for preventing lung cancer, more research, both preclinical and clinical, is required. Because methylation of DNA might be specific in type of a cell, heterogeneity of cancer cell is an important concern during sample analysis that may serve as a confounding influence when doing epigenomics in bulk. In recent years, single-cell transcriptomics has demonstrated its capacity to distinguish between the molecular structures of various intratumor cell populations. Since this technology is becoming more important in precision cancer therapy, our present knowledge of cancer biology is evolving. We expect that the newly advanced and refined methylation of single-cell technique will play a crucial role in intra-tumoral dissection of heterogeneity and deliver responses to questions concerning tumour initiation, progression, and therapy response. Less-intrusive techniques, together with developments in molecular technology, have showed promise for precision oncology. Lung cancer epigenetics research may therefore make use of a wide range of human bodily fluids, not only resection materials and biopsies, but also serum, sputum, plasma, saliva, and bronchoalveolar lavage. Accessibility isn't the only perk of these sources; they also let you gather data in a logical order. Cell-free DNA in the blood has recently been revealed to include methylation patterns that may be indicative of the origin of tumours and/or the development of diseases.

In conclusion, the development of cutting-edge technology and the expansion of the types and numbers of clinical trials that may be used for epigenomic research will be vital in furthering our understanding of the biology of lung cancer. The increasing clarity of this knowledge at the single-cell level bodes well for the progress of novel analytical and therapeutic approaches to this particularly tenacious kind of cancer.

References

1. Lillington GA. Lung cancer. Curr Opin Pulm Med. 2004;10(4):239–41.
2. Aberle DR, Brown K. Lung cancer screening with CT. Clin Chest Med. 2008;29(1):1–14. v
3. Abu Rous F, et al. Lung cancer treatment advances in 2022. Cancer Investig. 2023;41(1):12–24.
4. Amann A, et al. Lung cancer biomarkers in exhaled breath. Expert Rev Mol Diagn. 2011;11(2): 207–17.
5. Arifin AJ, Palma DA. The changing landscape of pneumonitis in non-small cell lung cancer. Lung Cancer. 2022;171:1–2.
6. Bastarrika G, Pueyo JC, Mulshine JL. Radiologic screening for lung cancer. Expert Rev Anticancer Ther. 2002;2(4):385–92.
7. Bearz A, et al. Target therapies in lung cancer. J Biomed Biotechnol. 2011;2011:921231.
8. Beattie EJ Jr. Lung cancer. CA Cancer J Clin. 1974;24(2):96–9.
9. Beattie EJ. Lung cancer. World J Surg. 1981;5(5):661–2.
10. Belani CP, et al. Women and lung cancer: epidemiology, tumor biology, and emerging trends in clinical research. Lung Cancer. 2007;55(1):15–23.
11. Cagle PT, Chirieac LR. Advances in treatment of lung cancer with targeted therapy. Arch Pathol Lab Med. 2012;136(5):504–9.
12. Cochrane A, Alvarez JM. Upstaging of lung cancer and waiting times for surgery. Heart Lung Circ. 2019;28(3):364–5.
13. Dong Y, et al. Research Progress on the relationship between blood lipids and lung cancer risk and prognosis. Zhongguo Fei Ai Za Zhi. 2020;23(9):824–9.
14. Donington JS, Le QT, Wakelee HA. Lung cancer in women: exploring sex differences in susceptibility, biology, and therapeutic response. Clin Lung Cancer. 2006;8(1):22–9.
15. Ellis J. The impact of lung cancer on patients and carers. Chron Respir Dis. 2012;9(1):39–47.
16. Endo C, Sakurada A, Kondo T. Early central airways lung cancer. Gen Thorac Cardiovasc Surg. 2012;60(9):557–60.
17. Epler GR. Screening for lung cancer. Is it worthwhile? Postgrad Med. 1990;87(6):181–6.
18. Erasmus JJ, Truong MT. Imaging of lung cancer: update on screening, staging, and therapy. Radiol Clin N Am. 2018;56(3):xv–xvi.
19. Evans M. Lung cancer: needs assessment, treatment and therapies. Br J Nurs. 2013;22(17): S15-6. s18, s20-2

20. Garelli E, et al. Abscopal effect in lung cancer: three case reports and a concise review. Immunotherapy. 2019;11(17):1445–61.
21. Ge X, et al. Research Progress of circular RNA in lung cancer. Zhongguo Fei Ai Za Zhi. 2020;23(12):1095–100.
22. Ginsberg MS. Letter from the guest editor: lung cancer. Semin Roentgenol. 2011;46(3):169.
23. Goya T, et al. Lung cancer. Gan To Kagaku Ryoho. 1999;26(1):49–53.
24. Grannis FW Jr. Minimizing over-diagnosis in lung cancer screening. J Surg Oncol. 2013;108 (5):289–93.
25. Han X, Ma S. Current situation of clinical feature and gene phenotype of young adult lung cancer. Zhongguo Fei Ai Za Zhi. 2020;23(5):388–92.
26. Hashemi ZS, et al. Lung cancer and miRNAs: a possible remedy for anti-metastatic, therapeutic and diagnostic applications. Expert Rev Respir Med. 2017;11(2):147–57.
27. Katzman D, Wu S, Sterman DH. Immunological aspects of cryoablation of non-small cell lung cancer: a comprehensive review. J Thorac Oncol. 2018;13(5):624–35.
28. Lanuti M. Surgical management of lung cancer involving the chest wall. Thorac Surg Clin. 2017;27(2):195–9.
29. Liao M. Some features of lung cancer in China. Lung Cancer. 1993;10(1-2):107–16.
30. Loizidou A, Lim E. Is small cell lung cancer a surgical disease at the present time? Thorac Surg Clin. 2021;31(3):317–21.
31. Lovly CM. Expanding horizons for treatment of early-stage lung cancer. N Engl J Med. 2022;386(21):2050–1.
32. Malyla V, et al. Recent advances in experimental animal models of lung cancer. Future Med Chem. 2020;12(7):567–70.
33. Martini N. Operable lung cancer. CA Cancer J Clin. 1993;43(4):201–14.
34. Minna JD, Roth JA, Gazdar AF. Focus on lung cancer. Cancer Cell. 2002;1(1):49–52.
35. Mizutani H, Gemma A. Lung cancer. Gan To Kagaku Ryoho. 2009;36(2):171–5.
36. Nanguzgambo AB, et al. Immunochemistry and lung cancer: application in diagnosis, prognosis and targeted therapy. Oncology. 2011;80(3-4):247–56.
37. Neville A. Lung cancer. Clin Evid. 2003;10:1804–23.
38. Neville A. Lung cancer. Clin Evid. 2005;14:1903–20.
39. North CM, Christiani DC. Women and lung cancer: what is new? Semin Thorac Cardiovasc Surg. 2013;25(2):87–94.
40. O'Keeffe P, Patel J. Women and lung cancer. Semin Oncol Nurs. 2008;24(1):3–8.
41. Ostrowski M, Marjański T, Rzyman W. Low-dose computed tomography screening reduces lung cancer mortality. Adv Med Sci. 2018;63(2):230–6.
42. Partridge MR. Lung cancer. Br J Hosp Med. 1990;43(6):413–21.
43. Pett SB Jr, Wernly JA, Akl BF. Lung cancer—current concepts and controversies. West J Med. 1986;145(1):52–64.
44. Piperi C, et al. Epigenetic effects of lung cancer predisposing factors impact on clinical diagnosis and prognosis. J Cell Mol Med. 2008;12(5a):1495–501.
45. Port JL, Kent M, Altorki NK. Early lung cancer detection and treatment strategies. Surg Oncol. 2002;11(4):191–9.
46. Quinn S. Lung cancer: the role of the nurse in treatment and prevention. Nurs Stand. 1999;13 (41):49–54. quiz 55
47. Rivera MP, Stover DE. Gender and lung cancer. Clin Chest Med. 2004;25(2):391–400.
48. Salehi-Rad R, et al. The biology of lung cancer: development of more effective methods for prevention, diagnosis, and treatment. Clin Chest Med. 2020;41(1):25–38.
49. Sandler JE, Kaumaya M, Halmos B. Biomarker use in lung cancer management: expanding horizons. Biomark Med. 2018;12(4):315–20.
50. Schiller JH. Lung cancer: therapeutic modalities and cytoprotection. Lung. 1998;176(3): 145–64.
51. Sellars RE, Zimmerman PV. Lung cancer. Med J Aust. 1997;167(2):99–104.
52. Song J, Zhang A. Screening for lung cancer. Clin J Oncol Nurs. 2014;18(5):601.

53. Sotto-Mayor R. Lung cancer in women: a different entity? Rev Port Pneumol. 2006;12(5):545–61.
54. Stahel RA. Biology of lung cancer. Lung Cancer. 1994;10(Suppl 1):S59–65.
55. Subramanian J, Govindan R. Lung cancer in 'Never-smokers': a unique entity. Oncology (Williston Park). 2010;24(1):29–35.
56. Sugarbaker DJ, Dasilva MC. Diagnostic workup of lung cancer. Surg Oncol Clin N Am. 2011;20(4):667–79.
57. Sundaram B, Kazerooni EA, Preface. Lung cancer is an important public health care issue. Radiol Clin N Am. 2012;50(5):xi.
58. Sung HJ, Cho JY. Biomarkers for the lung cancer diagnosis and their advances in proteomics. BMB Rep. 2008;41(9):615–25.
59. Talasaz A. Lung cancer and a bold new vision. Future Oncol. 2020;16(12):701–3.
60. Theegarten D, Hager T. Pathology of lung cancer. Radiologe. 2016;56(9):777–85.
61. Torok S, et al. Lung cancer in never smokers. Future Oncol. 2011;7(10):1195–211.
62. Tyczynski JE, Bray F, Parkin DM. Lung cancer in Europe in 2000: epidemiology, prevention, and early detection. Lancet Oncol. 2003;4(1):45–55.
63. Vavalà T, et al. An examination of two dichotomies: women with lung cancer and living with lung cancer as a chronic disease. Respirology. 2020;25(Suppl 2):24–36.
64. Veronesi G, et al. When is surgery indicated for small-cell lung cancer? Lung Cancer. 2015;90(3):582–9.
65. Wang CY, et al. Mechanisms of lung cancer caused by cooking fumes exposure: a minor review (△). Chin Med Sci J. 2017;32(3):193–7.
66. Wang Y, Zhou Y, Miao L. A review of drug therapy of lung cancer with interstitial lung disease. Zhongguo Fei Ai Za Zhi. 2020;23(4):286–93.
67. Wang Y, et al. Clinical implication of microrna for lung cancer. Cancer Biother Radiopharm. 2013;28(4):261–7.
68. Welcker K. Gender differences in lung cancer. Zentralbl Chir. 2015;140(3):260–5.
69. Wu K, et al. Next-generation sequencing for lung cancer. Future Oncol. 2013;9(9):1323–36.
70. Yao Y, et al. The effects and management of viral pneumonia on lung cancer patients. Zhongguo Fei Ai Za Zhi. 2020;23(4):255–60.
71. Yu H, et al. The clonal evolution and therapeutic approaches of lung cancer. Cell Biochem Biophys. 2014;70(1):63–71.
72. Yu M, Tan J, Wang J. Research Progress of single cell sequencing in lung cancer. Zhongguo Fei Ai Za Zhi. 2021;24(4):279–83.
73. Zagonel V, et al. Lung cancer in the elderly. Cancer Treat Rev. 1994;20(4):315–29.
74. Zarredar H, et al. Critical microRNAs in lung cancer: recent advances and potential applications. Anti Cancer Agents Med Chem. 2018;18(14):1991–2005.
75. Zhang L. Preface of interventional therapy for lung cancer. Zhongguo Fei Ai Za Zhi. 2020;23(6):407–8.
76. Zhou C. Blood-based tumor markers in lung cancer. Zhongguo Fei Ai Za Zhi. 2015;18(12):770–80.
77. Zhou H, Suo J, Zhu J. Therapeutic relevance of human microbiota and lung cancer. Zhongguo Fei Ai Za Zhi. 2019;22(7):464–9.
78. Zhou J, et al. Research Progress of tumor-associated neutrophils and lung cancer. Zhongguo Fei Ai Za Zhi. 2019;22(11):727–31.
79. Aberle MF, McLeskey SW. Biology of lung cancer with implications for new therapies. Oncol Nurs Forum. 2003;30(2):273–80.
80. Adjei AA. Lung cancer-celebrating progress and acknowledging challenges. J Thorac Oncol. 2013;8(11):1350–1.
81. Bunn PA Jr. Worldwide overview of the current status of lung cancer diagnosis and treatment. Arch Pathol Lab Med. 2012;136(12):1478–81.
82. Chen T, Yang Y. Role of circular RNA in diagnosis, development and durg resistance of lung cancer. Zhongguo Fei Ai Za Zhi. 2019;22(8):532–6.

Epigenetics of Pulmonary Tuberculosis

8

Madan Mohan Gupta, Ritu Gilhotra, Deepika Deopa, Asif Ahmad Bhat, Riya Thapa, Neelam Singla, Rashi Kulshrestha, and Gaurav Gupta

Abstract

Tuberculosis (TB) is a very contagious, chronic disease caused by acid-fast bacilli that are notoriously difficult to eradicate from the host's surroundings. For it to work, the host organism's innate and adaptive immune systems must be intact. There are several sensors for recognising patterns: when foreign pathogens or their by-products are detected by immune cells, which then trigger an immunological response. To make the host more susceptible to infection and prime the immune system to fight off the invading virus, epigenetic modification is essential. By altering the expression of genes, it alters the host cell's genetic makeup. Researchers have looked at the role of histone acetylation, ncRNA modification, methylation of DNA, and miRNA modification in TB pathophysiology to halt its development. While there has been a lot of study, many questions remain unanswered. In this chapter, we will go through the immunopathophysiological causes of TB, the basics of epigenetics, and how epigenetic research is now being used to understand the disease's pathophysiology and progression.

M. M. Gupta
School of Pharmacy, Faculty of Medical Sciences, The University of the West Indies, Saint Augustine, Trinidad and Tobago

R. Gilhotra · A. A. Bhat · R. Thapa · N. Singla · R. Kulshrestha · G. Gupta (✉)
School of Pharmacy, Suresh Gyan Vihar University, Jaipur, India

D. Deopa
School of Pharmacy, Suresh Gyan Vihar University, Jaipur, India

Six Sigma Institute of Technology and Science, Rudrapur, Uttarakhand, India

G. Gupta et al. (eds.), *Targeting Epigenetics in Inflammatory Lung Diseases*, https://doi.org/10.1007/978-981-99-4780-5_8

8.1 Introduction

Mycobacterium tuberculosis (Mtb) is a main cause of death globally, and it disproportionately affects the poor. Infection rates for Mtb have risen to one-third of the world's population or has done so in the past [1]. Twenty-five percent of all fatalities worldwide are caused by infectious illnesses, second only to cardiovascular (CV) disease in terms of mortality rates. Almost two-thirds of the 8.6 million people who get Mtb each year are male, according the World Health Organization (WHO) [2]. Co-infection with HIV accounts for 320,000 of the 1.3 million annual deaths among patients. There are now 84 positive Mtb smear tests per 100,000 patients each year in India. [3–5].

The two main goals of Mtb pharmacological treatment are lesion sterilisation and disease transmission prevention [6]. Mainstays of first-line treatment programmes include isoniazid, rifampicin, pyrazinamide, and ethambutol; these drugs may be used alone or in combination. There has been a 2–3% rise in infections of Mtb in the USA in 2010, as was projected [7]. Also, one-fourth of all newly confirmed cases of Mtb may be found in Africa, and 25% of all deaths caused by Mtb infections occur there. Multiple mutations, even though they are often uncommon and lead to worse results when treated with single medication [8]. Hence, the problem is addressed when many drugs are used together. Since they work in different ways, the efficacy of the treatments is increased while also addressing the problem of resistance [9]. For illnesses that do not respond to many medications, however, this is not the case. Second- and third-line drugs, which are used in addition to first-line medicines, are more toxic, costly, and ineffectual [10]. Newer drugs like bedaquiline and delamanid are effective against Mtb infections that have developed a resistance to many drugs. Another method for stopping the spread of Mtb is a vaccine developed from *Mycobacterium bovis* strain, which has been used since the 1930s. The BCG vaccine boosts immunity against the disease quickly [11]. Many issues, including social stigma, marginalisation, and less adherence to therapy due to the extended period, impede the management of Mtb infections. The risk of infection and co-occurring diseases is further increased by immune failure, such as HIV infection [12] (Figure 8.1).

A variety of novel strategies are now in development as a means of addressing the difficulties with the existing treatment. Interferon (IFN), imiquimod, interleukin (IL)-12, granulocyte-macrophage colony-stimulating factor (GM-CSF), and levamisole are some other drugs being evaluated as anti-Mtb therapy [13]. Additionally, research is being conducted on a vaccine derived from the rapidly reproducing saprophytic Mycobacterium vaccae as a novel defence against Mtb infections; however, first results have been contradictory [14]. The World Health Organization (WHO) has promoted "directly observed treatment, short-course (DOTS) "monitoring since the 1990s due to its enhanced treatment adherence and effectiveness. One of the primary contributors to relapse and drug resistance, noncompliance with prescribed therapy is the focus of this strategy [15].

Recently, individualised treatment regimens have been developed for conditions like M. tuberculosis. Making individualised host-directed medicines may benefit

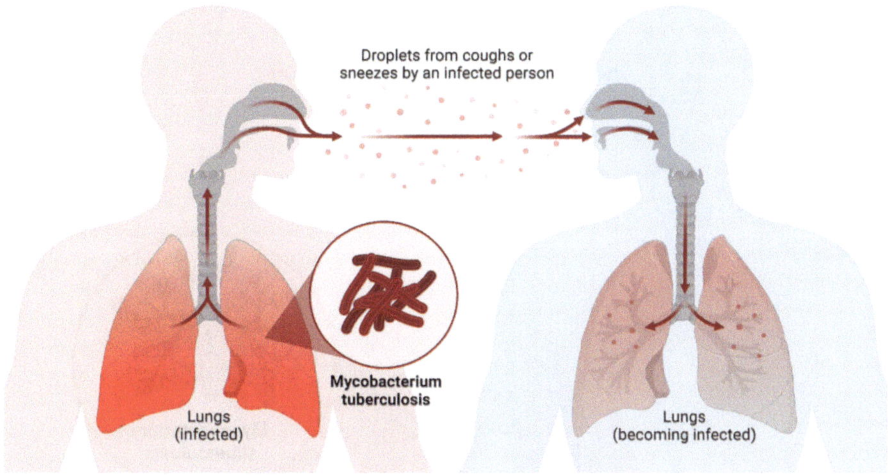

Fig. 8.1 Spread of tuberculosis

from understanding epigenetic pathways [16]. In a few recent studies, epigenetic changes including as miRNA-mediated DNA methylation, and histone modification control have been connected to interaction of Mtb pathogen with host cells. Investigating how cytokines like interferon-alpha, IL-12, and tumour necrosis factor-alpha affect the methylation and acetylation of histone proteins may lead to the creation of novel therapeutic strategies [17]. It will take some time until the field of epi-therapeutics, which focuses on treating illnesses by modifying epigenetic changes, is deemed completely established. We describe the Mtb infections immunopathophysiology, the epigenetic alteration brought on by the pathogen of Mtb, and the outcomes of recent in vitro and in vivo studies that have paved the way for further study into the use of epigenetics in the creation of potent treatments [18, 19].

8.2 Immuno-Pathophysiology of Mycobacterium Tuberculosis

The TB virus is widely dispersed in the atmosphere as an aerosol and is often acquired via inhalation. How an infection manifests inside a host hinges on both the effectiveness of the host's immune system and virulence of the pathogen [20]. The immunological-pathophysiological process may be used to explain this, since it involves both the adaptive and innate immune responses. If the pathogen of Mtb can overcome the host's physical defences, the innate immune system will be put at risk [21]. When Mtb invades a host, the innate immune system is activated and becomes the body's first line of defence. Macrophages are the first immune system cells to encounter the pathogen during an Mtb infection [22]. Invading pathogens secrete Pathogen-Associated Molecular Pattern (PAMP)s when entering a host body. Such patterns include conserved motifs that activate host defence mechanisms by

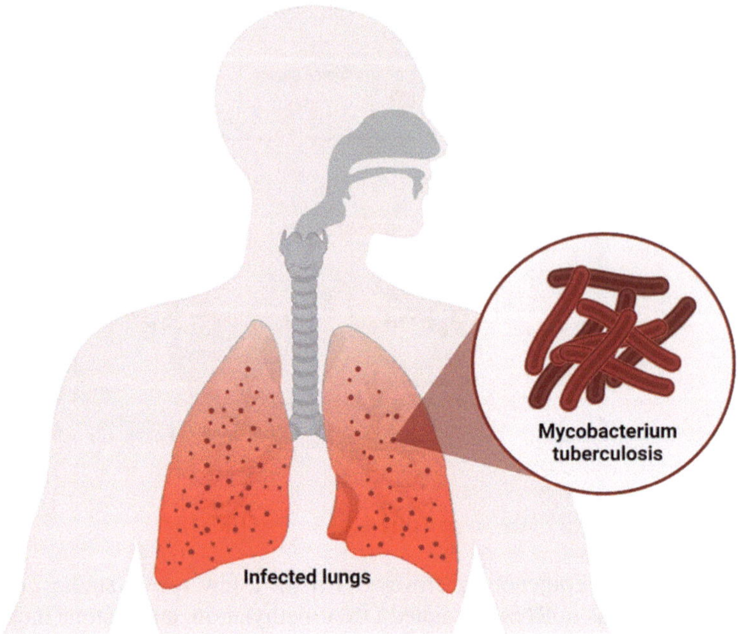

Fig. 8.2 Person infected with Mycobacterium tuberculosis

binding to receptors like as toll-like receptor (TLR)s. Hence, the immune system utilises these molecular patterns to initiate an inborn response of immune system against the attacking Mtb [22, 23] (Fig. 8.2).

Pattern recognition receptors are used to find additional ligands on the pathogen's surface. Surface protein A, lectin, and mannose, CD14 are some of these ligands. The most prevalent groups of pattern recognition receptor (PRR)s are TLRs, C-type lectin receptors (CLRs), RIG-I-like receptors (RLRs)s, complement receptors (1, 3, and 4), mannose receptors (MR), scavenger receptors (CR), CD43 and CD14, while additional PRRs play a role, the TLR 1–4 and 9 are especially important in the pathogenesis of Mtb [24–26]. Both non-immune cells and immune cells express the TLRs 1–4 and 9, correspondingly. Moreover, when host cells recognise molecular patterns linked to pathogens or danger, PRRs start the inflammasome pathway via regulating caspase-1. Mtb also robustly stimulates pertussis toxin (PTX)-3, a 42-KDa soluble PRR intricate in acute immunological responses to infection [27]. T and B cells are examples of adaptive immune responses in infections of Mtb. The recruitment of activated macrophages, granulocytes, and NK-cells with the aim of eliminating the pathogen facilitates the formation of the granuloma that is distinctive of Mtb. To make matters worse, immune cells stimulated by Mtb generate inflammatory cytokines including IFN-g and tumor necrosis factor (TNF-α) [18, 28].

With time, active macrophages transformed into epithelioid macrophages in response to persistent IFN activation [29]. They work together to form the granuloma, which encases the lung infection. The granuloma includes large quantities of

TNF, which inhibit infection, in addition to the core caseating necrosis, which is surrounded by fibroblasts, lymphocytes, and epithelial macrophages [30]. Moreover, 1-α hydroxylase overexpression at the granuloma site leads to calcification of the granuloma. If Mtb is still present inside the granuloma, it's called latent. Once the host's immune system is conceded, Mtb might become active again, causing a secondary case of tuberculosis [31]. Tregs are also present, and they employ IL-10, IL-35, and transforming growth factor (TGF) to suppress the immune system's capacity to destroy mycobacteria. Hence, the immuno-pathogenesis of Mtb is affected by Mtb-host immune cell interaction [32]. Moreover, Mtb bacillus makes use of the host immune system's signalling pathways to take advantage of the double-duty nature of PE proteins, hence facilitating the pathogen's growth and survival within the host [33]. The PE6 proteins further the disease by facilitating the proliferation and intracellular survival of pathogens via binding cell iron. Vitamin D has a part in the pathophysiology of infections of Mtb, because it is required to produce defensins and cathelicidins, that help in the removal of the bacteria [34, 35].

8.3 Tuberculosis and Epigenetic Modification and Regulations

C. H. Waddington first developed the concept of epigenetics in the 1940s to explain a method for manipulating genes without changing the cellular sequence [36]. The term "epigenetic regulation" refers to a little modification in gene countenance brought on by chromosomal changes deprived of altering the coding DNA's nucleotide sequence [37]. The Mtb pathogen seizes control of the host and rewrites the epigenome for its own defence via histone changes, miRNA-mediated regulation of genes and methylation of DNA [38]. These changes are essential for the host immunomodulation caused by Mtb. The TB bacilli impede epigenetic controls, particularly those facilitated by methylation of DNA [39]. Using diverse cohorts, different tissue types, and transcriptome analyses, it has been shown that epigenetic modifications linked to Mtb cause oxidative stress [40]. Senescence is triggered as a result, which accelerates cell ageing. Since these modifications caused by bacteria may be reversed, they provide a viable therapeutic target. IFN causes the production of human leukocyte antigen-DR (HLA-DR)/mRNA, which is suppressed by the bacilli Mtb together with the fractional reduction of Mtb-infected macrophages express CIITA [41]. However, Mtb infection, despite the expression of IFN regulatory factor-1 mRNA, remains unaffected by this condition. The transcriptional profile of genes involved in the immune system and the contact among the infectious pathogen and the host is regulated by epigenetic processes [42]. According to a recent study, Mtb strains, which are resistant, induce sub-optimal immune activation because it improves the bacilli's ability to survive within the cells through overexpression of genes and triggering the lipid metabolism of the host. In contrast, sensitive Mtb strains cause immune activation sufficient for the host in order to cure the illness [43, 44].

Both T cells and macrophages, two types of immune cells, play a significant role in the Mtb-induced changes. When it comes to epigenetic pathogenesis, Mtb relies heavily on both T helper and T effector cells [45]. The CD4 promotor leads to epigenetic inhibition of mature CD8+ T lymphocytes. Tregs may also produce DNA methylation and histone modifications on a variety of factors of transcription, including FOXP3 [46]. The expression of MHC II and the subsequent activity of CIITA are profoundly prejudiced by DNA methylation and other epigenetic changes. Additionally, macrophages are critical to the pathogenesis of Mtb. They may undergo epigenetic changes if the histones in their host cells are methylated [47].

8.3.1 Modifications of Histones

The nucleosome octamer is formed after 1.7 DNA rounds due to histone proteins, also known as chromatin remodelling proteins. These histone proteins oversee keeping the structural integrity of chromatin under check [48]. There are eight distinct categories of histone changes that occur after translation, totalling 60 positions. Proline isomerization, acetylation, deamination, methylation, and threonine and tyrosine phosphorylation are all examples [49]. All these modifications occur covalently, and they are regulated by a wide variety of enzymes including kinases, phosphatases, histone deacetylase (HDAC), histone acetyltransferase (HAT), histone methyltransferases (HMTs), and histone demethylases (HDMs) [50].

8.3.1.1 Methylation of Histones

Histones may be methylated at lysine or arginine, respectively, to switch on or off genes. The three kinds of methylation—monomeric, dimeric, and trimeric—have their own distinct consequences [51]. Ninety percent of methylation in human somatic cells occurs at CpG cytosines. Methylation at CpG nucleotides inhibits gene expression by blocking transcription factor binding to its specific binding site. This is accomplished by the recruitment of co-repressors. Inhibiting chromatin remodelling into repressor forms is one effect of this [52]. Rv1988 is a histone demethylase that exclusively targets the arginine at position 42 on histone H3, and it is a mycobacterial histone methyltransferase secretary protein that localises with chromatin in the host nucleus [53]. The transcription of host genes is greatly influenced by this activity. It is well known that Rv1988 blocks expression of genes by modifying histone H3 at location 42. Rv2966c, an Mtb protein has shown the same trends. This kind of protein diminishes the host defence initially deployed in response to infection by dampening the effect of protective-function genes [54]. Another research found that Mtb was able to avoid death by having an increase in H4K20me1 caused by the histone methyl transferase SET8. These methylation changes, depending on the kind, may limit gene expression, modify host histones via influencing cell signalling, etc. When the Mtb pathogen suppresses the KDM6B gene, for instance, the host histone protein H3K27 gets hypermethylated [55–57].

8.3.1.2 Acetylation of Histones

When a lysine position in DNA is modified with an acetyl group, the DNA is expressed differently. Any substances that may add an acetyl group to host proteins are classified as HATs [58]. To modify proteins besides histones via acetylation, the antigenic proteins of Mtb act as HATs. Nuclear Factor-kappa B (NF-kB) p65 is a protein that has been altered in a way other than via acetylation of histones [59]. One host enzyme that drives the response of inflammation to the Mtb is matrix metalloproteinase (MMP). MMP proteins acetylation allows Mtb bacilli to persist intracellularly in the host tissue. Macrophages and monocytes infected with Mtb secrete them alongside uninfected stromal cells [60]. Epigenetic pathways are key factors in regulating the activity of MMPs in infectious and non-infectious diseases like Mtb. The effects of epigenetic regulation on MMP-1/-3 synthesis, namely histone acetylation and its association with transcriptional activation, are explored and shown [61]. In addition, class 1 HDAC synthesis is inhibited by infection of macrophages with M. tuberculosis. HDACs are known to have a detrimental effect on gene expression. Therefore, MMP overexpression during Mtb infection destroys lung tissue, and elevating MMP expression necessitates the action of the HDAC and HATs proteins [62]. The breakdown of tissue caused by MMPs is a major contributor to immunopathology and epigenetic changes in the host. One histone alteration responsible for the observed shifts in gene transcription at different stages of development is methylation of arginine residues on histone H3 [63].

HDACs include HDAC1, HDAC3, HDAC2, and sirtuins, with HDAC3 playing a pivotal role in Mtb infection. Another evidence [62] suggests that Mtb employs HDAC1 to stifle IL12B expression [64]. By increasing the distances between nucleosomes, activation of chromatin is caused by HAT-induced acetylation of histone tails and suppresses host gene expression. Several additional proteins, including lipoarabinomannan protein LpqH (LpqH) and early secretory antigenic target 6 (ESAT-6), alter histones to inhibit the synthesis and presentation of antigens via major histocompatibility complex class II (MHC-II) [65]. In addition to preventing the production of IFN genes such as CIITA trans-activator, CD64, and, HLA-DR, Mtb infections inhibit the normal activation of the JAK-STAT1 signalling cascade. Hypoacetylation of Histone at the CIITA promoter is downregulated by INF genes, as seen with Rv3763, another mycobacterial protein [66]. It has been shown that Rv2966c interacts with the epigenomic macrophage to modify methylation of non-CpG at definite locations. Upholding homeostasis for epigenetic modifications on histones is crucial for accurate regulation of gene expression [67, 68].

8.3.2 Modification in Non-Coding RNAs' Expression

The types of proteins that are made because of translational activities are determined by the coding sequences that are already present on RNA templates. As the whole transcriptome and mammalian genome were sequenced and published, it was found that not all RNA arrangements were replicated in the proteins. This led to their categorization as nc-RNAs [69]. Several different infectious diseases and immunological abnormalities are caused by improper regulation of miRNAs, since miRNAs are principally responsible for controlling bacterial infections [70]. The cellular activities regulated by these nc-RNAs, which are typically approximately 22 nucleotides in length, include DNA methylation, modifications of histone, and other processes that influence about one-third of all genes in mammalian organisms. These miRNAs act as natural gene silencers, specifically aiming at mRNA to prevent translation from happening and therefore reducing gene expression [71, 72]. These miRNAs regulate crucial physiological pathways including cell proliferation, angiogenesis, and invasion in conjunction with other complex epigenetic processes like dynamics of genome organisation and chromatin structures in the nucleus. These gene activities are regulated by the RNA-induced silencing and mature miRNAs complex [73]. The dendritic cells immune response of Mtb is mediated in part by microRNAs. Recent research looked at the impact that microRNAs play in individuals with Mtb infections and tuberculosis. Which found an elevation of greater than 59 miRNAs in the serum of TB [74]. The infection of host macrophages with Mtb also causes them to create a modified form of miRNA. The miRNA modulates many signalling pathways during Mtb infection, including autophagy and apoptosis [75]. When overexpressed, miR-708-5p and miR-1178, for example, negatively regulate TLR-4 and inhibit the generation of mediators of inflammation including TNF-a, IL-1, IL-6, and IFN-γ. Upregulation of MiR-125a inhibits NF-kB pathway activity, which in turn inhibits tumor necrosis factor receptor-associated factor (TRAF)6's ability to negatively regulate NF-kB and the generation of proinflammatory cytokines [76, 77]. In response to Mtb infection, the TLR2/MyD88/NF-kB pathways are activated, leading to the production of miR-27b, which has a multiplicative effect on NF-kB and the activation of proinflammatory genes while upregulating p53 and favourably regulating death of cell. To reduce the immune response of T cells, miR-381-3p in Mtb-infected DCs causes downregulation of CD1c expression [78–80].

8.3.3 Modifications in Methylation of DNA

Mtb transfers a methyl group from the C5 position of the cytosine base to produce 5-methylcytosine with the assistance of DNA methyltransferase enzymes. In addition to being needed for processes like as differentiation, development, and reprogramming, it is necessary for the gene silencing that underpins the genesis of illnesses such as TB [81, 82]. DNA methylation has been shown to be a significant epigenetic target, with overexpression of DNA methylation Mtb immunity-related genes resulting in a diminished immune response to Mtb. Rv3263, which encodes the adenine methyltransferase, has been shown to be the sole identifiable source of methylation modification in the Mtb strain H37Rv [83, 84].

Methylation-controlled regulatory mechanisms have been implicated in the lineage-specific characteristics found in several Mtb strains. Due to this, it is well knowledge that Mtb modifies the host's epigenome [85]. An investigation using genetic ontology analysis showed that genes linked to the differentially methylated region (DMR) play a part in many immuno-biological processes, such as the activation and control of immune cells, the cellular response to interferon, and cytotoxicity. In contrast to histone modification, this epigenetic alteration is far less likely to be reversed, resulting in a prolonged silencing of the gene in question. Changes in DNA methylation have observable impacts on Mtb-infected DCs [82, 86]. Distal regulatory or enhancer regions, as well as low CpG density sites, show substantial changes between infected and uninfected DCs. Around 40% of miRNAs exhibited differential expression in DCs that had been infected with Mtb [59, 87]. Increased methylation of the IL-17 promoter is much more extreme than that of other receptors. These changes in methylation pattern may also be influenced by the host's genotype as well as the Mtb strain type. ESAT-6 has a crucial function in inhibiting methylation induced by IFN and acetylation of H3K4 at the Class II transactivator (CIITA) pl gene [88, 89].

In addition, it was revealed that ESAT-6 mediates TLR-2-dependent type IV CIITA expression downregulation but TLR-2-independent type I CIITA expression inhibition. Higher levels of methylation were seen in Mtb-infected monocytes, which also showed decreased IL-10 and greater IL-12 levels [9, 14]. This indicates that the monocytes' capacity to prevent lung harm from occurring due to severe inflammation has been impaired. Mtb's enhanced intracellular survival (Eis) protein increases the likelihood of infected macrophage survival by acetylating histone proteins. Hence, Mtb bacilli susceptibility is linked to methylation status, which varies across different populations [23, 36].

8.4 Treatments for Tuberculosis That Aim at Epigenetic Alterations in the Mycobacterium

Evidence suggests that systemic pharmacological HDAC inhibition enhances the Anti-Mtb activity by promoting the differentiation of macrophages into a more effective bactericidal phenotype and decreasing the cytokines production that cause inflammation [44]. An extensively studied in vivo model, HDAC inhibition significantly reduces the mycobacterial burden in zebrafish embryos infected by Mycobacterium marium, and HDAC inhibition of the class IIa leads to a considerable bacterial growth reduction. Suberoylandilide hydroxamic acid (SAHA), marketed under the brand names vorinostat and trichostatin A (TSA), is a kind of HDAC inhibitor that has generated an excitement because of its capability to greatly enhance clearance of bacteria by regulating the epigenetic Mtb-induced alterations [1, 87]. Although they do decrease inflammation, they also inhibit macrophages' capacity to destroy germs. This is achieved by lowering levels of nitric oxide (NO) and reactive oxygen species (ROS) in the exposed macrophages [22]. In combination with vitamin D, the non-specific HDAC inhibitor phenylbutyrate dual-targets host and pathogen. It is known that SAHA causes host alveolar macrophages and monocyte-derived macrophages to produce more IL-1 and less IL-10, in addition to improving GM-CSF and IFN-γ coproduction. These first-generation HDAC inhibitors target a wide variety of HDACs and have a far-reaching effect. Nevertheless, animals treated with tubustatin, a specific HDAC-6 inhibitor, had improved bacteraemia resolution, decreased organ damage, and a more controlled stress response due to an increase in neutrophil and monocyte counts and a reversal of lymphopenia [3, 11, 20]. It is pickier and requires more precise measures. Unlike broad-spectrum HDAC inhibitors, tubustatin enhances the microbicidal response induced by TLRs by macrophages, leading to greater intracellular microbial clearance. These inhibitors not only prevent HDAC from working, but they also artificially stimulate the expression of Nur77, an orphan nuclear receptor. The tricarboxylic acid (TCA) cycle in macrophages is modulated, as is well known, to improve anti-inflammatory activity. It has been shown that macrophages with a deficit in Nur77 are more likely to exhibit the proinflammatory phenotype, leading to enhanced release of cytokines such as IL-12, IL-6, and IFN-γ [67, 74, 78]. Mtb controls cytokine production to weaken the host immune system. Hence, therapeutically targeting HDACs to reduce their capacity to control transcription should be very useful. Bactericidal actions of HDAC inhibitors may be partially explained by their ability to regulate epigenetic processes. The upregulation of anti-Mtb activity by macrophages requires HDAC-6 [56]. Critically, it is preserved exclusively in macrophages which were derived from monocytes in those infected with a resistant Mtb. HDAC-6 is also essential for infected macrophages to keep their phagosomes, which contain Mtb, at a low pH. Since then, it has become an important new histone deacetylase inhibitor (HDT) target against Mtb [63]. Both HDAC-6 inhibitor (MC2780) and tubustatin A markedly suppress Mtb formation by increasing TNF, IL-12, and IFN and decreasing IL-10, respectively. The innate immune response is delayed because of the increased recruitment of macrophages, DCs, and neutrophils [81]. The HDAC3 inhibitor retroviral group fusion protein (RGFP)966 has been demonstrated to modify secretion of TNF and IL-6 by Mtb-infected macrophages, two pro-inflammatory cytokines. The production of nitric oxide synthase (NOS)2, cysteine-aspartic protease (CASP)1, and the M1 marker CD86 are all prompted. Inhibiting the epigenetic process of acetylation during M. tuberculosis infection, it has been demonstrated to be an important HDT in TB and, in combination with vitamin D3, induces a twofold rise in cyclic adenosine monophosphate (cAMP) [30, 89].

The Sirtuin (SIRT) or sirtuin family of NAD + -dependent HDACs is widely acknowledged to have a role in the pathophysiology of age-related illnesses. Inhibitors of both sirtuins and HDACs may be studied for their ability to reduce inflammation since they target the same proteins [2]. In vitro studies have shown that Cambinol, a selective inhibitor of SIRT-1/2, leads to decreased production of anti-inflammatory cytokines, offering protection against endotoxic and toxic shock [14]. As a result, SIRT-1/2 could have tremendous treatment possibilities for Mtb. The HAT inhibitor anacardic acid may be used to create isonicotinoylhydrazones when paired with the anti-tuberculosis drug isoniazid [90]. The Mtb-inhibiting activity of these isonicotinoylhydrazones is quite potent. Highly conserved structural patterns called bromodomains are associated with proteins that change chromatin, such as histone deacetylases. Specific targeting and novel post-translational modifications to histones during infection and inflammation are made possible by bromodomain inhibitors [91]. Bromodomain and Extra-Terminal domain (BET) family members mediate communication between transcription and histone acetylation. These drugs are expected to mitigate inflammation throughout life-threatening contaminations by activating CpG low promoters and permitting for regulation of the maximum effective lipopolysaccharide (LPS) effects [92]. There is hope for a more targeted approach using these medications thanks to the discovery of the bromodomain inhibitor JQ1-BET. I-BET, a distinct kind of systemic histone, prevents LPS-induced TNF and TLR4-stimulated mice macrophages from producing inflammatory cytokines such as IL-1b, IL-6, and IFN-a [93]. Since it creates the cell wall containing poly-L glutamate, the enzyme glutamate synthase is linked with Mtb's cause of illness. In fact, when phosphorothioate-modified antisense oligodeoxyribonucleotides were employed to target the mRNA of this enzyme, a 1.25-log drop in growth of Mtb and a 24% decrease in cell-wall L-glutamate were observed [94]. An eightfold increase in isoniazid sensitivity was seen in Mtb patients treated with a mixture of three Phosphorothioate Oligodeoxynucleotide (PS-ODN)s, each of which is an RNA molecule modified with 50 and 30 hairpins and targets mycolyl transferase transcripts. For infected macrophages, the ALD gene is the target that was effectively suppressed by the novel antisense mRNA targeting chemical 20-OMe PGOs, demonstrating high biological activity. Another thiocationic lipid-based formulation that has revealed in vitro action against Mtb is PAOs [95].

The DNA methyltransferase 1-azacytidine (DNMT1-aza) linkage is irreversible and leads to the destruction of the enzyme, drastically decreasing the methylation process. Andrew DiNardo et al. are investigating the efficacy of injectable azacytidine in the therapy of Mtb in a phase Ib/IIa open label, clinical research which is non-randomized (NCT03941496) [96]. Fifty participants will receive subcutaneous (SQ) azacytidine for 25 days at increasing doses from 5 mg/m^2 to 75 mg/m^2 over the course of 5 days. The main purposes of the research were to evaluate the prevalence and severity of adverse events and to determine whether they were mediated by epigenetic mechanisms. There has been no announcement yet about the trial's verdict [97]. Like azacytidine, the diphosphate version of the methylation inhibitor zebularine restores expression to silenced genes that have integrated into the DNA, therefore blocking DNA methylation. It has not been shown that it is effective in treating tuberculosis, although [98].

The acetylation of H3K27 at the target site is activated transcription of the target genes when the Cas9 is complexed with an acetyltransferase. When used in conjunction with LSD-1, the CRISPR-Cas9 system selectively targets DNA regions that boosts the expression of several genes. Latest studies on this method have shown promising therapeutic potential. Therapeutic use of Cas9 epigenetic effectors is a possibility because of their potential to selectively introduce, remove, or modify a wide range of epigenetic modifications while keeping tabs on the potential consequences of these manipulations [90, 92, 98].

8.5 Future Perspective

Histone modifications, changes in noncoding RNA synthesis, DNA methylation, and miRNA alterations are all examples of epigenetic processes that are intricately linked to one another. The induced epigenetic alterations in Mtb may be easily reversed by administering a drug that affects epigenetic processes. There have been promising new developments in "epi-drugs" for the epigenetic therapy of several tumours in recent years. Epigenetics has been understudied compared to genomic research despite its obvious importance in disease aetiology and outcome. When a host needs protection against infection, by Mtb-primed immune cells, a comprehensive knowledge of the pathogen's transcriptomic, proteomic, genomic, epigenetic, and vulnerabilities is required. The links between cavitation-inducing enzymes like MMP and epigenetic processes like acetylation of histone may be the subject of future study. Now, scientists have investigated whether the anti-Mtb drug isoniazid, a tried-and-true treatment, may also affect epigenetics by, say, increasing the rate of isonicotinylation at certain histone sites. Because of the alterations induced by isoniazid, cancer development is a possible side effect. To this end, further epigenetic research may shed light on several more pathways of side effects caused by doctors administering anti-TB drugs. The Mtb pathogen uses many epigenetic progressions to multiply inside the host, some of which may be therapeutic targets. As was said before, new HDT strategies for Mtb might be developed with more investigation into the PE6 protein mechanism.

Further research on the epigenetic developments in the pathogen of Mtb may use a wide range of methods, including chromatin immunoprecipitation, whole-genome bisulphite sequencing pyrosequencing, miRNA appearance, and polymerase chain reaction (PCR). The results of these epigenetics of the host research provide light on the workings of innate and adaptive immunity, as well as the interaction of immunosuppressive states with the Mtb pathogen, and may help in the development of future vaccines and treatment strategies. Another possible biomarker and important diagnostic tool are a blood test for epigenetically altered products. Drugs that reduce methylation, such as azacitidine, may be used to quiet Mtb-related genes specifically. Drugs like azacytidine are effective in targeting epigenetics; nevertheless, their toxicity is a major drawback. It is generally agreed that drugs containing aza or sulphur are poisonous and hence unsuitable. To avoid the potential dangers of azacytidine, other epigenetic alterations, such as histone acetylation or methylation, might be targeted instead. Hence, many other targets in addition to medicines pursuing the current goals need to be studied for a new epigenetic-specific treatment to address the disadvantages of the current therapy strategy.

8.6 Conclusion

Just a few environmental factors, such as, population density, malnutrition, genetics, the host's immunological state, may influence how severe an infection is. It is difficult to pinpoint any one of these aspects as the root cause of anything since they are all interrelated. Investigating epigenetic pathways is essential to understanding the genesis of Mtb. Many epigenetic processes, including as histone alterations, changes in the production of non-coding RNAs, changes to methylation of DNA, and microRNA, are interesting therapeutic targets because of the host environment's involvement in producing epigenetic alterations. They may also monitor a condition's rate of progression or how well a therapy is functioning. Moreover, "how epigenetics" affects the activation of the infection may be a determining factor in how latent Mtb infection is managed in infected individuals.

References

1. Abhimanyu, et al. Reversing post-infectious epigenetic-mediated immune suppression. Front Immunol. 2021;12:688132.
2. Abo-Kadoum MA, et al. Mycobacterium tuberculosis PE17 (Rv1646) promotes host cell apoptosis via host chromatin remodeling mediated by reduced H3K9me3 occupancy. Microb Pathog. 2021;159:105147.
3. Ala M, Ala M. Metformin for cardiovascular protection, inflammatory bowel disease, osteoporosis, periodontitis, polycystic ovarian syndrome, neurodegeneration, cancer, inflammation and senescence: what is next? ACS Pharmacol Transl Sci. 2021;4(6):1747–70.
4. Alizadeh Z, et al. Role of epigenetics in the pathogenesis of asthma. Iran J Allergy Asthma Immunol. 2017;16(2):82–91.
5. Almatroudi A. Non-coding RNAs in tuberculosis epidemiology: platforms and approaches for investigating the genome's dark matter. Int J Mol Sci. 2022;23(8):4430.
6. Arts RJW, et al. Immunometabolic pathways in BCG-induced trained immunity. Cell Rep. 2016;17(10):2562–71.
7. Arts RJW, et al. BCG vaccination protects against experimental viral infection in humans through the induction of cytokines associated with trained immunity. Cell Host Microbe. 2018;23(1):89–100.
8. Aspatwar A, et al. Tuberculosis vaccine BCG: the magical effect of the old vaccine in the fight against the COVID-19 pandemic. Int Rev Immunol. 2022;41(2):283–96.
9. Avci E, et al. Epigenetic mechanisms in parenchymal lung diseases: bystanders or therapeutic targets? Int J Mol Sci. 2022;23(1):4237.
10. Batista LAF, et al. Tuberculosis: A granulomatous disease mediated by epigenetic factors. Tuberculosis (Edinb). 2020;123:101943.
11. Borbora SM, Rajmani RS, Balaji KN. PRMT5 epigenetically regulates the E3 ubiquitin ligase ITCH to influence lipid accumulation during mycobacterial infection. PLoS Pathog. 2022;18 (6):e1010095.
12. Borges KCM, et al. Tuberculosis, BCG vaccination, and COVID-19: are they connected? Mini Rev Med Chem. 2022;22(12):1631–47.
13. Bryant JM, et al. Stepwise pathogenic evolution of Mycobacterium abscessus. Science. 2021;372(6541):eabb8699.
14. Chai Q, Lu Z, Liu CH. Host defense mechanisms against Mycobacterium tuberculosis. Cell Mol Life Sci. 2020;77(10):1859–78.
15. Chen L, et al. Ifnar gene variants influence gut microbial production of palmitoleic acid and host immune responses to tuberculosis. Nat Metab. 2022;4(3):359–73.
16. Cho T, et al. A review of the BCG vaccine and other approaches toward tuberculosis eradication. Hum Vaccin Immunother. 2021;17(8):2454–70.
17. Correa-Macedo W, et al. Alveolar macrophages from persons living with HIV show impaired epigenetic response to Mycobacterium tuberculosis. J Clin Invest. 2021;131(22):e148013.

18. Correia-Neves M, et al. Immunological hyporesponsiveness in tuberculosis: the role of mycobacterial glycolipids. Front Immunol. 2022;13:1035122.
19. Covián C, et al. BCG-induced cross-protection and development of trained immunity: implication for vaccine design. Front Immunol. 2019;10:2806.
20. Crimi E, et al. Clinical epigenetics and multidrug-resistant bacterial infections: host remodelling in critical illness. Epigenetics. 2020;15(10):1021–34.
21. Degenhardt F, Ellinghaus D, Juzenas S, Lerga-Jaso J, Wendorff M, Maya-Miles D, Uellendahl-Werth F, ElAbd H, Rühlemann MC, Arora J, Özer O, Lenning OB, Myhre R, Vadla MS, Wacker EM, Wienbrandt L, Blandino Ortiz A, de Salazar A, Garrido Chercoles A, Palom A, Ruiz A, Garcia-Fernandez AE, Blanco-Grau A, Mantovani A, Zanella A, Holten AR, Mayer A, Bandera A, Cherubini A, Protti A, Aghemo A, Gerussi A, Ramirez A, Braun A, Nebel A, Barreira A, Lleo A, Teles A, Kildal AB, Biondi A, Caballero-Garralda A, Ganna A, Gori A, Glück A, Lind A, Tanck A, Hinney A, Carreras Nolla A, Fracanzani AL, Peschuck A, Cavallero A, Dyrhol-Riise AM, Ruello A, Julià A, Muscatello A, Pesenti A, Voza A, Rando-Segura A, Solier A, Schmidt A, Cortes B, Mateos B, Nafria-Jimenez B, Schaefer B, Jensen B, Bellinghausen C, Maj C, Ferrando C, de la Horra C, Quereda C, Skurk C, Thibeault C, Scollo C, Herr C, Spinner CD, Gassner C, Lange C, Hu C, Paccapelo C, Lehmann C, Angelini C, Cappadona C, Azuure C, COVICAT study group, Aachen Study (COVAS), Bianco C, Cea C, Sancho C, Hoff DAL, Galimberti D, Prati D, Haschka D, Jiménez D, Pestaña D, Toapanta D, Muñiz-Diaz E, Azzolini E, Sandoval E, Binatti E, Scarpini E, Helbig ET, Casalone E, Urrechaga E, Paraboschi EM, Pontali E, Reverter E, Calderón EJ, Navas E, Solligård E, Contro E, Arana-Arri E, Aziz F, Garcia F, García Sánchez F, Ceriotti F, Martinelli-Boneschi F, Peyvandi F, Kurth F, Blasi F, Malvestiti F, Medrano FJ, Mesonero F, Rodriguez-Frias F, Hanses F, Müller F, Hemmrich-Stanisak G, Bellani G, Grasselli G, Pezzoli G, Costantino G, Albano G, Cardamone G, Bellelli G, Citerio G, Foti G, Lamorte G, Matullo G, Baselli G, Kurihara H, Neb H, My I, Kurth I, Hernández I, Pink I, de Rojas I, Galván-Femenia I, Holter JC, Afset JE, Heyckendorf J, Kässens J, Damås JK, Rybniker J, Altmüller J, Ampuero J, Martín J, Erdmann J, Banales JM, Badia JR, Dopazo J, Schneider J, Bergan J, Barretina J, Walter J, Hernández Quero J, Goikoetxea J, Delgado J, Guerrero JM, Fazaal J, Kraft J, Schröder J, Risnes K, Banasik K, Müller KE, Gaede KI, Garcia-Etxebarria K, Tonby K, Heggelund L, Izquierdo-Sanchez L, Bettini LR, Sumoy L, Sander LE, Lippert LJ, Terranova L, Nkambule L, Knopp L, Gustad LT, Garbarino L, Santoro L, Téllez L, Roade L, Ostadreza M, Intxausti M, Kogevinas M, Riveiro-Barciela M, Berger MM, Schaefer M, Niemi MEK, Gutiérrez-Stampa MA, Carrabba M, Figuera Basso ME, Valsecchi MG, Hernandez-Tejero M, Vehreschild MJGT, Manunta M, Acosta-Herrera M, D'Angiò M, Baldini M, Cazzaniga M, Grimsrud MM, Cornberg M, Nöthen MM, Marquié M, Castoldi M, Cordioli M, Cecconi M, D'Amato M, Augustin M, Tomasi M, Boada M, Dreher M, Seilmaier MJ, Joannidis M, Wittig M, Mazzocco M, Ciccarelli M, Rodríguez-Gandía M, Bocciolone M, Miozzo M, Imaz Ayo N, Blay N, Chueca N, Montano N, Braun N, Ludwig N, Marx N, Martínez N, Norwegian SARS-CoV-2 Study group, Cornely OA, Witzke O, Palmieri O, Pa Study Group, Faverio P, Preatoni P, Bonfanti P, Omodei P, Tentorio P, Castro P, Rodrigues PM, España PP, Hoffmann P, Rosenstiel P, Schommers P, Suwalski P, de Pablo R, Ferrer R, Bals R, Gualtierotti R, Gallego-Durán R, Nieto R, Carpani R, Morilla R, Badalamenti S, Haider S, Ciesek S, May S, Bombace S, Marsal S, Pigazzini S, Klein S, Pelusi S, Wilfling S, Bosari S, Volland S, Brunak S, Raychaudhuri S, Schreiber S, Heilmann-Heimbach S, Aliberti S, Ripke S, Dudman S, Wesse T, Zheng T, STORM Study group, The Humanitas Task Force, The Humanitas Gavazzeni Task Force, Bahmer T, Eggermann T, Illig T, Brenner T, Pumarola T, Feldt T, Folseraas T, Gonzalez Cejudo T, Landmesser U, Protzer U, Hehr U, Rimoldi V, Monzani V, Skogen V, Keitel V, Kopfnagel V, Friaza V, Andrade V, Moreno V, Albrecht W, Peter W, Poller W, Farre X, Yi X, Wang X, Khodamoradi Y, Karadeniz Z, Latiano A, Goerg S, Bacher P, Koehler P, Tran F, Zoller H, Schulte EC, Heidecker B, Ludwig KU, Fernández J, Romero-Gómez M, Albillos A, Invernizzi P, Buti M, Duga S, Bujanda L, Hov JR, Lenz TL, Asselta R, de Cid R, Valenti L, Karlsen TH, Cáceres M, Franke A. Detailed stratified GWAS analysis for severe COVID-19 in

four European populations. Hum Mol Genet. 2022;31(23):3945–66. https://doi.org/10.1093/hmg/ddac158. PMID: 35848942; PMCID: PMC9703941

22. DiNardo AR, et al. Tuberculosis endotypes to guide stratified host-directed therapy. Med. 2021;2(3):217–32.
23. DiNardo AR, et al. DNA hypermethylation during tuberculosis dampens host immune responsiveness. J Clin Invest. 2020;130(6):3113–23.
24. Divangahi M, et al. Trained immunity, tolerance, priming and differentiation: distinct immunological processes. Nat Immunol. 2021;22(1):2–6.
25. Esterhuyse MM, et al. Epigenetics and proteomics join transcriptomics in the quest for tuberculosis biomarkers. mBio. 2015;6(5):e01187–15.
26. Farhadi J, et al. Epigenetics and Behçet's disease: DNA methylation specially highlighted. Iran J Allergy Asthma Immunol. 2019;18(5):462–72.
27. Fatima S, et al. Epigenetic code during mycobacterial infections: therapeutic implications for tuberculosis. FEBS J. 2022;289(14):4172–91.
28. Ferluga J, et al. Natural and trained innate immunity against Mycobacterium tuberculosis. Immunobiology. 2020;225(3):151951.
29. Foster M, et al. BCG-induced protection against Mycobacterium tuberculosis infection: evidence, mechanisms, and implications for next-generation vaccines. Immunol Rev. 2021;301(1): 122–44.
30. Gauba K, et al. Immunomodulation by epigenome alterations in Mycobacterium tuberculosis infection. Tuberculosis (Edinb). 2021;128:102077.
31. Giamarellos-Bourboulis EJ, et al. Activate: Randomized Clinical Trial of BCG Vaccination against Infection in the Elderly. Cell. 2020;183(2):315–323.e9.
32. Gonzalez-Perez M, et al. The BCG vaccine for COVID-19: first verdict and future directions. Front Immunol. 2021;12:632478.
33. Groslambert J, Prokhorova E, Ahel I. ADP-ribosylation of DNA and RNA. DNA Repair (Amst). 2021;105:103144.
34. Hashemian SM, et al. Non-coding RNAs and exosomes: their role in the pathogenesis of sepsis. Mol Ther Nucleic Acids. 2020;21:51–74.
35. Herrington CS, Poulsom R, Coates PJ. Recent advances in pathology: the 2020 annual review issue of the journal of pathology. J Pathol. 2020;250(5):475–9.
36. Jeyakumar SM. Micronutrient deficiency in pulmonary tuberculosis - perspective on hepatic drug metabolism and pharmacokinetic variability of first-line anti- tuberculosis drugs: special reference to Fat-soluble vitamins A, D, & E and nutri-epigenetics. Drug Metab Lett. 2021;14(3): 166–76.
37. Kathirvel M, Mahadevan S. The role of epigenetics in tuberculosis infection. Epigenomics. 2016;8(4):537–49.
38. Kaufmann E, et al. BCG educates hematopoietic stem cells to generate protective innate immunity against tuberculosis. Cell. 2018;172(1–2):176–190.e19.
39. Kaul D. Coronin-1A epigenomics governs mycobacterial persistence in tuberculosis. FEMS Microbiol Lett. 2008;278(1):10–4.
40. Khadela A, et al. Epigenetics in tuberculosis: immunomodulation of host immune response. Vaccines (Basel). 2022;10(10):1740.
41. Khader SA, et al. Targeting innate immunity for tuberculosis vaccination. J Clin Invest. 2019;129(9):3482–91.
42. Koeken V, et al. Trained innate immunity and resistance to Mycobacterium tuberculosis infection. Clin Microbiol Infect. 2019;25(12):1468–72.
43. Kulesza J, et al. BCG and SARS-CoV-2-What Have We Learned? Vaccines (Basel). 2022;10 (10):1641.
44. Kumar R, et al. Immunometabolism of phagocytes during Mycobacterium tuberculosis infection. Front Mol Biosci. 2019;6:105.
45. Lerm M, Dockrell HM. Addressing diversity in tuberculosis using multidimensional approaches. J Intern Med. 2018; https://doi.org/10.1111/joim.12776.

46. Lerm M, Netea MG. Trained immunity: a new avenue for tuberculosis vaccine development. J Intern Med. 2016;279(4):337–46.
47. Li L, et al. Diagnosis of pulmonary nodules by DNA methylation analysis in bronchoalveolar lavage fluids. Clin Epigenetics. 2021;13(1):185.
48. Lin D, et al. Decoding the spatial chromatin organization and dynamic epigenetic landscapes of macrophage cells during differentiation and immune activation. Nat Commun. 2022;13(1): 5857.
49. Liu Y, et al. Epigenetics in immune-mediated pulmonary diseases. Clin Rev Allergy Immunol. 2013;45(3):314–30.
50. Madden K, et al. Surveying the epigenetic landscape of tuberculosis in alveolar macrophages. Infect Immun. 2022;90(5):e0052221.
51. Maneechotesuwan K. Role of microRNA in severe asthma. Respir Investig. 2019;57(1):9–19.
52. Marimani M, Ahmad A, Duse A. The role of epigenetics, bacterial and host factors in progression of Mycobacterium tuberculosis infection. Tuberculosis (Edinb). 2018;113:200–14.
53. Martinez N, Kornfeld H. Diabetes and immunity to tuberculosis. Eur J Immunol. 2014;44(3): 617–26.
54. Merchant SA, Shaikh MJS, Nadkarni P. Tuberculosis conundrum - current and future scenarios: a proposed comprehensive approach combining laboratory, imaging, and computing advances. World J Radiol. 2022;14(6):114–36.
55. Mongan NP, Emes RD, Archer N. Detection and analysis of RNA methylation. F1000Res. 2019;8 https://doi.org/10.12688/f1000research.17956.1.
56. Moorlag S, et al. β-glucan induces protective trained immunity against mycobacterium tuberculosis infection: a key role for IL-1. Cell Rep. 2020;31(7):107634.
57. Moorlag S, et al. BCG vaccination induces long-term functional reprogramming of human neutrophils. Cell Rep. 2020;33(7):108387.
58. Mortaz E, et al. Epigenetics and chromatin remodeling play a role in lung disease. Tanaffos. 2011;10(4):7–16.
59. Naeem MA, et al. Stealth strategies of Mycobacterium tuberculosis for immune evasion. Curr Issues Mol Biol. 2021;41:597–616.
60. Okafor CN, Rewane A, Momodu II. Bacillus calmette guerin, in StatPearls. 2022, StatPearls Publishing Copyright © 2022. Treasure Island, FL: StatPearls Publishing LLC.
61. Pagán AJ, et al. mTOR-regulated mitochondrial metabolism limits mycobacterium-induced cytotoxicity. Cell. 2022;185(20):3720–3738.e13.
62. Pal R, Bisht MK, Mukhopadhyay S. Secretory proteins of Mycobacterium tuberculosis and their roles in modulation of host immune responses: focus on therapeutic targets. FEBS J. 2022;289(14):4146–71.
63. Pehrson I, et al. The spectrum of tuberculosis described as differential DNA methylation patterns in alveolar macrophages and alveolar T cells. Clin Epigenetics. 2022;14(1):175.
64. Pisu D, et al. Single cell analysis of M. tuberculosis phenotype and macrophage lineages in the infected lung. J Exp Med. 2021;218(9):e20210615.
65. Rawat BS, et al. Therapeutic potentials of immunometabolomic modulations induced by tuberculosis vaccination. Vaccines (Basel). 2022;10(12):2127.
66. Richard-Greenblatt M, Av-Gay Y. Epigenetic phosphorylation control of mycobacterium tuberculosis infection and persistence. Microbiol Spectr. 2017;5(2) https://doi.org/10.1128/microbiolspec.TBTB2-0005-2015.
67. Salaikumaran MR, Badiger VP, Burra V. 16S rRNA methyltransferases as novel drug targets against tuberculosis. Protein J. 2022;41(1):97–130.
68. Scully EP, Bryson BD. Unlocking the complexity of HIV and Mycobacterium tuberculosis coinfection. J Clin Invest. 2021;131(22):e154407.
69. Sengupta S, et al. Mycobacterium tuberculosis Phosphoribosyltransferase Promotes Bacterial Survival in Macrophages by Inducing Histone Hypermethylation in Autophagy-Related Genes. Front Cell Infect Microbiol. 2021;11:676456.

70. Silva-García O, Valdez-Alarcón JJ, Baizabal-Aguirre VM. Wnt/β-catenin signaling as a molecular target by pathogenic bacteria. Front Immunol. 2019;10:2135.
71. Singh AK, Netea MG, Bishai WR. BCG turns 100: its nontraditional uses against viruses, cancer, and immunologic diseases. J Clin Invest. 2021;131(11):e148291.
72. Singh AK, et al. Re-engineered BCG overexpressing cyclic di-AMP augments trained immunity and exhibits improved efficacy against bladder cancer. Nat Commun. 2022;13(1):878.
73. Soleimanpour S, et al. A century of attempts to develop an effective tuberculosis vaccine: Why they failed? Int Immunopharmacol. 2022;109:108791.
74. Song H, et al. Biological roles of RNA m(5)C modification and its implications in Cancer immunotherapy. Biomark Res. 2022;10(1):15.
75. Soto JA, et al. BCG vaccination induces cross-protective immunity against pathogenic microorganisms. Trends Immunol. 2022;43(4):322–35.
76. Sui J, et al. Epigenetic changes in Mycobacterium tuberculosis and its host provide potential targets or biomarkers for drug discovery and clinical diagnosis. Pharmacol Res. 2022;179: 106195.
77. Tarashi S, et al. The human microbiota in pulmonary tuberculosis: not so innocent bystanders. Tuberculosis (Edinb). 2018;113:215–21.
78. Tarashi S, et al. The inter-talk between Mycobacterium tuberculosis and the epigenetic mechanisms. Epigenomics. 2020;12(5):455–69.
79. Ter Steeg L, et al. Trained immunity as a preventive measure for surgical site infections. Clin Microbiol Rev. 2021;34(4):e0004921.
80. Testa D, et al. Allergic rhinitis and asthma assessment of risk factors in pediatric patients: a systematic review. Int J Pediatr Otorhinolaryngol. 2020;129:109759.
81. van Rensburg IC, Loxton AG. Transcriptomics: the key to biomarker discovery during tuberculosis? Biomark Med. 2015;9(5):483–95.
82. Vercelli D. Does epigenetics play a role in human asthma? Allergol Int. 2016;65(2):123–6.
83. Wang B, et al. Integrative analysis of pooled CRISPR genetic screens using MAGeCKFlute. Nat Protoc. 2019;14(3):756–80.
84. Warnat-Herresthal S, et al. Swarm Learning for decentralized and confidential clinical machine learning. Nature. 2021;594(7862):265–70.
85. Wu H, et al. Genome-wide DNA methylation profiling in differentiating Crohn's disease from intestinal tuberculosis. Genes Genomics. 2022;44(5):603–15.
86. Wu Z, et al. mbtD and celA1 association with ethambutol resistance in Mycobacterium tuberculosis: a multiomics analysis. Front Cell Infect Microbiol. 2022;12:959911.
87. Yang F, et al. The gut microbiota mediates protective immunity against tuberculosis via modulation of lncRNA. Gut Microbes. 2022;14(1):2029997.
88. Zhou J, et al. Trained immunity contributes to the prevention of Mycobacterium tuberculosis infection, a novel role of autophagy. Emerg Microbes Infect. 2021;10(1):578–88.
89. Zhu J, et al. Pulmonary tuberculosis associated with immune checkpoint inhibitors: a pharmacovigilance study. Thorax. 2022;77(7):721–3.
90. Angria N, et al. Expression of miRNA-29a-3p and IFN-γ as biomarkers in active and latent pulmonary tuberculosis. Ann Med Surg (Lond). 2022;83:104786.
91. Bhaskar A, et al. Host sirtuin 2 as an immunotherapeutic target against tuberculosis. elife. 2020;9:e55415.
92. Brezgin S, et al. Dead cas systems: types, principles, and applications. Int J Mol Sci. 2019;20 (23):6041.
93. Danjuma L, et al. Genomic plasticity between human and mycobacterial DNA: a review. Tuberculosis (Edinb). 2017;107:38–47.
94. Joshi L, Chelluri LK, Gaddam S. Mesenchymal stromal cell therapy in MDR/XDR tuberculosis: a concise review. Arch Immunol Ther Exp. 2015;63(6):427–33.
95. Kotze LA, et al. Establishment of a patient-derived, magnetic levitation-based, three-dimensional spheroid granuloma model for human tuberculosis. mSphere. 2021;6(4):e0055221.

96. Li F, et al. Metabolic plasticity and regulation of T cell exhaustion. Immunology. 2022;167(4): 482–94.
97. Niller HH, et al. Pathogenic mechanisms of intracellular bacteria. Curr Opin Infect Dis. 2017;30 (3):309–15.
98. Saccone D, Asani F, Bornman L. Regulation of the vitamin D receptor gene by environment, genetics and epigenetics. Gene. 2015;561(2):171–80.

Epigenetics of Idiopathic Pulmonary Fibrosis

9

Sumeet Kumar Singh, Sampat Singh Tanwar, Dhaneshvaree Patel, Poonam Yadav, Sonu Rajput, Anjali Sharma, Jasvinder Singh Bhatti, Amit Khurana, and Umashanker Navik

Abstract

Idiopathic pulmonary fibrosis (IPF) is a fatal variation of the interstitial pulmonary illness characterised by extracellular matrix deposition that leads to secretion of inflammatory cytokines and causes fibrosis in the lungs. Further progression of fibrosis leads to cancerous stage of lungs and death. IPF is the worst pathological condition to be focused on and explored because of its rising prevalence, poor prognosis, and inadequate treatment. Even though the disease's origin is still unknown, several genetic, environmental, and underlying pulmonary problems could set off a number of molecular pathways which is involved in the development of IPF. However, several genetic loci and genetic polymorphisms linked to IPF have been examined by genome-wide association studies and whole-genome sequencing. The newly found gene may clarify key elements in the identification, prognosis, and treatments of IPF. Additionally IPF can result from a variety of epigenetic alternations, including modification of histone, methylation of DNA, and non-coding RNA. This book chapter summarises the pathogenesis of pulmonary fibrosis, available treatment and the pathways that involved in IPF

Sumeet Kumar Singh, Sampat Singh Tanwar, Dhaneshvaree Patel, and Poonam Yadav are contributed equally to this work.

S. K. Singh · S. S. Tanwar · D. Patel · P. Yadav · S. Rajput · A. Sharma · U. Navik (✉)
Department of Pharmacology, Central University of Punjab, Bathinda, India

J. S. Bhatti
Department of Human Genetics and Molecular Medicine, Central University of Punjab, Bathinda, India

A. Khurana
Institute of Molecular Pathobiochemistry, Experimental Gene Therapy and Clinical Chemistry (IFMPEGKC), RWTH Aachen University Hospital, Aachen, Germany

G. Gupta et al. (eds.), *Targeting Epigenetics in Inflammatory Lung Diseases*,
https://doi.org/10.1007/978-981-99-4780-5_9

progression and may develop into lung cancer. Furthermore, this highlights epigenetic and molecular mechanism of IPF progression.

Keywords

Idiopathic pulmonary fibrosis · Pathogenesis · Molecular Pathways · Epigenetics · Treatment

9.1 Introduction

Idiopathic pulmonary fibrosis (IPF) is a long-term, persistent illness associated with lungs and marked by scarred lungs and the typical interstitial pneumonia histology. Further it is characterised by intensifying cough, dyspnoea, and worsening of life quality [1]. Extra-cellular matrix deposition causes fibrosis in which functional tissue is replaced by non-functional scarred tissue, affect lung function and often causes death [2]. Among the various type of lung diseases IPF is becoming the centre of attraction for many researchers, because its development is slow and it is resistant to traditional treatments as well. IPF has a 3-year median life expectancy and leads to dyspnoea and final respiratory collapse [3]. There is a considerable risk of death in many IPF patients who encounter acute episodes of respiratory deterioration [4]. Acute exacerbations are idiopathic acute worsening when there is no known reason for more than 50% of these acute episodes [5]. It is believed that acute IPF exacerbations signify an inherent speeding up of the underlying disease process, possibly brought on by a hidden stressor like a viral infection, micro aspiration, or ambient pollution [6]. The most prevalent form of idiopathic interstitial pneumonia is IPF. Although the condition has been regarded as rare, its frequency of occurrence is less than testicular, brain, and stomach cancers. Over time, the prevalence of IPF has increased with estimates for Europe and North America ranging from 2 to 18 cases per 100,000 persons per year. Little information about global variance is available, however, Asia and South America, where incidence is thought to be between 0 and 42 instances per 100,000 people annually, may have lower rates. People over the age of 50 are more susceptible to IPF, and men are more affected than women by the disease. The typical time to survive after diagnosis is between 2 and 4 years [2]. IPF occurs between 0.09 and 1.30 and 0.33 and 4.51 out of every 10,000 people, respectively [7]. In the US, it is reported that the occurrence of IPF is higher and increased rates of IPF-related hospital admissions and fatalities also point to a rising disease burden [3]. Identification of the underlying cause is necessary for the diagnosis: interstitial pneumonia histology pattern, typically with high-resolution computerised tomography scanning; lung biopsy perhaps necessary in some patients. The diagnostic process also requires ruling out any coexisting illnesses or other interstitial lung disorders [1]. Comprehensive management of the patient with IPF entails making a precise diagnosis using an interdisciplinary review that is meticulous; managing common comorbidities like depression, gastroesophageal reflux disease, obstructive sleep apnea; immunising against influenza and pneumococcal infection; providing education and structured exercise and through formal pulmonary rehabilitation classes; and, in cases of severe or worsening disease, determining whether or not lung transplantation is appropriate. All IPF patients should be examined for clinical trials of innovative therapeutic agents in the absence of effective treatments. Patients who do not meet the requirements for a clinical trial should be given the option of empiric therapy with acetylcysteine or proton pump inhibition [6]. Although pirfenidone, a new antifibrotic medication, has received approval for usage in some countries, there are no widely accepted treatments for IPF [6]. Pirfenidone and nintedanib are approved for the treatment of IPF because they can halt the spread of the illness, and functional decline, but they don't provide a cure and have tolerability problems [4]. Notably, prednisone and immunomodulatory drugs like azathioprine are ineffective in treating interstitial lung disease (Fig. 9.1). The active therapy arm of a recent randomised controlled study of prednisone, azathioprine, and the antioxidant acetylcysteine was terminated early due to safety and efficacy concerns. Development of drugs has focused on fibrosis and its proliferation as a result of this lack of efficacy, and a growing number of targeted treatments are currently undergoing early-phase clinical studies. In the coming 10 years, it appears likely that the therapeutic environment for IPF will undergo significant change [6]. There are more therapy options available for IPF. However, a combination of treatment approaches with various mechanisms of action may be necessary to target the large number of profibrotic cytokines and growth factors implicated in disease development [4]. Building on the advances in our understanding of IPF pathobiology, it is hoped that future research into the role of gene variants, epigenetic changes, and other molecular biomarkers reflecting disease activity and behaviour will help the development of innovative medications for individualised therapy of IPF, enable earlier and more definite diagnosis, and better disease phenotyping [5].

9.2 Involvement of Environmental Factor in the Pathophysiology of IPF

Earlier, it was considered that the primary cause of IPF is inflammation, which later develops into fibrosis. However, both anti-inflammatory therapy and immunosuppressive agent were unable to mitigate the IPF. Recent studies suggested that various environmental factors, genetic factors, and microbial infection lead to fibrosis development. It has been determined that progressive injury of the alveolar epithelium is the primary cause of an altered healing process in which numerous lung cells exhibit abnormal behaviours, resulting in the onset and maintenance of the fibrotic process [5]. Fibrotic foci are the principal histological characteristics of IPF which is triggered by alveolar epithelial damage, which also appears to stimulate fibroblast proliferation, differentiation of myofibroblast, and collagen accumulation. This extracellular matrix and collagen accumulation stiffens the lungs and obliterates

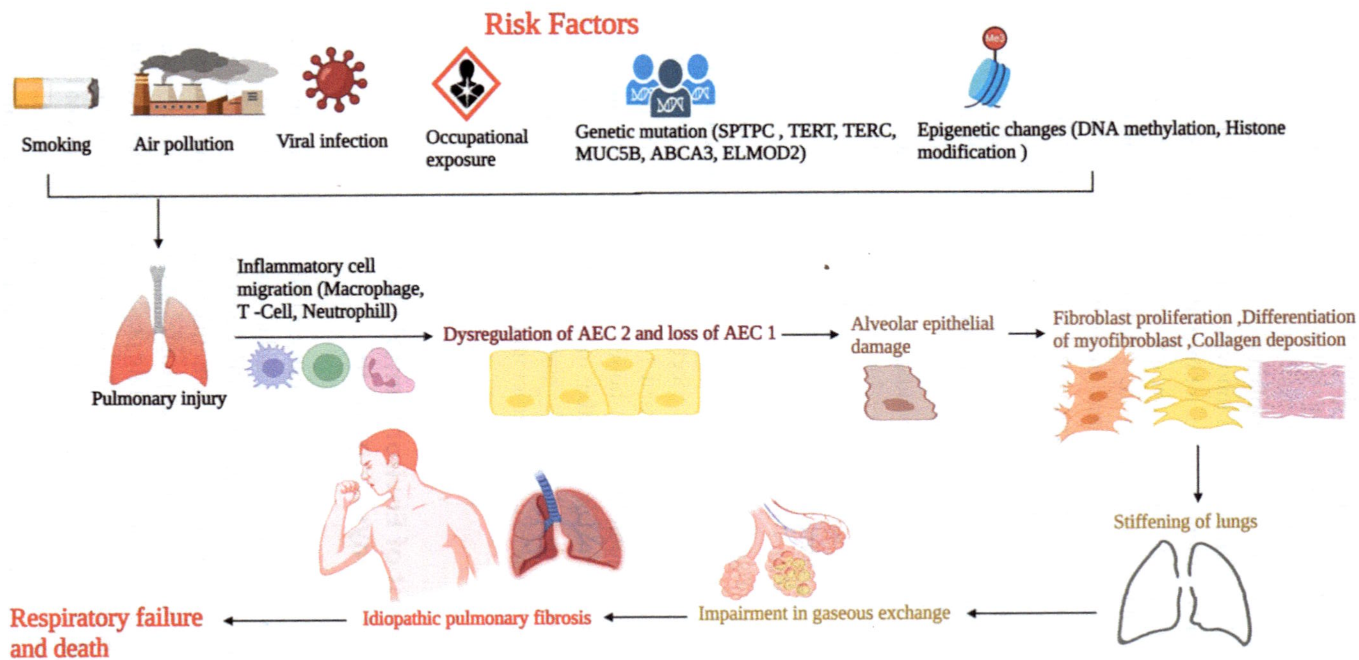

Fig. 9.1 Pathogenesis of idiopathic pulmonary fibrosis and various risk factors for pulmonary alterations that result in idiopathic pulmonary fibrosis, which develops into malignancy, respiratory failure, and death. Abbreviations: *SFTPC* surfactant protein C (SP-C) gene, *TERT* telomerase reverse transcriptase, *TERC* telomerase RNA component, *MUC5B* mucin 5B, *ABCA3* ATP-binding cassette sub-family A member 3, *ELMOD* 2 ELMO domain containing 2, *AEC 1* alveolar epithelial type-1 cells, *AEC 2* alveolar epithelial type-2 cells

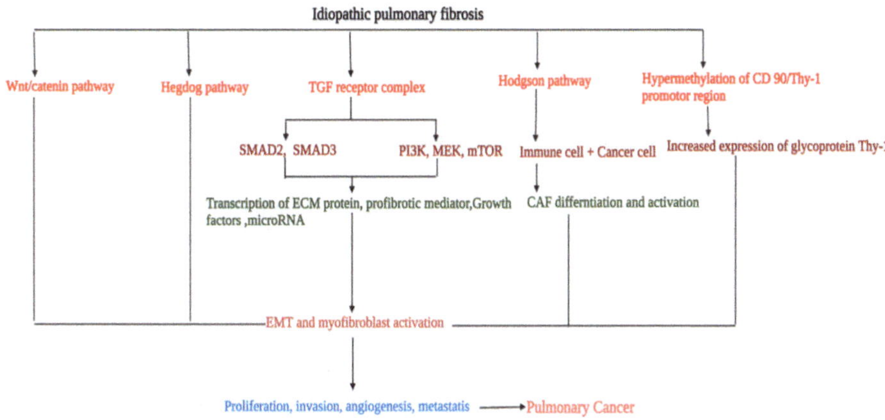

Fig. 9.2 Progression of idiopathic pulmonary fibrosis to pulmonary cancer: Idiopathic pulmonary fibrosis can proceed to lung cancer as a result of the activation of several molecular pathways that result in genetic alterations. Abbreviations: *TGF* transforming growth factor, *CD 90/Thy-1* cluster of differentiation 90, *SMAD 2* suppressor of mothers against decapentaplegic 2, *SMAD 3* suppressor of mothers against decapentaplegic 3, *PI3K* phosphoinositide 3-kinase, *MEK* mitogen-activated protein kinase, *mTOR* mammalian target of rapamycin, *ECM* extracellular matrix, *CAF* cancer-associated fibroblast, *EMT* epithelial-mesenchymal transition

the delicate lace-like structure of the alveolar gaps, leading to impairment in gaseous exchange, and ultimately, respiratory failure and death [8]. Type 2 alveolar epithelial cells (AEC2) are the stem cells in the pulmonary cells; persistent dysregulation of AEC2 has been identified as the main mechanism behind the development of fibrosis. AEC2 also involve significantly in the regeneration of AEC2 after lungs injury. The dysregulated AEC2 and loss of AEC1 are observed in IPF tissue [2] (Fig. 9.2).

IPF susceptibility is likely influenced by several genetic characteristics, including gene variants and transcriptional changes that lead to epithelial integrity loss. Currently genome-wide association studies (GWAS) found that common genetic variations that are essential for the integrity of the epithelium have been defined as IPF risk factors. These studies suggested that telomere biology such as Telomerase reverse transcriptase (TERT), Telomerase RNA component (TERC), OB Fold-containing Protein 1 (OBFC1), host defence (mucin 5B (MUC5B), ATPase phospholipid transporting 11A, and toll-interacting protein (TOLLIP)), and cellular barrier function (desmoplakin, dipeptidyl peptidase 9) could play a role in the development of disease. Both the GWAS identified additional common variants linked to illness and established the promoter of the MUC5B gene as a risk factor. The functional gain of the promoter region of MUC5B, i.e., rs35705950, has been established as a major risk factor for the development of sporadic IPF [5, 9]. Apart from genetic factors, environmental exposures and microbes play significant roles in the progression of IPF. Cigarette smoking, wood dust, metal dust, coal dust, elements such as silicon and beryllium, asbestos, and radiation are some of the substances which are potential risk of IPF [2, 9] (Fig.9.1). Bacterial, fungal, and viral (hepatitis C, Epstein-Barr, Cytomegalovirus, herpesvirus) infections also lead to the development of IPF [9, 10].

9.3 IPF Association with Lung Cancer

There are various molecular and cellular mechanisms which are common in both such as endoplasmic reticulum stress, and fibroblast transition proliferation, activation, and oxidative stress, along with several epigenetic markers and genetics which make IPF patients more vulnerable to developing Lung cancer (LC) (Fig. 9.2) [11, 12]. According to earlier research, IPF patients had more sensitivity in developing primary lung cancer (22% of patients) in comparison to the normal public [13–17]. Similarly to this, individuals who receive lung transplants for IPF have also become vulnerable for primary lung cancer of more than 20 times than that of the general population [18, 19]. LC prevalence among patients with IPF reportedly ranges between 2.7 to 48% [13]. The study also shows the overall lung cancer occurrence in IPF. IPF cumulative incidence was reported by Ozawa et al. to be 3.3, 15.4, and 54.7% after one, five, and ten years of follow-up [14]. In a second experiment, individuals with IPF had cumulative lung cancer occurrence rates of 41% at 1 year and 82% at 3 years [20]. Additionally identified as complicating variables in emergence of LC in individuals having IPF are their age and smoking history. Finger clubbing is observed in nearly all individuals with IPF and lung cancer (95%) against around 60% of IPF patients who are alone.

Numerous research contrast IPF with pulmonary cancer to shed light on the aetiology of these two conditions, which have poor prognoses. Homogeneity, metastasis, and laterality of cancer are used to refute the parallels between IPF and cancer. Furthermore, IPF exists when both lungs are affected at once. However, it is predicated on the generally held notion that cancers usually typically grow in a single lung before colonising and metastasising various body part. Fibrosis in IPF patients initiate with the lower lobes and lung periphery, that are frequently the sites of lung malignancies from an anatomical perspective [21]. Additionally, altered cell-cell interactions, unchecked multiplication, and aberrant actuation of particular signal transduction pathways are characteristics of these two disorders [22, 23] (Fig. 9.3).

9.3.1 Pathways Involved in the Progression of IPF into LC

9.3.1.1 Transforming Growth Factor-β1

Profibrotic mediators that support the onset of IPF include platelet-derived growth factor (PDGF), transforming growth factor β-1(TGFβ-1), tumour necrosis factor (TNF), endothelin-1, chemokine connective tissue growth factor, and osteopontin. Since it controls the activity of myofibroblasts and the ensuing remodelling of extracellular matrix deposition (ECM), TGFβ-1 stands out among them as the chief regulator of fibrotic development [24]. Chemotactic factors include TGF β-1 which attracts macrophages and monocytes, causing the release of PDGF from these

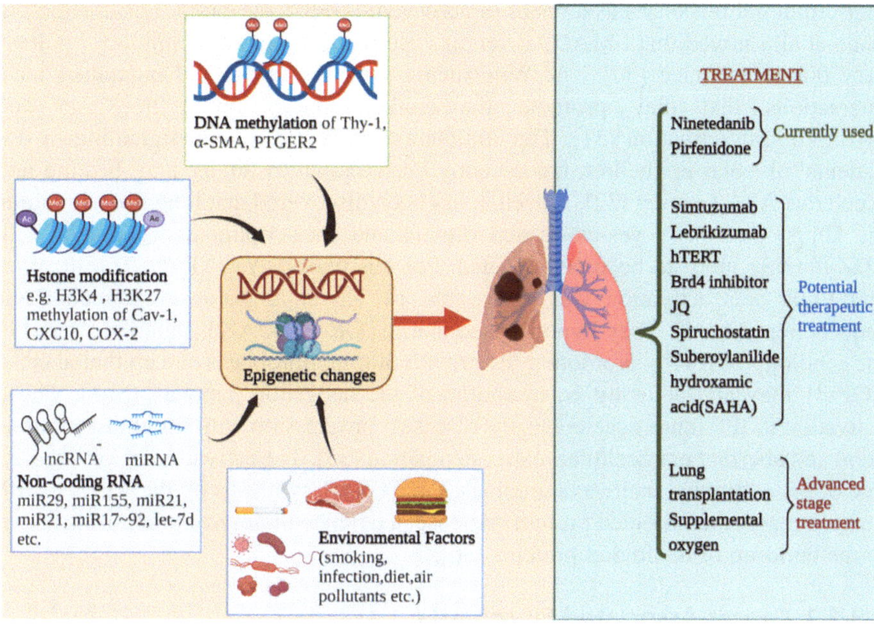

Fig. 9.3 A diagrammatic representation of various risk factors and epigenetic changes that includes DNA methylation, histone modification and non-coding RNA which lead to the development of idiopathic pulmonary fibrosis and outlining ongoing and potential therapeutic treatments for idiopathic pulmonary fibrosis. Abbreviations: *α-SMA* alpha smooth muscle actin, *PTGER2* prostaglandin E receptor 2, *H3K4* histone H3 lysine K4, *H3K27* histone H3 lysine K27, *Cav-1* caveolin 1, *CXC10* C-X-C chemokine ligand 10, *COX-2* cyclooxygenase-2, *miR* MicroRNA, *let-7d* lethal-7d microRNA, *hTERT* human telomerase reverse transcriptase

sorts of cells, interleukin-1 (IL-1), basic Fibroblast growth factors, and TNF. Because of this, expression of TGFβ-1 has been increased in macrophages and alveolar epithelial cells from the lung tissues of IPF patients [25]. Furthermore, in the bleomycin mice, an in vivo IPF model [26] was demonstrated in the research [27], which found that 9 out of 12 standardised lung tissue samples from patients with IPF showed high membrane PD-L1 expression in alveolar macrophages. These macrophages play a crucial part in the development of lung cancer and IPF. Additionally, they discovered that IPF patients had considerably greater serum levels of soluble PD-L1 (sPD-L1) than a healthy control group. Checkpoint inhibitors that target PD-1/PD-L1 unleash antitumor T lymphocytes in cancer biology, enabling their activation, proliferation, and tumour cell eradication. Activating the TGF β-1 receptor complex also activates downstream canonical (Suppressor of mothers against decapentaplegic (SMAD2 and 3)) and non-canonical (PI3K, Mitogen-activated protein kinase (MEK), mammalian target of rapamycin (mTOR), etc.) signalling cascades, that ultimately impact the transcription of ECM proteins, profibrotic mediators, growth factors, and microRNAs [28, 29]. Particularly, phosphorylated SMAD2 and SMAD3 are translocated into the nucleus to modulate transcriptional responses as a result of activated TGFβ-1 receptors. For this reason, Zhao et al. showed that SMAD3 deletion mice reduced bleomycin-induced pulmonary fibrosis in mice [30]. The Wnt/catenin system and TGFβ-1 signalling have interactions that may promote the epithelial-mesenchymal transition and myofibroblast activation [31]. The stimulation of Wnt/ß-catenin signalling in IPF patients' alveolar epithelium has recently been shown to hinder lung healing and accelerate AEC2 ageing [32]. Indeed, it has been discovered that lung fibrosis causes the Hippo pathway's yes-associated protein and transcriptional coactivator with PDZ-binding motif to become activated. For this reason, YAP/TAZ, TGF-ß, Wnt, and PI3K axis activation was discovered by Xu et al. using single-cell RNA sequencing [33, 34]. They hypothesised that mTOR/PI3K/AKT signalling aided in the aetiology of IPF by promoting the growth of lung fibroblasts and epithelial cells. TGF-ß1 may consequently be in charge of the hedgehog pathway (Shh), whose activation in IPF lungs accelerates the disease's progression with fibroblast proliferation and aberrant extracellular matrix deposition [35]. The activation of v6 integrin, one of the additional pathways implicated in TGF-ß1 synthesis by the AECs of IPF lungs, triggers the unfolded protein response, a defence biological process attributed to the build-up of misfolded proteins [36].

9.3.1.2 Cancer-Associated Fibroblasts

Cancer-associated fibroblasts (CAFs) function the same as myofibroblasts in IPF and they are significant elements in the tumour microenvironment and are largely attracted to and activated by cytokines produced by cancer cells and immune cells that have infiltrated the tumour [37, 38]. Due to their significance in the tumour microenvironment's pathways of proliferation, invasion, inflammation, angiogenesis, and metastasis, CAFs, which make up the majority of the cells in tumour stroma, significantly contribute to the biology of tumours. Additionally, stromal cells, which are both malignant and non-cancerous, produce different growth factors, such as FGF, TGF-β1, TNF, EGF, IL-1, vascular endothelial growth factor, and IL-6 [39, 40], which promote CAF trans-differentiation and activation, which also has the effect of boosting inflammation and carcinogenesis. After accounting for everything, it is possible to draw comparisons between fibrosis and LC, and among the various growth factors, TGF-β1 represents the key driving signalling for both fibroblasts and CAFs' trans-differentiation, that is important for tumour development and therapeutic resistance [41, 42]. Numerous in vitro studies conducted in recent years have demonstrated that TGF-β triggered epithelial-mesenchymal transition into non-cancerous epithelial cells by activating the mTOR and PI3K pathways resulting in cancer cell's resistance to apoptosis, encouraging CAFs to transdifferentiate [43]. Furthermore, CAFs exhibit great heterogeneity, which is most likely resulting from their liability to various tumour-secreted components, which prompt them to communicate various molecular markers. Although they are diverse, the molecular processes that activate CAFs and involved in the progression of cancer may be shared by different malignancies. In addition to TGF-β, epidermal growth factor receptor, the Wnt/ß-catenin, Hippo pathways, and JAK/STAT are other important molecular pathways for this function [44].

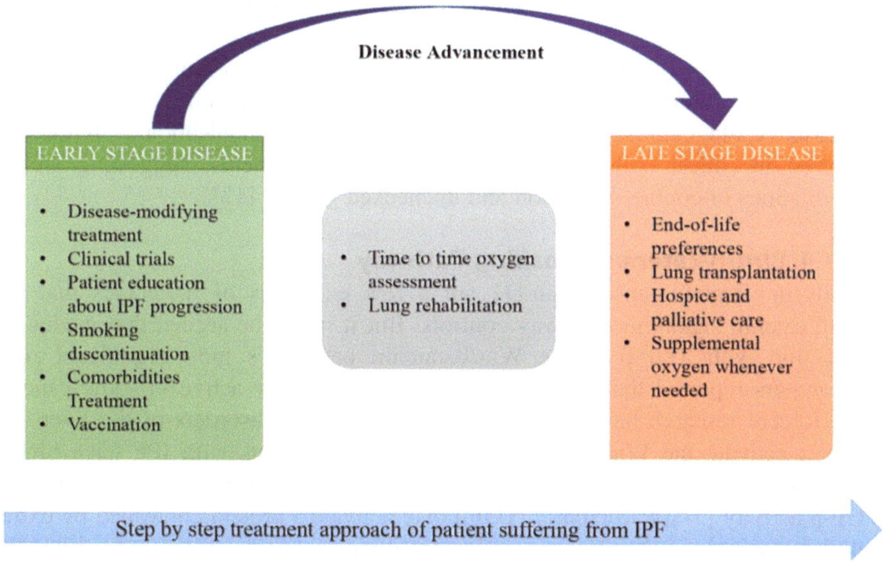

Fig. 9.4 Idiopathic pulmonary fibrosis management

By interacting with or activating a number of additional downstream pathways, the profibrotic stimuli enhance the degree of TGFß expression that fosters activation of myofibroblast further which leads to the aberrant ECM accumulation and progression of IPF (such as the canonical SMAD3 pathway or non-canonical pathway and growth factors). The LC counterpart to the myofibroblasts, the CAFs, can come from different sources, such as cancer-associated adipocytes, and resident fibroblasts, which can originate from circulating progenitors in the bone marrow and are prevalent in tumour stroma [35] (Fig. 9.4).

9.3.1.3 Abnormal Cell-Cell Communication

Cxs (connexions) family made intracellular channels that connect cells metabolically and electrically. The coordination of cell growth and tissue repair depends on Cxs [45]. Cx43, one of them, is engaged in wound healing and tissue repair healing and is the one that is most prevalent on fibroblast membranes. Repression of Cx43 at wound sites aids in the healing of damaged skin tissues by promoting keratinocyte and fibroblast migration and proliferation. Therefore, Cx43 downregulation is linked with greater levels of TGF-β expression, collagen formation, and accelerated myofibroblast differentiation, all of which may aid in the promotion of healing. These modifications help explain how aberrant healing and fibrosis are characterised by a loss of control over fibroblast growth, a decline in control of fibroblast proliferation that distinguishes fibrosis and aberrant repair. The discovery that keloids and hypertrophic scars exhibit lower levels of Cx43 expression than normal skin tissues does corroborate this claim [46]. Despite the fact that human lung carcinoma cell lines expressing enhanced level of Cx43 demonstrated lower proliferation, decrease expression of Cxs are usually linked to the onset of cancer and the breakdown of intercellular communication [47]. Primary lung fibroblasts from IPF patients were found to have decreased intercellular communication as well as decreased Cx43 expression [48]. Limited cell-cell communication is frequently observed in cancer cells and fibroblasts from IPF patients, suggesting shared abnormalities of contact inhibition and unchecked growth [12].

9.3.1.4 Wnt/β-catenin Signalling Pathway

Matrilysin, laminin, and cyclin-D1 are a few examples of molecules that the Wnt/ß-catenin signalling pathway controls. But it might be argued that mediating interaction with TGF-β is the Wnt/ß-catenin pathway's most significant role. Researchers reported that this pathway is inappropriately active in some cancers [49]. Recent research has shown that fibroproliferative disorders of the liver and kidneys activate the Wnt/ß-catenin pathway [50]. Patients with IPF have highly active Wnt/ß-catenin pathways in their lung tissues, which may be a reflection of TGF-β activity [51]. Particularly, TGFβ may activate a protein controlled by extracellular signals. Apoptosis and proliferation are regulated by the phosphatidylinositol 3-kinase (PI3K)/Akt pathway, which is activated by the kinases 1 and 2 and target genes of this pathway. The functions of PI3K in myofibroblast development and proliferation have been demonstrated after TGF-β stimulation [52]. When the PI3K pathway is activated in cancer cells, the regulatory mechanisms that regulate cell proliferation break down. Blockers of the PI3K pathway have been utilised as therapeutic targets, and researchers are examining how they affect the growth and survival of tumours in different types of cancer [53].

On the other hand, unnatural actions of these kinases have been linked to a variety of cancers' growth, development, and spread [54]. Common carcinogenic and fibrogenic mediators include TGF-β, Platelet-derived growth factor(PDGF), Vascular endothelial growth factor(VEGF), and Fibroblast growth factors (FGF). VEGF, in particular, may stimulate ERK1/2 and PI3K in a manner that directly or indirectly promotes cell survival and proliferation. Accordingly, it was discovered that endothelial progenitor cells (EPCs) from patients with IPF had increased VEGF mRNA expression levels [55].

9.3.1.5 Hepatocyte Growth Factor

Mesenchymal cells produce hepatocyte growth factor (HGF), which has been found to be an effective mitogen for mature hepatocytes. The c-Met proto-oncogene product known as the HGF receptor is mostly expressed in different kinds of epithelial cells. IPF patients' bronchoalveolar lavage fluid and serum have greater amounts of HGF than serum from healthy individuals [56, 57]. In the IPF patients lung tissue, hyperplastic AECs and macrophages also show elevated levels of HGF [58]. In addition, Trkb treatment stopped mice from developing lung fibrosis in the bleomycin model [59]. In light of these findings, treatment with HGF could offer a new way for the inhibition of lung injury with in pulmonary fibrosis patients [60].

9.4 Several Aberrant Genes Expression Are Involved in the Pathogenesis of IPF

Both sporadic and familial forms of pulmonary fibrosis are correlated with genetic variations. Rare and frequent genetic variations are linked to five genes associated with telomere biology—TERT, TERC, DKC1 (Dyskerin pseudouridine synthase 1), TINF2 (TERF1-interacting nuclear factor 2), RTEL1 (Regulator of telomere elongation helicase 1), and PARN (Poly(A)-specific ribonuclease) as well as FIP (feline infectious peritonitis)-related mutations connected to alveolar stability SFTPA1, SFTPC, SFTPA2, NAF1, and ATP-binding cassette-type 3 have also been discovered through familial studies. Common variations, which are minor allele frequencies of >5%, also seem to affect the risk of FIP [61–65].

The promoter region of MUC5B has the most frequently repeated risk mutation (rs35705950), which was first discovered and has been tightly linked with IPF and familial IPF in a combined linkage and association analysis. In pulmonary fibrosis, there have been two substantial GWAS of IPF individuals (both familial and sporadic) with controls. Along with confirming TERT's function at 5p15, MUC5B at 11p15, and the 3q26 area close to TERC, the seven novel linked loci were discovered by GWAS, including DSP (6q24), FAM13A(4q22), ATP11A (13q34), OBFC1 (10q24), DPP9(19q13), and the chromosomal regions 7q22 and 15q14-15, respectively, those who have nominally coupled with others. In contrast, common variations typically have a smaller effect magnitude, but they occur more frequently, particularly together, and may increase the percentage of disease risk [66].

9.4.1 Surfactant Protein-C (SFTPC)

According to numerous reports, SFTPC is linked with IPF. The gene SFTPC, which has gene ID 6440, is found on the short arm of chromosome 8 and codes for a 197 amino acid precursor protein through six exons. The first example of newborn ILD having a heterozygous change from A to G in the intron 4 splice donor site was described by Nogee et al. in 2001 [67]. This nucleotide mutation resulted in the protein precursor's C-terminal region losing 37 amino acids and bypassing exon 4. As a result, the lung was deficient in mature surfactant protein C (SP-C) tissue and fluid from bronchoalveolar lavage. Afterwards, two more Leu188 to Gln (L188Q), a heterozygous mutation, and Ile73 were found in the SFTPC gene and correspond to Thr (I73T) [68, 69]. Prosurfactant protein C (pro-SP-C) builds up within alveoli and altered protein intracellular trafficking are potential effects of these two mutations. The mature 35-residue hydrophobic protein (SP-C), which is encoded by the SFTPC, is created by many rounds of proteolytic cleavage of the pro-SP-C. SP-C is subsequently secreted into the alveolar space. The SP-C is crucial for preserving alveolar stability, along with surfactant phospholipids and other surfactant proteins [70]. The aberrant SP-C proteins produced by SFTPC mutations can activate apoptotic pathways, instigate ER stress, and block the ubiquitin/proteasome system [71]. When SFTPCL188Q was transfected into cultured type II alveolar epithelial cells (AECII), aberrant lamellar structures, ER stress, and an unfolded protein response appeared [72]. Pro-SP-BRICHOS C's domain, a unique domain identified in 12 protein families with a variety of functions and illness correlations, has been revealed to include certain fatal mutations [73]. When SP-C is made from the pro-SP-C, the BRICHOS domain acts as a chaperone. It is essential for the maturation of pro-SP-C, the precise folding procedure, and the correct packaging before it joins other surfactant components in the lamellar bodies for exocytosis. This mechanism may explain why infant/adult respiratory distress is not the same as pulmonary fibrosis, which is a condition caused by unfolded pro-SP-C but not a deficit. It has also been demonstrated that pulmonary fibrosis and the process of collagen deposition in the alveolus wall both include epithelial-to-mesenchymal transition (EMT). In A549 cells, targeted expression of SFTPCL188Q led to reduced zonula occludin-I and E-cadherin expression enhanced smooth mesenchymal marker expression muscle actin, demonstrating the connection between the variations of EMT and SFTPC [74]. In humans, SP-C makes up 10% of the protein components of pulmonary surfactants together with SP-A, SP-B, and SP-D, with lipids making up the remaining 90%. The main function of surfactants is to reduce surface tension at the air-water interface and prevent alveolar collapse. The surfactant-related genetic and acquired disorders attracted attention in the aetiology study of IPF since they had been linked in the pathophysiology of IPF together with surfactant modifications and alveolar type II cell death [75].

9.4.2 Telomerase Reverse Transcriptase and Telomerase RNA Component

Chromosome ends have telomeres, which are repeating nucleotide sequences that prevent the chromosomes from gradually shrinking when cells replicate normally. Telomere length is restored by telomerases' two primary components, telomerase RNA (encoded by TERC) and telomerase reverse transcriptase (encoded by TERT). Oral leukoplakia, unusual skin darkening, and dystrophic nails are the hallmarks of the rare inherited telomere shortening condition known as dyskeratosis congenita (DKC). About 20% of individuals have lung fibrosis, and DKC complications including bone marrow failure can also occur. DKC served as the setting for the initial discovery of telomerase component mutations. More recent studies have found associations between FIP and a number of telomerase maintenance pathway genes, including those implicated in telomere stabilisation (DKC1, PARN, and RTELI) and catalytic activity (TERT and TERC) [76].

In peripheral blood and the lung, these pathogenic mutations promote telomere shortening by impairing telomerase activity. TERT variations have so far been found in 15% of FIP cases and 1–3% of sporadic cases, making them the uncommon variants associated with pulmonary fibrosis. TERT, RTEL1, and PARN variations were previously identified by a study associated with occasional IPF being linked to FIP, further supporting the pathophysiology of IPF in which telomere dysfunction plays a role as well as emphasising the genetic similarities between FIP and IPF sporadic. Further research has linked IPF to telomere dysfunction as there is evidence that telomere length is not the only connected to rare variant mutations in telomerase. A study discovered that without known mutations for TERT or TERC, short telomeres were present in 25% of sporadic IPF participants and 24% of familial IPF participants. Additionally, all participants in this study who had pulmonary fibrosis and those with TERC or TERT mutations also showed short telomeres. Uncertainty exists regarding the exact processes through which lung disease is brought on by telomere abnormalities. Epithelial cell senescence and a reduced ability to respond to epithelial damage have both been linked to defects in telomere maintenance. Telomere shortening happens throughout subsequent cycles of cell division and eventually triggers DNA damage pathways that result in apoptosis or senescence. Although early senescence can disrupt lung epithelial homeostasis and cause a remodelling response that leads to fibrotic lesions, cellular senescence is sometimes acceptable [66].

9.4.3 Pulmonary-Surfactant Associated Proteins (SFTPA2) Mutations

Researchers thought that mutations in the other surfactant proteins (A, B, and D) might also be discovered after discovering SFTPC mutations in FIP. The discovery of two families with FIP caused by mutations in SFTPA2 was reported by Wang et al. in 2009. Neighbouring genes encode two SPA isoforms (SFTPA1 and SFTPA2) and Wang et al. used whole genome linkage on a family of 15 individuals who had FIP, bronchoalveolar cell carcinoma, or unspecified lung disease and were connected to an area around SFTPA1 and SFTPA2 [63].

9.4.4 Mucin 5B (MUC5B)

A genome-wide linkage study was conducted in 2011 and discovered a locus on chromosome 11 that was strongly connected to the possibility of IPF [77]. After resequencing this region, a frequent single nucleotide polymorphism (rs35705950) in the promoter of the mucin 5B (Muc5B) gene was discovered, and it was connected to a six- to eightfold increased risk for IPF. Additional independent cohorts have now verified the link between this MUC5B promoter polymorphism and IPF, mostly made up of Caucasians [78]. It's interesting to note that the MUC5B SNP appears to be frequent in cases of FIP and sporadic IPF [77]. The results reveal that rs35705950 did not raise the risk of scleroderma-related interstitial lung diseases or sarcoidosis. However, rs35705950 was discovered to be uncommon in the Korean population. This connection between rs35705950 and IPF was verified in a cohort of Mexican patients [79]. Similar to this, rs35705950 was uncommon in IPF patients in a Chinese community, while various MUC5B polymorphisms were linked to the condition [80].

MUC5B expression was uniformly higher in patients of IPF lungs compared with controls, despite the MUC5B SNP being present, despite the rs35705950 MUC5B SNP being linked with increased MUC5B mRNA expression in control people's lungs. This result is in line with individuals with IPF having more MUC5B-expressing cells in the distal airways [81]. In the Framingham cohort, MUC5B rs35705950 has also been identified as a risk factor for asymptomatic interstitial lung abnormalities detected on CT scans in persons over 50. Uncommonly, those with the minor allele of rs35705950 are more likely to experience the syndrome, despite the fact that IPF patients who have the risk allele appear to have increased mortality in comparison to non-carriers. According to past animal studies, MUC5B may alter airway host defence [82], despite the fact that it is not yet known how MUC5B influences fibrotic remodelling [83].

9.4.5 Toll-Like Receptor

TLR signalling disruption has been identified in patients with IPF as a bridge between innate and adaptive immune responses [84]. Yet, the precise role of these signalling cascades in the fibroproliferative response is still largely unknown. There are ten functioning TLRs in humans, each of which has a unique receptor-ligand relationship. To recognise distinct external and intracellular signals, the [85] TLRs are either localised to the cell membrane (TLR1, 2, 4, 5, 6) or endosomal compartments (TLR3, 7, 8, 9), correspondingly. The majority of TLRs signal via a MyD88-dependent pathway, which ultimately leads to NF-κB activation and transcription of proinflammatory cytokine genes. A MyD88-independent mechanism underlies TLR3 and alternative TLR4 signalling, with TRIF recruitment ultimately leading to IRF3 or IRF7 transcription of type I interferon (IFN) genes [85].

Below is a description of the genetic risk mutations influencing TLR family signalling that are linked to IPF. GWAS results revealed three typical variations (rs111521887, rs5743894, rs574389) which were found in the toll-interacting protein gene connected with IPF. There has been conflicting research regarding whether these variations are providing separate associations with IPF or are in linkage disequilibrium. However, the IL-1 receptor-associated kinase inhibition (IRAK) phosphorylation of TLRs, notably TLR2 and TLR4, have been shown to limit TLR activity [86]. It is reported that IPF epithelia have increased levels of TLR2 and TLR4 activation, possibly as a result of continuous exposure to pathogenic microorganisms [86]. It has been demonstrated that decreased toll-interacting protein (TOLLIP) expression causes macrophages to secrete more pro-inflammatory cytokines as a result of TLR activation [87]. Through TLR4-dependent signalling, TOLLIP stimulates the production of IL-10, which protects mice from a fibrosis model caused by bleomycin. Additionally, TOLLIP inhibits TGF-β signalling by destroying TGF-ß1 receptors via interactions with SMAD7. These findings collectively imply that TOLLIP expression may protect against IPF by reducing pro-inflammatory and profibrotic pathways [88]. The rs1278769 variation in ATP11A discovered in the GWAS, however, less functionally characterised, might affect TLR4 signalling. It has been demonstrated that the gene 13 ATP11A, which codes for a phospholipid flippase, increases MyD88-dependent NF-kB activation and the generation of proinflammatory cytokines by inhibiting TLR4 endocytosis [89].

Dysregulated TLR9 activation encourages fibroblast-to-myofibroblast differentiation and has been associated with a more aggressive IPF phenotype. In most cases, microbial genetic material contains hypomethylated CpG nucleotides [90]. Since TLR9 also recognises mitochondrial DNA released from wounded cells, it's conceivable that the cell deterioration and non-apoptotic cell death observed in the IPF distal airway may likewise play a substantial role in TLR9-mediated fibrosis. IL1RN gene (rs408392 and rs419598) was considerably connected to the IPF condition [91]. These risk alleles cause a decrease in IL-1Ra expression, which results in unrestricted IL-1R activation in innate immune cells and airway/alveolar epithelial cells. In contrast to the conventional bleomycin model, adenoviral-mediated IL-1 overexpression is employed to study the role of IL-1R signalling in animal models of pulmonary fibrosis [92, 93].

IPF patients' alveolar macrophages exhibit an increased IL-1:IL-1Ra ratio indicating this pro-inflammatory condition. It is reported that the IL-1R/MyD88 axis reduces fibrosis in bleomycin and silica-induced fibrosis through genetic and pharmaceutical targeting. Both non-pathogenic and pathogenic bacterial signals are directly responsive to many of these hyperactive TLR signalling pathways, including TLR2 and TLR4 [94].

9.4.6 Adenosine Triphosphate-Binding Cassette Transporter A3(ABCA3)

The protein ABCA3 has been discovered in children with ILD or newborns with respiratory distress syndrome. The transfer of lipids across plasma membranes is facilitated by this transporter protein, which is mostly expressed in the AECII. The ABCA3 gene results in the production of a 1704 amino acid protein which can be found on the surfactant secretory vesicle on the limiting membrane of lamellar bodies, indicating that it may be involved in the metabolism and transport of surfactant. The ABCA3 gene has a large range of variations, and the genetic defects in surfactant metabolism that result in neonatal respiratory distress and paediatric ILD have been linked to 150 different mutations [95–97].

9.5 Epigenetic and Genetic Abnormalities

For the majority of cancers, documented pathogenic mechanisms include hypomethylation of oncogenes and methylation of tumour suppressor genes. Patients with IPF have recently been found to exhibit epigenetic responses to environmental exposures, such as smoking and nutritional variables, as well as ageing. Additionally, current research has shown that individuals with IPF experience global methylation pattern changes that are similar to those in lung cancer patients [98]. Hypermethylation of the CD90/Thy-1 promoter region reduces the expression of the glycoprotein Thy-1, which is generally produced by fibroblasts, in the context of IPF [99, 100]. In addition, the conversion of fibroblasts into myofibroblasts and the invasive nature of malignancies are linked to the absence of this molecule in IPF patients. This suggests a unique therapeutic approach for this condition: pharmacological inhibition of Thy-1 gene methylation may restore Thy-1 production. In addition, particular gene changes have been implicated in the occurrence and progression of cancer [101]. Similarly, the oncogene p53 was expressed in almost half of the cases of IPF, along with fragile histidine triads, microsatellite instability, and loss of heterozygosity, commonly in the peripheral honeycombed lung areas that are unique to idiopathic pulmonary fibrosis [101–104]. Also, familial IPF has been linked to mutations that are typically linked to the occurrence and progression of cancer, such as those that influence telomere shortening and telomerase production [105, 106]. Recent studies have looked at circulating and cell-free DNA as a cancer biomarker for diagnosis and prognosis. In these studies, individuals with cancer and IPF had higher free circulating DNA concentrations than those with other fibrotic lung illnesses. The pathophysiology of both disorders was linked to aberrant expression levels of mRNA in addition to circulating DNA. According to these findings, short non-protein-coding RNAs control genes relevant to carcinogenesis, which are implicated in the growth, invasion, and metastasis of cancer cells [98, 107, 108]. According to recent studies, 10% of mRNAs are expressed abnormally in IPF patients. Among them, miR-21 and miR-155 were upregulated, whereas lethal-7 (let-7), miR-29, miR-30, and miR-200 were downregulated. These alterations correlated with gene families involved in apoptosis, ECM modulation, fibrosis, and the formation of epithelial-to-mesenchymal transition (EMT). Some of these mRNAs might affect TGF-β expression and be affected by it, hastening the functional loss in IPF patients [109–111].

9.6 Epigenetic Regulation of Gene Expression

The chromatin regulation, expression of genes, and particular cell type activity which participates in disease pathophysiology are mediated by various environmental exposures which have been translated by epigenetic processes that is linked to the incidence of disease. Environmental factors lead to change in non-inherent epigenetic marks, which are more dynamic than inheritable epigenetic marks [112]. Methylation of DNA, modification of histones, and long-chain non-coding RNAs are the processes which are part of the epigenome (Table 9.1).

Epigenetic factors are influenced by environmental exposure, ageing, and the diet of an individual. The progression of IPF depends upon a complex interaction between various environmental and genetic factors [113]. Inhalation of several environmental factors like metal dust, wood dust, industries, and dust from textile, and cigarette smoking affects the genetic and epigenetic marks. Several studies show

Table 9.1 Various epigenetic modifications and their consequences on genes and protein translation

S. no.	Epigenetic process	Modification	Consequences
1.	DNA methylation of CpG dinucleotide	Hyper methylation	Gene silencing
		Hypo-methylation	Active transcription
2.	Histone methylation	H3K4 trimethylation	Transcriptional activation
		H3K27 trimethylation	Gene silencing
3.	Histone acetylation	Histone tail acetylation by HATs	Active gene transcription
		Histone tail deacetylation by HDACs	Gene silencing
4.	miRNA leading mRNA degradation	Binding to 3′ UTR of mRNA	Inhibition of protein translation

that modulation of epigenetic factors regulates the expressions of genes which are involved in the occurrence of IPF.

The link between DNA methylation and histone alterations is an emerging model for epigenetic control of gene expression. One illustration of these relationships is the association of the regulatory component DNMT (DNA methyltransferases) 3L with the N-terminal of the histone H3 tail, which is linked in sequence to the mammalian new genetic methyltransferases like DNMT3A and DNMT3B. These data support the notion that the histone H3 tail's unmethylated lysine 4 serves as a chromatin regulator of DNA methylation. DNMT3L identifies unmethylated histone H3 tails at lysine 4 and triggers de novo methylation of DNA by recruiting or activating DNA methyltransferase3A2. DNA methyltransferases similarly favour targeting nucleosome-bound DNA [114].

It is already evident that epigenetic marks which are related to the progression of IPF are also being regulated by environmental exposure [115] such as wood dust, textile dust, metal dust, silica, cigarette smoking, and many more [112, 115]. Out of all the environmental exposures, cigarette smoking is one of the major causes which has been linked to the development of IPF. Some studies related with case control have shown a link between cigarette smoking [116], occupational exposure, and the progression of IPF [115]. Various epigenome-wide association studies have confirmed that cigarette smoke can alter the DNA methylation at multiple CpG sites, and it can also regulate the miRNA expression in epithelial cells of bronchioles [117]. Cigarette smoking also involves in the methylation of *Wnt7a*, which is a specific promoter of a gene that is involved in progression of IPF [112]. Diet and ageing are also an equal contributors in the alteration of epigenomes. The pathobiology of ageing includes abnormal telomere shortening and dysfunction which is also linked to the development of IPF [10]. Other ageing-associated alterations, like the dysfunctioning of mitochondria and changed proteostasis enhance cell senescence. Persistent deposition of senescent cells is found in IPF lungs and it is responsible for stem cell dysfunctioning and produces pro-inflammatory cytokines such as IL-6, TGF-β, and metalloproteases which may cause fibrosis [118]. Exposure to fine particulate matter 2.5 (PM2.5), pesticides, and fungicides can also lead to the methylation of DNA [115]. The initiation and progression of IPF can also be a result of viral infection. A meta-analysis including 634 IPF patients, 653 control and 19 virus species (such as Epstein-Barr Virus, CHV, Human herpesvirus), suggested that viral infection is strongly associated with higher risk of IPF development .[118]. Many case control studies have suggested that the exposure of cigarette smoke can lead to the development and progression of IPF, however, there is still need of evident result to confirm the linkage and exposure of smoking and epigenomic of IPF (Fig. 9.3).

9.6.1 DNA Methylation

In numerous targeted studies, it is already shown that modulation of epigenetics controls the expression of genes which are involved in the development and progression of IPF. Likewise, Thy-1(CD90) is a glycosylphosphatidylinositol-linked membranous (outer) glycoprotein and is a major regulator of cell-matrix and cell-cell interactions [119] but in IPF patients, its expression is absent in fibroblastic foci and myofibroblasts [112]. Thy-1 suppresses myofibroblast differentiation and its expression is downregulated by hypermethylation of promoter gene DNA in rat lung fibroblast [119] and also by histone modification [115]. The expression of the α-smooth muscle actin (α-SMA) gene varies depending on the degree of methylation of three CpG islands in the promoter of the α-SMA gene in fibroblasts, myofibroblasts, and alveolar epithelial type II cells. Inhibition of DNMT by siRNA induces the expression of α-SMA while its overexpression downregulates the α-SMA expression [120]. p14 (ARF) and prostaglandin E receptor 2 gene (PTGER2) are two examples of genes involved in fibroblast apoptosis whose expression was reduced in IPF lung fibroblasts due to hypermethylation of their promoters. It has been seen that prostaglandin E2 increases the DNMT3a activity, which leads to global hypermethylation and also results in the upregulation of those genes which are responsible for the suppression of cell proliferation in pulmonary fibroblast [121]. In IPF patients, TP53INP1 protein, which is a cell stress response protein induced by p53, is the most hypomethylated and upregulated gene [121].

9.6.2 Histone Modification

Histone deacetylases are the enzymes that deacetylase specific non-histone and histone proteins at particular locations [122]. In the process of deacetylation, histone deacetylases remove the acetyl group from lysine and retain the positive charge of histone, which results in the development of dense chromatin, hence inhibiting gene expression. On the other hand, histone acetyltransferases, i.e., HATs are another enzymes that acetylate the amino acids having conserved lysine by transferring the acetyl group from acetyl coenzyme A to form ε-N-acetyl-lysine, that attach with positive charged histone lysine and increases hydrophobicity. This binding leads to the emergence of a loose chromatin structure that promotes the transcription of uncoiled DNA [123]. Defects in the acetylation of histone can lead to the repression of expression of two genes which are antifibrotic, COX2 (cyclooxygenase-2), and chemokine IP-10 [115, 124]. C-X-C motif chemokine 10 is an antifibrotic molecule, produced by lung fibroblast, as a counterregulatory factor. It has been shown that the methylation of histone H3 lysine 27 by histone lysine methyltransferases (EZH2, G9a, G9a-like protein) represses the CXCL10 gene [124].

Recent studies demonstrated the suppression of the caveolin (Cav-1) gene by TGF-β leads to fibroblast proliferation and attains anti-apoptotic tendency. Cav-1 is constitutively repressed in IPF and the repression of the Cav-1 gene is regulated by histone modification (trimethylation of H3 lysine 4) [125]. It has been also reported that class III histone deacetylase, sirtuin, encourages proteasomal degradation of p21 to lessen the senescence brought on by TGF-β in human bronchial epithelial cells [121].

9.6.3 Non-Coding RNA Regulation

After exploring global miRNA profiles in lung tissue of IPF patients, epigenetic regulators of fibroproliferation, miR-29, and let-7d have been recognised while miR-155, miR-154, and miR-21 were found to be upregulated [114, 121]. The analysis of the lung fibrosis model of murine shows that let-7d has a protective role through the restriction of SMAD-3-dependent EMT. According to a subsequent investigation, in IPF lungs, miR-154 was one of the miRs that was most highly elevated. miR-154 appears to be a viable therapeutic target since transfection of primary fibroblasts with it led to substantial increases in cell migration and proliferation by activating a Wnt pathway [121]. A study reported that in IPF patients' lung tissue, the miR17~92 clusters were reduced, while DNA methylation of its promoters' region by DNMT1 was increased [112, 126]. miR17~92 is crucial in the development of lung epithelial cells and targets essential fibrotic genes [126]. A short non-coding RNA, miR-29 has also shown antifibrotic action in lung fibrosis, and it is majorly expressed during the fibrosis of various tissues. miR-21 has been demonstrated to be increased in the lungs of individuals with IPF and an experimental model of pulmonary fibrosis, primarily in fibroblastic foci [121].

9.7 Treatment of Idiopathic Pulmonary Fibrosis

Since delayed access to care and treatment is independently related to a higher risk of mortality, it is indicated that patients with known or suspected IPF be sent as soon as possible to a facility with experience in providing care for the condition [127]. The step-by-step treatment approach should be followed in the management of patients with IPF (Fig. 9.4). There is currently no cure for IPF, although nintedanib and pirfenidone are two drugs that work to delay the disease's development and reduce mortality. The patient's tolerance for side effects will determine which medicine is preferred because, based on the available evidence, none is more effective than the other [128]. Pirfenidone, a pyridone, is an orally available, small molecule having multiple effects such as anti-inflammatory, antioxidant and antifibrotic [8, 123]. Pirfenidone inhibits ECM production and fibrogenesis by suppressing the transcription of profibrotic and procollagen factors, such as TGF-β and PGDF, and it also suppresses the production of reactive oxygen species [123]. Pirfenidone also inhibits the proliferation of fibroblasts [128].

Nintedanib, a tyrosine kinase inhibitor, was developed for use in cancer treatment because it inhibits the proangiogenic factors [8]. Nintedanib targets platelet-derived growth factor receptors, receptor, and vascular endothelial growth factor receptor, that are highly upregulated in the lung tissue of IPF patients [123]. In a phase 2 study of patients with IPF, nintedanib showed a dose-dependent tendency to slow the loss of lung function and lower the frequency of acute exacerbations. In two additional phase 3 studies, nintedanib was compared to a placebo. For inclusion, the forced vital capacity had to be larger than or equal to 50% of the expected value, and the predicted lung carbon monoxide diffusion capacity had to be between 30 and 79%. In both studies, the relative fall in forced vital capacity after 52 weeks was 47.9 and 55.1% lower in the treatment group compared to the control group [2]. Both the above drugs have the same effect on the rate of decrement in forced vital capacity. Till date, none of the drugs, whether it is nintedanib or pirfenidone, has shown a survival benefit in clinical trials. However, these drugs have proven their effect on decreasing the mortality of IPF patients [2]. Various studies had been done using different drugs to treat IPF. However, none of them proved to be an effective and a reliable treatment therapy. Drugs, such as ambrisentan, everolimus, prednisolone, azathioprine, acetylcysteine, and warfarin had been identified as potentially harmful therapies in clinical trials. However, some therapies were potentially ineffective such as bosentan, imatinib, macitentan, and sildenafil [2]. Further, several drugs were tried and under clinical trials unfortunately no success has been achieved (Table 9.2).

9.7.1 Symptom-Focused Therapy

Due to the significant symptom load of IPF, including dyspnoea and cough, adjunctive symptom-based therapy is crucial. Corticosteroids and opiates might help in reducing chronic cough, anxiety, dyspnoea respectively [2]. Further, studies have demonstrated that pulmonary rehabilitation can reduce these symptoms, enhancing the quality of life. Moreover, LTOT (long-term oxygen therapy) is a necessary treatment in patients with IPF suffering from disease progression [128]. To treat low oxygen levels in IPF patients, supplemental oxygen therapy should be considered [2].

Table 9.2 Various drugs which have been terminated due to adverse events and ineffectiveness during clinical trials on IPF patients

S. no.	Compound	Category	Study phase	Results	References
1.	Warfarin	anticoagulant	Phase III	Increased mortality, study terminated	[129]
2.	Everolimus	mTOR inhibitor	Phase II	High adverse events and increased disease progression	[130]
3.	Bosentan	Dual endothelin receptor antagonist	Phase III	No significant changes in treatment group	[131]
4.	Ambrisentan	Type A endothelin receptor antagonist	Phase III	Terminated due to lack of efficacy and increased disease progression	[132]
5.	Macitentan	Dual endothelin receptor antagonist	Phase II	No significant changes in treatment and control group	[133]
6.	Sildenafil	PDE-5 inhibitor	Phase II	No significant changes in treatment and control group	[134]
7.	Etanercept	Recombinant soluble human TNF-α receptor	Phase II	No significant difference in primary end points	[135]
8.	Imatinib	Tyrosine kinase inhibitor	Phase II	No difference in survival or lung function between treatment groups	[136]

9.8 Future Perspective of IPF Treatments

There are several approaches that should be considered in the treatment of IPF such as alveolar epithelial injury prevention, targeting the coagulation cascade, boosting epithelial proliferation and collagen synthesis, preventing fibroblast proliferation and extracellular matrix direct targeting, inhibiting TGF-β and other profibrotic cytokines, and increasing epithelial resistance to injury [8]. The drugs which are in advanced stages in clinical trials are Simtuzumab (monoclonal antibody against lysyl oxidase-like 2), Lebrikizumab (monoclonal antibody against IL-13), and STX-100 [8]. An approach to combating telomere shortening, a known contributor to epithelial senescence is an appealing therapeutic approach for the treatment of IPF. hTERT administered by using an adenoviral vector has shown the potential treatment to express enzyme telomerase in other human diseases. Additionally, by inhibiting the development of senescent cells as a result of mitochondrial malfunction, treating lung fibrosis may be possible by enhancing mitochondrial biogenesis or mitophagy by utilising activators of PINK1 or SIRT3 [118]. Brd4 inhibitor JQ1 and HDAC inhibitors such as Spiruchostatin A (SpA), and Suberoylanilide hydroxamic acid (SAHA) have shown the reversal of profibrotic phenotype in primary fibroblasts and bleomycin mouse model [137].

9.9 Conclusion

In the last 15 years, the perspective of IPF has been changed enormously. Earlier IPF was considered idiopathic, but series of findings such as genetic polymorphisms, ageing, epigenetic alterations put the light on various pathogenesis through which IPF is characterised, that are ECM deposition, inflammation, and fibrosis. Several studies have also shown that alteration in epigenetics leads to the progression of IPF. Some animal studies have shown effectiveness in mitigating the IPF by targeting these epigenetic alterations. So far, the management of IPF relies on two newly approved drugs, that are pirfenidone and nintedanib as they have shown effect in reducing the IPF severity, yet these two are unable to prove efficiency in reducing mortality of IPF patients. Some drugs, like Simtuzumab, which is a monoclonal antibody, Lebrikizumab, monoclonal antibody against IL-13, and STX-100 are in advanced stage of clinical trials. Apart from these drugs, various other drugs were proved ineffective and hazardous during clinical trials. Other approaches should also be considered in the therapy of IPF, such as alveolar epithelial injury prevention, preventing fibroblast proliferation, ECM accumulation, and inhibiting profibrotic and TGF-β. Targeting telomere shortening by administration of hTERT has potential to combat IPF. Due to the significant symptom load of idiopathic pulmonary fibrosis, including dyspnoea and cough, adjunctive symptom-based therapy is crucial. Consider probable contributory comorbidities in individuals with persistent cough, such as gastroesophageal reflux disease. Corticosteroids and opiates might help in reducing chronic cough, anxiety, dyspnoea, respectively. Supplemental Oxygen for long-term is also a necessary treatment in patients with IPF suffering disease progression. To treat low oxygen level in IPF patients, supplemental oxygen therapy should be considered. Altogether, IPF is a disease of concern as till date no effective treatment is available and targeting root cause of disease such as epigenetic alterations is a crucial approach to mitigate the disease.

Acknowledgements Author would like to acknowledge infrastructure facility and Research Seed Money provided by the Central University of Punjab, Bathinda, and UGC Start-Up grant from UGC-BSR and PECFAR fellowship by IGSTC. PY would like to acknowledge the SRF fellowship received from ICMR.

Conflict of Interest None.

References

1. Martinez FJ, Collard HR, Pardo A, Raghu G, Richeldi L, Selman M, et al. Idiopathic pulmonary fibrosis. Nat Rev Dis Primers. 2017;3:17074. https://doi.org/10.1038/nrdp.2017.74.
2. Richeldi L, Collard HR, Jones MG. Idiopathic pulmonary fibrosis. Lancet. 2017;389(10082):1941–52. https://doi.org/10.1016/s0140-6736(17)30866-8.
3. Lederer DJ, Martinez FJ. Idiopathic pulmonary fibrosis. N Engl J Med. 2018;378(19):1811–23. https://doi.org/10.1056/NEJMra1705751.
4. Spagnolo P, Kropski JA, Jones MG, Lee JS, Rossi G, Karampitsakos T, et al. Idiopathic pulmonary fibrosis: disease mechanisms and drug development. Pharmacol Ther. 2021;222:107798. https://doi.org/10.1016/j.pharmthera.2020.107798.
5. Sgalla G, Iovene B, Calvello M, Ori M, Varone F, Richeldi L. Idiopathic pulmonary fibrosis: pathogenesis and management. Respir Res. 2018;19(1):32. https://doi.org/10.1186/s12931-018-0730-2.
6. Wolters PJ, Collard HR, Jones KD. Pathogenesis of idiopathic pulmonary fibrosis. Annu Rev Pathol. 2014;9:157–79. https://doi.org/10.1146/annurev-pathol-012513-104706.
7. Maher TM, Bendstrup E, Dron L, Langley J, Smith G, Khalid JM, et al. Global incidence and prevalence of idiopathic pulmonary fibrosis. Respir Res. 2021;22(1):197. https://doi.org/10.1186/s12931-021-01791-z.
8. Woodcock HV, Maher TM. The treatment of idiopathic pulmonary fibrosis. F1000Prime Rep. 2014;6:16. https://doi.org/10.12703/P6-16.
9. Mustafin RN. Molecular genetics of idiopathic pulmonary fibrosis. Vavilovskii Zhurnal Genet Selektsii. 2022;26(3):308–18. https://doi.org/10.18699/VJGB-22-37.
10. Leung J, Cho Y, Lockey RF, Kolliputi N. The role of aging in idiopathic pulmonary fibrosis. Lung. 2015;193(4):605–10. https://doi.org/10.1007/s00408-015-9729-3.
11. Samarelli AV, Masciale V, Aramini B, Coló GP, Tonelli R, Marchioni A, et al. Molecular mechanisms and cellular contribution from lung fibrosis to lung cancer development. Int J Mol Sci. 2021;22(22) https://doi.org/10.3390/ijms222212179.
12. Kinoshita T, Goto T. Molecular mechanisms of pulmonary fibrogenesis and its progression to lung cancer: a review. Int J Mol Sci. 2019;20(6) https://doi.org/10.3390/ijms20061461.
13. Karampitsakos T, Tzilas V, Tringidou R, Steiropoulos P, Aidinis V, Papiris SA, et al. Lung cancer in patients with idiopathic pulmonary fibrosis. Pulm Pharmacol Ther. 2017;45:1–10. https://doi.org/10.1016/j.pupt.2017.03.016.
14. Ozawa Y, Suda T, Naito T, Enomoto N, Hashimoto D, Fujisawa T, et al. Cumulative incidence of and predictive factors for lung cancer in IPF. Respirology. 2009;14(5):723–8. https://doi.org/10.1111/j.1440-1843.2009.01547.x.
15. Goto T, Maeshima A, Akanabe K, Oyamada Y, Kato R. Acute exacerbation of idiopathic pulmonary fibrosis of microscopic usual interstitial pneumonia pattern after lung cancer surgery<en-aut-mei> </en-aut-mei>. Ann Thorac Cardiovasc Surg. 2011;17(6):573–6. https://doi.org/10.5761/atcs.cr.10.01619.
16. Goto T, Maeshima A, Oyamada Y, Kato R. Idiopathic pulmonary fibrosis as a prognostic factor in non-small cell lung cancer. Int J Clin Oncol. 2014;19(2):266–73. https://doi.org/10.1007/s10147-013-0566-1.
17. Goto T. Measuring surgery outcomes of lung cancer patients with concomitant pulmonary fibrosis: a review of the literature. Cancers. 2018;10(7):223.
18. Hendriks LE, Drent M, van Haren EH, Verschakelen JA, Verleden GM. Lung cancer in idiopathic pulmonary fibrosis patients diagnosed during or after lung transplantation. Respir Med Case Rep. 2012;5:37–9. https://doi.org/10.1016/j.rmedc.2011.10.003.
19. Daniels CE, Jett JR. Does interstitial lung disease predispose to lung cancer? Curr Opin Pulm Med. 2005;11(5):431–7. https://doi.org/10.1097/01.mcp.0000170521.71497.ba.

20. Tomassetti S, Gurioli C, Ryu JH, Decker PA, Ravaglia C, Tantalocco P, et al. The impact of lung cancer on survival of idiopathic pulmonary fibrosis. Chest. 2015;147(1):157–64. https://doi.org/10.1378/chest.14-0359.
21. Antoniou KM, Tomassetti S, Tsitoura E, Vancheri C. Idiopathic pulmonary fibrosis and lung cancer: a clinical and pathogenesis update. Curr Opin Pulm Med. 2015;21(6):626–33. https://doi.org/10.1097/mcp.0000000000000217.
22. Vancheri C, Failla M, Crimi N, Raghu G. Idiopathic pulmonary fibrosis: a disease with similarities and links to cancer biology. Eur Respir J. 2010;35(3):496–504. https://doi.org/10.1183/09031936.00077309.
23. Vancheri C. Common pathways in idiopathic pulmonary fibrosis and cancer. Eur Respir Rev. 2013;22(129):265–72. https://doi.org/10.1183/09059180.00003613.
24. El Agha E, Moiseenko A, Kheirollahi V, De Langhe S, Crnkovic S, Kwapiszewska G, et al. Two-way conversion between lipogenic and myogenic fibroblastic phenotypes marks the progression and resolution of lung fibrosis. Cell Stem Cell. 2017;20(2):261–73.e3. https://doi.org/10.1016/j.stem.2016.10.004.
25. Luo YH, Wang C, Xu WT, Zhang Y, Zhang T, Xue H, et al. 18β-glycyrrhetinic acid has anti-cancer effects via inducing apoptosis and G2/M cell cycle arrest, and inhibiting migration of A549 lung cancer cells. Onco Targets Ther. 2021;14:5131–44. https://doi.org/10.2147/ott.S322852.
26. Pallante P, Malapelle U, Nacchio M, Sgariglia R, Galati D, Capitelli L, et al. Liquid biopsy is a promising tool for genetic testing in idiopathic pulmonary fibrosis. Diagnostics. 2021;11(7):1202.
27. Jovanovic D, Roksandic Milenkovic M, Kotur Stevuljevic J, Markovic J, Ceriman V, Kontic M, et al. Membrane PD-L1 expression and soluble PD-L1 plasma levels in idiopathic pulmonary fibrosis-a pilot study. J Thorac Dis. 2018;10(12):6660–9. https://doi.org/10.21037/jtd.2018.11.16.
28. Marriott S, Baskir RS, Gaskill C, Menon S, Carrier EJ, Williams J, et al. ABCG2pos lung mesenchymal stem cells are a novel pericyte subpopulation that contributes to fibrotic remodeling. Am J Physiol Cell Physiol. 2014;307(8):C684–98. https://doi.org/10.1152/ajpcell.00114.2014.
29. Karki S, Surolia R, Hock TD, Guroji P, Zolak JS, Duggal R, et al. Wilms' tumor 1 (Wt1) regulates pleural mesothelial cell plasticity and transition into myofibroblasts in idiopathic pulmonary fibrosis. FASEB J. 2014;28(3):1122–31. https://doi.org/10.1096/fj.13-236828.
30. Rock JR, Barkauskas CE, Cronce MJ, Xue Y, Harris JR, Liang J, et al. Multiple stromal populations contribute to pulmonary fibrosis without evidence for epithelial to mesenchymal transition. Proc Natl Acad Sci U S A. 2011;108(52):E1475–83. https://doi.org/10.1073/pnas.1117988108.
31. Strieter RM, Keeley EC, Hughes MA, Burdick MD, Mehrad B. The role of circulating mesenchymal progenitor cells (fibrocytes) in the pathogenesis of pulmonary fibrosis. J Leuk Biol. 2009;86(5):1111–8. https://doi.org/10.1189/jlb.0309132.
32. Shiga K, Hara M, Nagasaki T, Sato T, Takahashi H, Takeyama H. Cancer-associated fibroblasts: their characteristics and their roles in tumor growth. Cancers (Basel). 2015;7(4):2443–58. https://doi.org/10.3390/cancers7040902.
33. Chen Q, Chen S, Zhao J, Zhou Y, Xu L. MicroRNA-126: a new and promising player in lung cancer. Oncol Lett. 2021;21(1):35. https://doi.org/10.3892/ol.2020.12296.
34. Domen A, Quatannens D, Zanivan S, Deben C, Van Audenaerde J, Smits E, et al. Cancer-associated fibroblasts as a common orchestrator of therapy resistance in lung and pancreatic cancer. Cancers (Basel). 2021;13(5) https://doi.org/10.3390/cancers13050987.
35. Bochet L, Lehuédé C, Dauvillier S, Wang YY, Dirat B, Laurent V, et al. Adipocyte-derived fibroblasts promote tumor progression and contribute to the desmoplastic reaction in breast cancer. Cancer Res. 2013;73(18):5657–68. https://doi.org/10.1158/0008-5472.Can-13-0530.
36. De P, Aske J, Dey N. Cancer-associated fibroblast functions as a road-block in cancer therapy. Cancers (Basel). 2021;13(20) https://doi.org/10.3390/cancers13205246.

37. Monteran L, Erez N. The dark side of fibroblasts: cancer-associated fibroblasts as mediators of immunosuppression in the tumor microenvironment. Front Immunol. 2019;1835
38. De Jaeghere EA, Denys HG, De Wever O. Fibroblasts fuel immune escape in the tumor microenvironment. Trends Cancer. 2019;5(11):704–23. https://doi.org/10.1016/j.trecan.2019.09.009.
39. Yazdani S, Bansal R, Prakash J. Drug targeting to myofibroblasts: implications for fibrosis and cancer. Adv Drug Deliv Rev. 2017;121:101–16. https://doi.org/10.1016/j.addr.2017.07.010.
40. Caja L, Dituri F, Mancarella S, Caballero-Diaz D, Moustakas A, Giannelli G, et al. TGF-β and the tissue microenvironment: relevance in fibrosis and cancer. Int J Mol Sci. 2018;19(5):1294.
41. Chen P-Y, Wei W-F, Wu H-Z, Fan L-S, Wang W. Cancer-associated fibroblast heterogeneity: a factor that cannot be ignored in immune microenvironment remodeling. Front Immunol. 2021;12:2760.
42. Ji X, Ji J, Shan F, Zhang Y, Chen Y, Lu X. Cancer-associated fibroblasts from NSCLC promote the radioresistance in lung cancer cell lines. Int J Clin Exp Med. 2015;8(5):7002–8.
43. Hua W, ten Dijke P, Kostidis S, Giera M, Hornsveld M. TGFβ-induced metabolic reprogramming during epithelial-to-mesenchymal transition in cancer. Cell Mol Life Sci. 2020;77(11):2103–23. https://doi.org/10.1007/s00018-019-03398-6.
44. Pereira BA, Vennin C, Papanicolaou M, Chambers CR, Herrmann D, Morton JP, et al. CAF subpopulations: a new reservoir of stromal targets in pancreatic cancer. Trends Cancer. 2019;5(11):724–41. https://doi.org/10.1016/j.trecan.2019.09.010.
45. Losa D, Chanson M, Crespin S. Connexins as therapeutic targets in lung disease. Expert Opin Ther Targets. 2011;15(8):989–1002. https://doi.org/10.1517/14728222.2011.584875.
46. Mori R, Power KT, Wang CM, Martin P, Becker DL. Acute downregulation of connexin43 at wound sites leads to a reduced inflammatory response, enhanced keratinocyte proliferation and wound fibroblast migration. J Cell Sci. 2006;119(24):5193–203. https://doi.org/10.1242/jcs.03320.
47. Cesen-Cummings K, Fernstrom MJ, Malkinson AM, Ruch RJ. Frequent reduction of gap junctional intercellular communication and connexin43 expression in human and mouse lung carcinoma cells. Carcinogenesis. 1998;19(1):61–7. https://doi.org/10.1093/carcin/19.1.61.
48. Trovato-Salinaro A, Trovato-Salinaro E, Failla M, Mastruzzo C, Tomaselli V, Gili E, et al. Altered intercellular communication in lung fibroblast cultures from patients with idiopathic pulmonary fibrosis. Respir Res. 2006;7(1):122. https://doi.org/10.1186/1465-9921-7-122.
49. Mazieres J, He B, You L, Xu Z, Jablons DM. Wnt signaling in lung cancer. Cancer Lett. 2005;222(1):1–10. https://doi.org/10.1016/j.canlet.2004.08.040.
50. Hoguin A, Rastogi A, Bowler C, Tirichine L. Genome-wide analysis of allele-specific expression of genes in the model diatom Phaeodactylum tricornutum. Sci Rep. 2021;11(1):2954. https://doi.org/10.1038/s41598-021-82529-1.
51. Caraci F, Gili E, Calafiore M, Failla M, La Rosa C, Crimi N, et al. TGF-beta1 targets the GSK-3beta/beta-catenin pathway via ERK activation in the transition of human lung fibroblasts into myofibroblasts. Pharmacol Res. 2008;57(4):274–82. https://doi.org/10.1016/j.phrs.2008.02.001.
52. Conte E, Fruciano M, Fagone E, Gili E, Caraci F, Iemmolo M, et al. Inhibition of PI3K prevents the proliferation and differentiation of human lung fibroblasts into myofibroblasts: the role of class I P110 isoforms. PLoS One. 2011;6(10):e24663.
53. Guerreiro AS, Fattet S, Kulesza DW, Atamer A, Elsing AN, Shalaby T, et al. A Sensitized RNA interference screen identifies a novel role for the PI3K p110γ isoform in medulloblastoma cell proliferation and chemoresistance. Mol Cancer Res. 2011;9(7):925–35. https://doi.org/10.1158/1541-7786.Mcr-10-0200.
54. Grimminger F, Schermuly RT, Ghofrani HA. Targeting non-malignant disorders with tyrosine kinase inhibitors. Nat Rev Drug Discov. 2010;9(12):956–70. https://doi.org/10.1038/nrd3297.
55. Malli F, Koutsokera A, Paraskeva E, Zakynthinos E, Papagianni M, Makris D, et al. Endothelial progenitor cells in the pathogenesis of idiopathic pulmonary fibrosis: an evolving concept. PLoS One. 2013;8(1):e53658. https://doi.org/10.1371/journal.pone.0053658.

56. Sakai T, Satoh K, Matsushima K, Shindo S, Abe S, Abe T, et al. Hepatocyte growth factor in bronchoalveolar lavage fluids and cells in patients with inflammatory chest diseases of the lower respiratory tract: detection by RIA and in situ hybridization. Am J Respir Cell Mol Biol. 1997;16(4):388–97. https://doi.org/10.1165/ajrcmb.16.4.9115749.
57. Maeda J, Ueki N, Hada T, Higashino K. Elevated serum hepatocyte growth factor/scatter factor levels in inflammatory lung disease. Am J Respir Crit Care Med. 1995;152(5):1587–91. https://doi.org/10.1164/ajrccm.152.5.7582299.
58. Shiratori M, Michalopoulos G, Shinozuka H, Singh G, Ogasawara H, Katyal SL. Hepatocyte growth factor stimulates DNA synthesis in alveolar epithelial type II cells in vitro. Am J Respir Cell Mol Biol. 1995;12(2):171–80. https://doi.org/10.1165/ajrcmb.12.2.7532419.
59. Yaekashiwa M, Nakayama S, Ohnuma K, Sakai T, Abe T, Satoh K, et al. Simultaneous or delayed administration of hepatocyte growth factor equally represses the fibrotic changes in murine lung injury induced by bleomycin. Am J Respir Crit Care Med. 1997;156(6):1937–44. https://doi.org/10.1164/ajrccm.156.6.9611057.
60. Panganiban RA, Day RM. Hepatocyte growth factor in lung repair and pulmonary fibrosis. Acta Pharmacol Sin. 2011;32(1):12–20. https://doi.org/10.1038/aps.2010.90.
61. Noth I, Zhang Y, Ma SF, Flores C, Barber M, Huang Y, et al. Genetic variants associated with idiopathic pulmonary fibrosis susceptibility and mortality: a genome-wide association study. Lancet Respir Med. 2013;1(4):309–17. https://doi.org/10.1016/s2213-2600(13)70045-6.
62. Nathan N, Giraud V, Picard C, Nunes H, Dastot-Le Moal F, Copin B, et al. Germline SFTPA1 mutation in familial idiopathic interstitial pneumonia and lung cancer. Hum Mol Genet. 2016;25(8):1457–67. https://doi.org/10.1093/hmg/ddw014.
63. Wang Y, Kuan PJ, Xing C, Cronkhite JT, Torres F, Rosenblatt RL, et al. Genetic defects in surfactant protein A2 are associated with pulmonary fibrosis and lung cancer. Am J Hum Genet. 2009;84(1):52–9. https://doi.org/10.1016/j.ajhg.2008.11.010.
64. Crossno PF, Polosukhin VV, Blackwell TS, Johnson JE, Markin C, Moore PE, et al. Identification of early interstitial lung disease in an individual with genetic variations in ABCA3 and SFTPC. Chest. 2010;137(4):969–73. https://doi.org/10.1378/chest.09-0790.
65. Stanley SE, Gable DL, Wagner CL, Carlile TM, Hanumanthu VS, Podlevsky JD, et al. Loss-of-function mutations in the RNA biogenesis factor NAF1 predispose to pulmonary fibrosis-emphysema. Sci Transl Med. 2016;8(351):351ra107. https://doi.org/10.1126/scitranslmed.aaf7837.
66. Kaur A, Mathai SK, Schwartz DA. Genetics in idiopathic pulmonary fibrosis pathogenesis, prognosis, and treatment. Front Med. 2017;4 https://doi.org/10.3389/fmed.2017.00154.
67. Nogee LM, Dunbar AE 3rd, Wert SE, Askin F, Hamvas A, Whitsett JA. A mutation in the surfactant protein C gene associated with familial interstitial lung disease. N Engl J Med. 2001;344(8):573–9. https://doi.org/10.1056/nejm200102223440805.
68. Thomas AQ, Lane KF, Phillips JA, Prince MA, Markin CR, Speer MC, et al. Heterozygosity for a surfactant protein C gene mutation associated with usual interstitial pneumonitis and cellular nonspecific interstitial pneumonitis in one kindred. Am J Respir Crit Care Med. 2002;165(9):1322–8.
69. Brasch F, Griese M, Tredano M, Johnen G, Ochs M, Rieger C, et al. Interstitial lung disease in a baby with a de novo mutation in the SFTPC gene. Eur Respir J. 2004;24(1):30–9. https://doi.org/10.1183/09031936.04.00000104.
70. Li J, Liepinsh E, Almlén A, Thyberg J, Curstedt T, Jörnvall H, et al. Structure and influence on stability and activity of the N-terminal propeptide part of lung surfactant protein C. FEBS J. 2006;273(5):926–35. https://doi.org/10.1111/j.1742-4658.2006.05124.x.
71. Maguire JA, Mulugeta S, Beers MF. Multiple ways to die: delineation of the unfolded protein response and apoptosis induced by Surfactant Protein C BRICHOS mutants. Int J Biochem Cell Biol. 2012;44(1):101–12. https://doi.org/10.1016/j.biocel.2011.10.003.
72. Lawson WE, Cheng D-S, Degryse AL, Tanjore H, Polosukhin VV, Xu XC, et al. Endoplasmic reticulum stress enhances fibrotic remodeling in the lungs. Proc Natl Acad Sci U S A. 2011;108(26):10562–7. https://doi.org/10.1073/pnas.1107559108.

73. Willander H, Hermansson E, Johansson J, Presto J. BRICHOS domain associated with lung fibrosis, dementia and cancer--a chaperone that prevents amyloid fibril formation? FEBS J. 2011;278(20):3893–904. https://doi.org/10.1111/j.1742-4658.2011.08209.x.
74. Tanjore H, Cheng DS, Degryse AL, Zoz DF, Abdolrasulnia R, Lawson WE, et al. Alveolar epithelial cells undergo epithelial-to-mesenchymal transition in response to endoplasmic reticulum stress. J Biol Chem. 2015;290(6):3277. https://doi.org/10.1074/jbc.A110.181164.
75. Zhou W, Wang Y. Candidate genes of idiopathic pulmonary fibrosis: current evidence and research. Appl Clin Genet. 2016;9:5–13. https://doi.org/10.2147/tacg.S61999.
76. Armanios M. Telomerase and idiopathic pulmonary fibrosis. Mutat Res. 2012;730(1–2):52–8. https://doi.org/10.1016/j.mrfmmm.2011.10.013.
77. Seibold MA, Wise AL, Speer MC, Steele MP, Brown KK, Loyd JE, et al. A common MUC5B promoter polymorphism and pulmonary fibrosis. N Engl J Med. 2011;364(16):1503–12. https://doi.org/10.1056/NEJMoa1013660.
78. Fingerlin TE, Murphy E, Zhang W, Peljto AL, Brown KK, Steele MP, et al. Genome-wide association study identifies multiple susceptibility loci for pulmonary fibrosis. Nat Genet. 2013;45(6):613–20. https://doi.org/10.1038/ng.2609.
79. Peljto AL, Selman M, Kim DS, Murphy E, Tucker L, Pardo A, et al. The MUC5B promoter polymorphism is associated with idiopathic pulmonary fibrosis in a Mexican cohort but is rare among Asian ancestries. Chest. 2015;147(2):460–4.
80. Wei R, Li C, Zhang M, Jones-Hall YL, Myers JL, Noth I, et al. Association between MUC5B and TERT polymorphisms and different interstitial lung disease phenotypes. Transl Res. 2014;163(5):494–502.
81. Seibold MA, Smith RW, Urbanek C, Groshong SD, Cosgrove GP, Brown KK, et al. The idiopathic pulmonary fibrosis honeycomb cyst contains a mucocilary pseudostratified epithelium. PLoS One. 2013;8(3):e58658. https://doi.org/10.1371/journal.pone.0058658.
82. Roy MG, Livraghi-Butrico A, Fletcher AA, McElwee MM, Evans SE, Boerner RM, et al. Muc5b is required for airway defence. Nature. 2014;505(7483):412–6. https://doi.org/10.1038/nature12807.
83. Yang IV, Fingerlin TE, Evans CM, Schwarz MI, Schwartz DA. MUC5B and idiopathic pulmonary fibrosis. Ann Am Thoracic Soc. 2015;12 Suppl 2(Suppl 2):S193–9. https://doi.org/10.1513/AnnalsATS.201503-110AW.
84. Karampitsakos T, Woolard T, Bouros D, Tzouvelekis A. Toll-like receptors in the pathogenesis of pulmonary fibrosis. Eur J Pharmacol. 2017;808:35–43. https://doi.org/10.1016/j.ejphar.2016.06.045.
85. Kawasaki T, Kawai T. Toll-like receptor signaling pathways. Front Immunol. 2014;5 https://doi.org/10.3389/fimmu.2014.00461.
86. Zhang G, Ghosh S. Negative regulation of toll-like receptor-mediated signaling by tollip*. J Biol Chem. 2002;277(9):7059–65. https://doi.org/10.1074/jbc.M109537200.
87. Didierlaurent A, Brissoni B, Velin D, Aebi N, Tardivel A, Käslin E, et al. Tollip regulates proinflammatory responses to interleukin-1 and lipopolysaccharide. Mol Cell Biol. 2006;26(3):735–42. https://doi.org/10.1128/MCB.26.3.735-742.2006.
88. Kowalski EJA, Li L. Toll-interacting protein in resolving and non-resolving inflammation. (1664–3224 (Print)).
89. van der Mark VA, Ghiboub M, Marsman C, Zhao J, van Dijk R, Hiralall JK, et al. Phospholipid flippases attenuate LPS-induced TLR4 signaling by mediating endocytic retrieval of Toll-like receptor 4. Cell Mol Life Sci. 2017;74(4):715–30. https://doi.org/10.1007/s00018-016-2360-5.
90. Trujillo G, Meneghin A, Flaherty KR, Sholl LM, Myers JL, Kazerooni EA, et al. TLR9 differentiates rapidly from slowly progressing forms of idiopathic pulmonary fibrosis. Sci Transl Med. 2010;2(57):57ra82. https://doi.org/10.1126/scitranslmed.3001510.
91. Korthagen NM, van Moorsel CHM, Kazemier KM, Ruven HJT, Grutters JC. IL1RN genetic variations and risk of IPF: a meta-analysis and mRNA expression study. Immunogenetics. 2012;64(5):371–7. https://doi.org/10.1007/s00251-012-0604-6.

92. Gasse P, Mary C, Guenon I, Noulin N, Charron S, Schnyder-Candrian S, et al. IL-1R1/MyD88 signaling and the inflammasome are essential in pulmonary inflammation and fibrosis in mice. J Clin Invest. 2007;117(12):3786–99. https://doi.org/10.1172/jci32285.
93. Borthwick LA. The IL-1 cytokine family and its role in inflammation and fibrosis in the lung. Semin Immunopathol. 2016;38(4):517–34. https://doi.org/10.1007/s00281-016-0559-z.
94. Michalski JE, Schwartz DA. Genetic risk factors for idiopathic pulmonary fibrosis: insights into immunopathogenesis. J Inflamm Res. 2020;13:1305–18. https://doi.org/10.2147/jir.S280958.
95. Bruder E, Hofmeister J, Aslanidis C, Hammer J, Bubendorf L, Schmitz G, et al. Ultrastructural and molecular analysis in fatal neonatal interstitial pneumonia caused by a novel ABCA3 mutation. Mod Pathol. 2007;20(10):1009–18. https://doi.org/10.1038/modpathol.3800928.
96. Campo I, Zorzetto M, Mariani F, Kadija Z, Morbini P, Dore R, et al. A large kindred of pulmonary fibrosis associated with a novel ABCA3 gene variant. Respir Res. 2014;15(1):43. https://doi.org/10.1186/1465-9921-15-43.
97. Ciantelli M, Ghirri P, Presi S, Sigali E, Vuerich M, Somaschini M, et al. Fatal respiratory failure in a full-term newborn with two ABCA3 gene mutations: a case report. J Perinatol. 2011;31(1):70–2. https://doi.org/10.1038/jp.2010.122.
98. Pandit KV, Milosevic J, Kaminski N. MicroRNAs in idiopathic pulmonary fibrosis. Transl Res. 2011;157(4):191–9. https://doi.org/10.1016/j.trsl.2011.01.012.
99. Sanders YY, Kumbla P, Hagood JS. Enhanced myofibroblastic differentiation and survival in Thy-1 (−) lung fibroblasts. Am J Respir Cell Mol Biol. 2007;36(2):226–35.
100. Sanders YY, Pardo A, Selman M, Nuovo GJ, Tollefsbol TO, Siegal GP, et al. Thy-1 promoter hypermethylation: a novel epigenetic pathogenic mechanism in pulmonary fibrosis. Am J Respir Cell Mol Biol. 2008;39(5):610–8.
101. Kuwano K, Kunitake R, Kawasaki M, Nomoto Y, Hagimoto N, Nakanishi Y, et al. P21Waf1/Cip1/Sdi1 and p53 expression in association with DNA strand breaks in idiopathic pulmonary fibrosis. Am J Respir Crit Care Med. 1996;154(2 Pt 1):477–83. https://doi.org/10.1164/ajrccm.154.2.8756825.
102. Hojo S, Fujita J, Yamadori I, Kamei T, Yoshinouchi T, Ohtsuki Y, et al. Heterogeneous point mutations of the p53 gene in pulmonary fibrosis. Eur Respir J. 1998;12(6):1404–8. https://doi.org/10.1183/09031936.98.12061404.
103. Uematsu K, Yoshimura A, Gemma A, Mochimaru H, Hosoya Y, Kunugi S, et al. Aberrations in the fragile histidine triad (FHIT) gene in idiopathic pulmonary fibrosis. Cancer Res. 2001;61(23):8527–33.
104. Demopoulos K, Arvanitis DA, Vassilakis DA, Siafakas NM, Spandidos DA. MYCL1, FHIT, SPARC, p16(INK4) and TP53 genes associated to lung cancer in idiopathic pulmonary fibrosis. J Cell Mol Med. 2002;6(2):215–22. https://doi.org/10.1111/j.1582-4934.2002.tb00188.x.
105. Rosales W, Carulla J, García J, Vargas D, Lizcano F. Role of histone demethylases in cardiomyocytes induced to hypertrophy. Biomed Res Int. 2016;2016:2634976. https://doi.org/10.1155/2016/2634976.
106. Correll KA, Edeen KE, Redente EF, Zemans RL, Edelman BL, Danhorn T, et al. TGF beta inhibits HGF, FGF7, and FGF10 expression in normal and IPF lung fibroblasts. Physiol Rep. 2018;6(16):e13794. https://doi.org/10.14814/phy2.13794.
107. Lovat F, Valeri N, Croce CM. MicroRNAs in the pathogenesis of Cancer. Semin Oncol. 2011;38(6):724–33. https://doi.org/10.1053/j.seminoncol.2011.08.006.
108. Oak SR, Murray L, Herath A, Sleeman M, Anderson I, Joshi AD, et al. A micro RNA processing defect in rapidly progressing idiopathic pulmonary fibrosis. PLoS One. 2011;6(6):e21253. https://doi.org/10.1371/journal.pone.0021253.
109. Corella D, Ordovas JM. Basic concepts in molecular biology related to genetics and epigenetics. Rev Española Cardiol (English Edition). 2017;70(9):744–53. https://doi.org/10.1016/j.rec.2017.05.011.

110. Roach KM, Feghali-Bostwick CA, Amrani Y, Bradding P. Lipoxin A4 attenuates constitutive and TGF-β1–dependent profibrotic activity in human lung myofibroblasts. J Immunol. 2015;195(6):2852–60.
111. Samara KD, Trachalaki A, Tsitoura E, Koutsopoulos AV, Lagoudaki ED, Lasithiotaki I, et al. Upregulation of citrullination pathway: from autoimmune to idiopathic lung fibrosis. Respir Res. 2017;18(1):218. https://doi.org/10.1186/s12931-017-0692-9.
112. Yang IV, Schwartz DA. Epigenetics of idiopathic pulmonary fibrosis. Transl Res. 2015;165 (1):48–60. https://doi.org/10.1016/j.trsl.2014.03.011.
113. Yang IV. Epigenomics of idiopathic pulmonary fibrosis. (1750-192X (Electronic)).
114. Yang IV, Schwartz DA. Epigenetic control of gene expression in the lung. Am J Respir Crit Care Med. 2011;183(10):1295–301. https://doi.org/10.1164/rccm.201010-1579PP.
115. Penz-Österreicher M, Österreicher CH, Trauner M. Fibrosis in autoimmune and cholestatic liver disease. Best Pract Res Clin Gastroenterol. 2011;25(2):245–58. https://doi.org/10.1016/j.bpg.2011.02.001.
116. Ding Q, Luckhardt T, Hecker L, Zhou Y, Liu G, Antony VB, deAndrade J, et al. New insights into the pathogenesis and treatment of idiopathic pulmonary fibrosis. (1179–1950 (Electronic)).
117. Zong D, Liu X, Li J, Ouyang R, Chen P. The role of cigarette smoke-induced epigenetic alterations in inflammation. Epigenetics Chromatin. 2019;12(1):65. https://doi.org/10.1186/s13072-019-0311-8.
118. Pardo A, Selman M. The interplay of the genetic architecture, aging, and environmental factors in the pathogenesis of idiopathic pulmonary fibrosis. Am J Respir Cell Mol Biol. 2021;64(2): 163–72. https://doi.org/10.1165/rcmb.2020-0373PS.
119. Sanders YY, Pardo A, Selman M, Nuovo GJ, Tollefsbol TO, Siegal GP, et al. Thy-1 promoter hypermethylation: a novel epigenetic pathogenic mechanism in pulmonary fibrosis. Am J Respir Cell Mol Biol. 2008;39(5):610–8. https://doi.org/10.1165/rcmb.2007-0322OC.
120. Hu B, Gharaee-Kermani M, Wu Z, Phan SH. Epigenetic regulation of myofibroblast differentiation by DNA methylation. (1525–2191 (Electronic)).
121. Tzouvelekis A, Kaminski N. Epigenetics in idiopathic pulmonary fibrosis. Biochem Cell Biol. 2015;93(2):159–70. https://doi.org/10.1139/bcb-2014-0126.
122. Seto E, Yoshida M. Erasers of histone acetylation: the histone deacetylase enzymes. (1943–0264 (Electronic)).
123. Sehgal M, Jakhete SM, Manekar AG, Sasikumar S. Specific epigenetic regulators serve as potential therapeutic targets in idiopathic pulmonary fibrosis. Heliyon. 2022;8(8):e09773. https://doi.org/10.1016/j.heliyon.2022.e09773.
124. Zhang M, Serna-Salas S, Damba T, Borghesan M, Demaria M, Moshage H. Hepatic stellate cell senescence in liver fibrosis: characteristics, mechanisms and perspectives. Mech Ageing Dev. 2021;199:111572. https://doi.org/10.1016/j.mad.2021.111572.
125. Sanders YY, Liu H, Scruggs AM, Duncan SR, Huang SK, Thannickal VJ. Epigenetic regulation of caveolin-1 gene expression in lung fibroblasts. Am J Respir Cell Mol Biol. 2017;56(1): 50–61. https://doi.org/10.1165/rcmb.2016-0034OC.
126. Dakhlallah D, Batte K, Wang Y, Cantemir-Stone CZ, Yan P, Nuovo G, et al. Epigenetic regulation of miR-17~92 contributes to the pathogenesis of pulmonary fibrosis. Am J Respir Crit Care Med. 2013;187(4):397–405. https://doi.org/10.1164/rccm.201205-0888OC.
127. Lamas DJ, Kawut Smu, Bagiella E, Philip N, Arcasoy SM, Lederer DJ. Delayed access and survival in idiopathic pulmonary fibrosis: a cohort study. (1535–4970 (Electronic)).
128. Abuserewa ST, Duff R, Becker G. Treatment of idiopathic pulmonary fibrosis. Cureus. 2021;13(5):e15360. https://doi.org/10.7759/cureus.15360.
129. Noth I, Anstrom KJ, Calvert SB, de Andrade J, Flaherty KR, Glazer C, Kaner RJ, et al. A placebo-controlled randomized trial of warfarin in idiopathic pulmonary fibrosis. (1535–4970 (Electronic)).
130. Malouf MA, Hopkins P, Snell G, Glanville AR. An investigator-driven study of everolimus in surgical lung biopsy confirmed idiopathic pulmonary fibrosis. (1440–1843 (Electronic)).

131. King TE, Jr., Behr J, Brown KK, du Bois RM, Lancaster L, de Andrade JA, Stähler G, et al. BUILD-1: a randomized placebo-controlled trial of bosentan in idiopathic pulmonary fibrosis. (1535–4970 (Electronic)).
132. Raghu G, Behr J, Brown KK, Egan JJ, Kawut SM, Flaherty KR, Martinez FJ, et al. Treatment of idiopathic pulmonary fibrosis with ambrisentan: a parallel, randomized trial. (1539–3704 (Electronic)).
133. Raghu G, Million-Rousseau R, Morganti A, Perchenet L, Behr J. Macitentan for the treatment of idiopathic pulmonary fibrosis: the randomised controlled MUSIC trial. (1399–3003 (Electronic)).
134. Zisman Da, Schwarz M, Anstrom KJ, Collard HR, Flaherty KR, Hunninghake GW. A controlled trial of sildenafil in advanced idiopathic pulmonary fibrosis. (1533–4406 (Electronic)).
135. Raghu G, Brown Kk, Costabel U, Cottin V, du Bois RM, Lasky JA, Thomeer M, et al. Treatment of idiopathic pulmonary fibrosis with etanercept: an exploratory, placebo-controlled trial. (1535–4970 (Electronic)).
136. Daniels CE, Lasky Ja, Limper AH, Mieras K, Gabor E, Schroeder DR. Imatinib treatment for idiopathic pulmonary fibrosis: Randomized placebo-controlled trial results. (1535–4970 (Electronic)).
137. Helling BA, Yang IV. Epigenetics in lung fibrosis: from pathobiology to treatment perspective. Curr Opin Pulm Med. 2015;21(5):454–62. https://doi.org/10.1097/MCP.0000000000000191.

Epigenetics of Influenza: The Host-Virus Interaction.

10

Muhammad Mustafa, Muhammad Shahid Nadeem, Abeer Asif, and Imran Kazmi

Abstract

With advancements in science and technology, an equivalent growth in diseases has also been observed. A disease that has remained prevalent for decades, showing an almost incomparable persistence and adaptability is influenza. With increasing research on DNA, we now know that external influences, like influenza, can successfully alter gene expression with lasting effects through epigenetics. In this chapter, the possible role that both influenza and its host play on an epigenetic level to alter each other's gene expression will be explored along with the possible effects these epigenetic changes have on virulence, the immune system, and other cellular mechanisms.

10.1 Introduction

While cells in a multicellular organism have the same genetic sequences, their phenotypes may differ greatly. The core of epi-(above)-genetics is this nongenetic cellular memory, which accumulates developmental and environmental signals [1]. The term "epigenetics" was coined to describe the poorly understood processes that led to the development of a fertilised zygote into a complex, mature organism. Due to the realization that all cells in an organism carry the same DNA and an improved understanding of gene expression mechanisms, the definition was revised to place more emphasis on the ways in which heritable traits can be connected to

M. Mustafa
KAM School of Life Sciences, Forman Christian College (A Chartered University), Lahore, Pakistan

M. S. Nadeem (✉) · A. Asif · I. Kazmi
Department of Biochemistry, Faculty of Science, King Abdulaziz University, Jeddah, Saudi Arabia

G. Gupta et al. (eds.), *Targeting Epigenetics in Inflammatory Lung Diseases*,
https://doi.org/10.1007/978-981-99-4780-5_10

modification of DNA or the structural and regulatory proteins bound to it, rather than changes in nucleotide sequence. It may be beneficial to return to the definition of epigenetics as it originally existed, according to recent research into how these systems operate throughout embryonic development [2]. The network of functional linkages between the influenza virus and its hosts depends on epigenetic changes [2]. The influenza virus, of the Orthomyxoviridae family, is composed of influenza subtypes A, B, and C and has existed since ancient times, wreaking havoc through history, and killing millions [3, 4]. This spherical virus is encapsulated in a capsid and protected by a mix of surface proteins; neuraminidase (N), responsible for the destruction of infected cells to release the viral particles letting them infect more cells, and hemagglutinin (H), allows identification and aggregation of the viral fragments, and holds eight segments of single-stranded RNA (ssRNA) [5]. Eighteen hemagglutinin and nine neuraminidase proteins exist in influenza A, with H1/2/3 and N1/2 commonly known to impact human hosts [6]. This single-stranded RNA is what transfers itself into the host's nucleus through endocytosis, replicates using nuclear replicative machinery, and releases itself to infect neighbouring respiratory epithelial cells after an almost 6-h maturation period [7]. This exchange of nucleic acids offers an intimate point of contact, that many studies now confirm induces a change in gene regulation of both the host and the virus, engendering an amalgam of modifications [6, 8].

On the other hand, the chromatin structure of the eukaryotic genome is composed of a nucleosome, made up of an amalgamation of histones: H4, H3, H2B, and H2A [9]. Two of each of these histones come together to engender an octamer around which the 147 base pair (bp) DNA is wound, giving rise to a nucleosome [9]. Alongside the H1 histone, the nucleosome can coil to form a compact structure, which is then further altered through posttranslational changes to regulate cellular mechanisms [10].

10.1.1 An Introduction to Epigenetics

The DNA of an organism or an individual is pre-ordained at birth and remains consistent throughout their lifetime, or so we thought. Epigenetics is a new study that explores the various effects or alterations that may take place due to various factors like stress, diet, or exercise. These changes not only affect the way DNA/genes are expressed within the body but also may alter the genes of future offspring. These changes usually evolve because of modifications in histones (phosphorylation, methylation, ubiquitination, or acetylation), DNA methylation, or through non-coding microRNAs (miRNAs). Phosphorylation refers to the addition of phosphate molecules to activate (or deactivate) enzymes; methylation refers to the addition of methyl groups to alter gene expression; ubiquitination includes the addition of ubiquitin molecules which allows the degradation of specific protein molecules; and lastly, acetylation includes the addition of acetyl molecules to allow transcription [11]. These epigenetic changes may be positive or negative according to the change they induce and their long-term influence. For example, Colorectal cancer patients who had experienced severe famine in the Dutch famine (1944–1945) were less likely to develop tumours labelled with a CpG island methylator phenotype (CIMP) [12]. In addition, increased B6 vitamin intake was proportional to methylation of MutL homolog 1's (MLH1), a gene involved in repairing DNA damage, promoter in men [13]. Additionally, even regular exercise has been reported to induce major changes in DNA methylation, affecting vulnerability to disease and metabolic phenotype [14].

Recently, research has shown that even external factors like diseases can lead to significant epigenetic changes. Viral particles hijack the cell's molecular mechanism and affect both the host's immune response to the virus, as well as the virus' pathogenicity. This same perspective can also be reversed to see the effect of the host on the virus' genome, histone modifications, and other structural changes.

10.1.2 Epigenetics and Viruses

The influence of virus on epigenetics is a serious concern as it may lead to increased vulnerability of the host, or additional pathogenicity of the virus. Many diseases like cancer, influenza, hepatitis, tuberculosis, etc. have been reported to play their part in influencing the genome through epigenetics. For example, in the human immunodeficiency virus (HIV) significant changes in gene promoter regions were observed that resulted in alterations in H3K9ac and H3K4me3 signalling consequently playing a role in HIV latency. Moreover, for instance an increase in methyl groups in the Class II major histocompatibility complex transactivator (CIITA) gene promoter's CpG islands results in a more persistent case of hepatitis B virus [15]. As another example, tuberculosis (TB) methylation profiles showed a distinct difference in macrophage methylation between active and latent TB [16]. Similarly, here we will highlight the epigenetic changes enforced by and on to the influenza virus by the host.

10.2 Influenza and its Epigenetic Changes on the Host

The influenza virus is an RNA-based virus with no histones of its own. However, its interactions with its host leave an imprint, some of which are epigenetic. Here we will explore the epigenetic changes that the influenza virus inflicts on its host. These effects may range from changes to the host's immune system and cellular mechanism, to changes that help the virus' infection into the host and pathogenicity. The influenza virus infects the respiratory tract, causing a robust pro-inflammatory response but also compromising response of host immune system, increasing vulnerability to secondary infections [17, 18]. Opportunistic bacterial infections following the primary influenza virus are a common complication that accounts for a majority of deaths during pandemics [18]. Research has shown that there may be a higher risk of a secondary bacterial infection even after viral clearance. These illnesses often manifest themselves within a week after contracting the influenza virus [19]. Influenza virus infection, especially H5N1 and pathogenic H1N1, acute

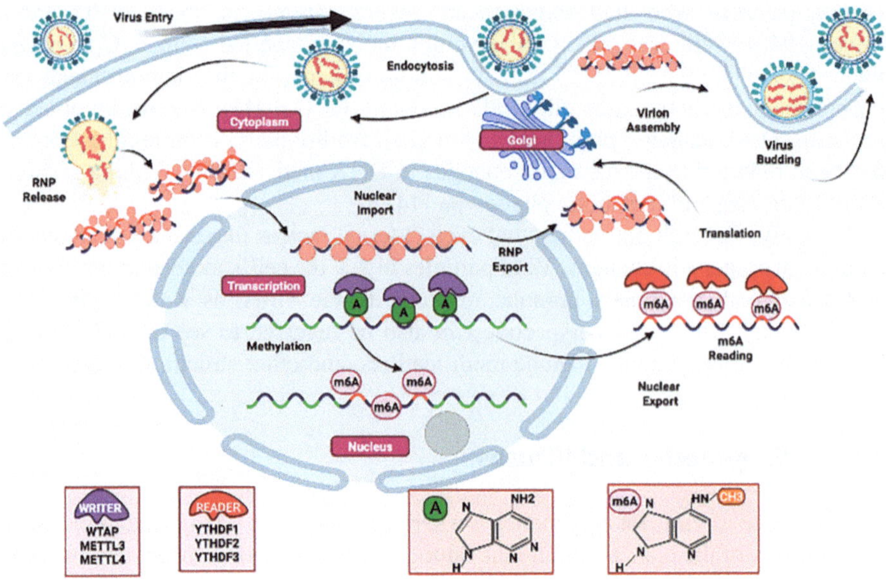

Fig. 10.1 Mechanisms of immunity's epigenetic remodelling during influenza virus contamination [8]

inflammation brought on by increased cytokine release has been seen. It has been demonstrated that increased cytokine release during infection of influenza virus caused by inflammatory gene promoter DNA by change in methylation. Additionally, the pathogenicity of the influenza virus strain and its variation has a significant impact on the change in methylation level. With highly pathogenic H5N1 influenza viruses, studies have seen the most substantial alterations in the DNA methylation of inflammatory genes [20]. Fig. 10.1 shows the epigenetic modifications of the immunity that occur due to influenza virus A.

10.2.1 Influenza's Effects on the Body's Inbuilt Immune System

The body's immune system is at the forefront of any battle that takes place against intruders. Therefore, it is expected that an intruder would be most likely to affect it in some way. A study reported in 2018 had concluded that although they found little epigenetic alterations in the DNA methylation, they did find a decrease in acetylation of histones, which meant that the host's transcriptional system was impaired [11]. Moreover, they found a unique player—lysine 79 of histone 3 (H3K79)—which plays a role in regulating the cell cycle and repairing damaged DNA. H3K79 when methylated by disruptor of telomeric silencing 1-like (Dot1L), a histone methyltransferase enzyme, instigated interferon signals, leading to metabolic changes, that worked towards counteracting the viral infection [21]. Therefore, through decreased methylation of H3K79, the antiviral response is subdued and viral replication increases [11]. Moreover, the influenza virus is known to quell antigen presentation in host by epigenetically downregulating antigen presentation genes [22]. This downregulation was found to possibly vary according to the type of virus: Middle East respiratory syndrome coronavirus (MERS-CoV), relies on DNA methylation, while the H5N1 influenza infection relies on a mix of epigenetic mechanisms to affect antigen presentation [22]. The influenza virus has long-term repercussions too as it damages neutrophil's—a common white blood cell involved in fighting infections—performance and decreases the host's defence against secondary bacterial infections after the virus' exposure [23]. Its immunosuppressive condition is achieved through the activation of IFN-B (beta interferon) and chemokine's transcriptional regulation [23]. As part of the innate immunity, the Interferon Regulatory Factor 7, 3, and transcription factors NF-kB of activated B cells are triggered when host pathogen recognition receptors detect influenza A virus (IAV) infection intracellularly. The pattern recognition receptors (PRRs) comprise melanoma differentiation-associated gene 5, retinoic acid-inducible gene-I protein, and Toll-like receptors (TLRs). The type I and type III IFNs expression is then induced by these active transcription factors. The interaction between IFNs released by infected cells and their receptors activates the JAK-STAT signalling pathway, which controls the production of numerous genes that stimulate IFN as shown in Fig. 10.2 [8, 24].

10.2.2 Influenza's Effect on Host's Cellular Mechanism

Upon infection, influenza carries out a series of events that damage and initiate chaos in the cell's routine. One of these instigated changes includes a disruption of the host's mRNA splicing function [25]. Studies have shown that following an infection, hundreds of genes in the host reveal changed alternative splicing [25]. A possible reason behind this may be its manipulation of the splicing machinery of the host, or perhaps that the non-structural protein 1 (NS1) protein of the influenza virus shows a tendency to bind to intronic regions, contributing to a higher intron retention and altered splicing [26]. In addition, it has also been shown that defective termination of the transcription process takes place post-infection [27–29]. Influenza's unwanted interactions with its host do not stop there. The virus goes on to interfere with the host's long non-coding RNAs, specifically upregulating PSMB8-AS1 in infected cells [30]. This increase is in direct proportion to the increase in viral genes—NS1 and NP—and protein expression—NS1, PB1(polymerase basic protein 1), and NP (nucleoprotein) [30]. Overall, it has been reported in an experiment that 1913 long noncoding RNAs were altered in A549 cells after they were inoculated with the influenza A virus [30].

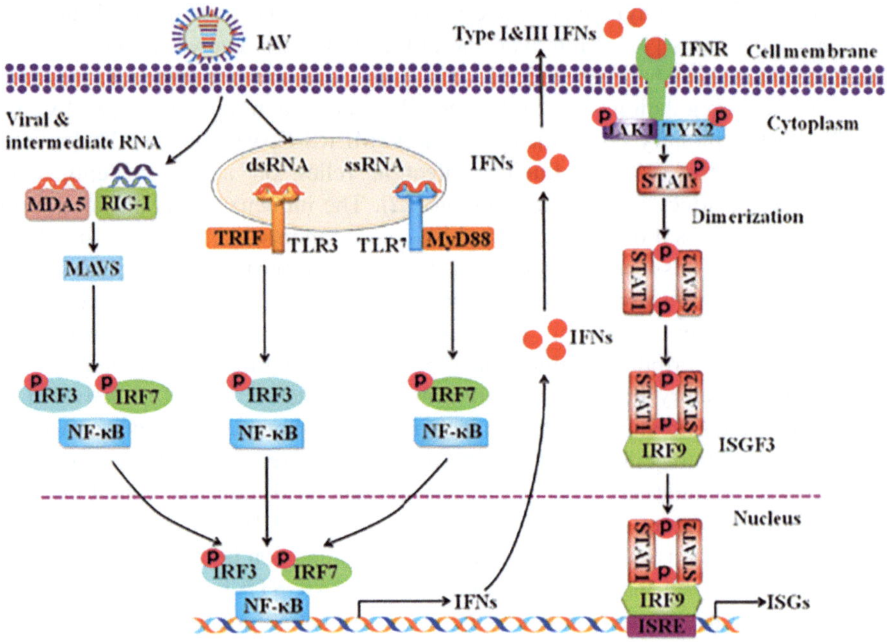

Fig. 10.2 Diagram of the innate immune system's defence against infection by the influenza A virus (IAV) [24]

10.2.3 Epigenetics Effect on Pathogenesis and Progression

Influenza, without a doubt induces epigenetic changes into the host. Some of these changes inevitably result in favour of the infectivity of virus. In a study observing mouse lung cells infected with influenza, they uncovered that by deleting the histone methylation gene it is possible to affect replication and pathology of the virus [31]. Additionally, the study highlights that the influenza virus interferes with the gene silencing mechanism, regulated by histone methyltransferase, to activate controlled genes [31]. These genes are unique because once activated they allow viral replication to occur [31]. Additionally, the influenza virus, specifically H3N2 subtype, contains a special non-structural protein 1 (NS1) protein which works as a histone mimic, similar to H3 histone tail of host, hampering the cell's antiviral response [32]. NS1 does this by binding to human PAF1 transcription elongation complex (hPAF1C), responsible for decreasing the force of influenza infection, and suppressing its expression leading to increased susceptibility and reduced antiviral response [32].

It has often been demonstrated that H1N1 influenza patients who experience a catastrophic illness outcome have CRP (C-reactive protein) levels that are noticeably higher. The findings of this study indicate that serum CRP, together with other biomarkers, may be used to predict the complications of H1N1 influenza [33]. A baseline rise in numerous biomarkers related to inflammation, coagulation, or immunological function in individuals with infection of H1N1 pdm09 virus of different severity clearly predicted a higher risk of illness development. Interventions intended to lower these baseline increases could potentially have an impact on how a disease develops [34]. A novel technique for diagnosis of early influenza has been made possible by the findings of miRNA and its distinct flu sufferers' expression profiles. Additionally, we showed a correlation between throat swab miRNA indicators and influenza virus infection. miR-449b-5p, miR-205-5p, miR-181a-5p, miR-34c-3p, miR-34b-5p, miR-30c-5p, and miR-29a-3p are some examples of these biomarkers. They could be utilised to distinguish patients having infA and infB from unaffected healthy individuals. Additionally, they can be used to diagnose H1N1 and H3N2 infections. We anticipate that in the near future, this non-invasive method employing the miRNAs from throat swabs will be a reliable method for diagnosing influenza [35]. Infection with the influenza virus usually has more severe side effects in people over 65 (the elderly). Immunosenescence makes people more susceptible to viral infections and reduces the effectiveness of immunization. Designing preventative and immunomodulatory techniques to lower mortality and morbidity in the old age requires a thorough understanding of age-related immune system variations [36].

10.3 Epigenetic Changes Induced by Host on Influenza Virus

The influenza virus is significantly dependent on post-translational modifications, ubiquitination, acetylation, phosphorylation, and SUMOylation (attachment of SUMO proteins to lysine residues in proteins), to maintain normal viral protein structure and function [37]. Like the long list of changes that the influenza virus imparts onto the host, the host too plays a role in impacting the infecting virus. The host holds special enzymes called acetyl transferases which shift an acetyl group from an acetyl-CoA to a histone's lysine amino acids to relax and open coiled chromatin structure and allow transcription. The host body uses PCAF (P300/CBP-associated factor) and GCN5 (general control non-derepressible 5), two special acetyltransferases, to attach acetyl groups on the virus' Lys-90 and Lys-31 on the nucleoprotein (NP) which then impacts the polymerase activity of the virus [38]. Interestingly, the study mentioned that by silencing PCAF, viral polymerase became more active but with the silencing of GCN5, the viral polymerase activity decreased; showing that acetylation of both lysine residues has an opposite effect on the virus' replication [38]. In another study, they found that H3K4 (histone 3, lysine) methylation decreases playing a possible role in body's defence against virus. Moreover, they found that seasonal viral flu caused fewer changes in DNA methylation as compared to the virus in hens or A549 cells [39].

The host's fight against influenza can also instigate antiviral response but also stifle the virus' normal mechanism. For instance, on viral entry, demethylation of the IL-6 promoter leads to increased IL-6 expression resulting in cytokine secretion and the launch of an immune response [40]. Additionally, an influenza infection instigates the expression of miRNA-29 through epigenetic changes, resulting in an increase in DNA methyltransferase (DNMT) expression that induces cyclooxygenase-2 (COX-2) expression and an accumulation of COX2-derived prostaglandin E2 (PGE2) [41]. This then promotes an inflammatory cascade, playing a role in the body's antiviral programme [41]. Moreover, in an influenza infection study on mice, eosinophils, blood defence white blood cells, show a decreased activity alongside a viral recognition protein transcription and T-box transcription factor (Tbx21) promoter's CpG methylation [42]. In an example of a more direct impact of the host on the virus, epigenetic modifications in the Interleukin 32 (IL32) promoter increase its transcription, effectively hampering viral replication [43]. This is done when the influenza virus instigates the CpG demethylation at the CREB, cAMP-response element binding protein binding point, increasing the binding of IL32 promoter and CREB, and increasing IL32 transcription [43]. Therefore, the host succeeds in not only launching its own immune response but also hinders the viral virulence.

10.4 Summary

Due to the constraints of space, the emphasis of this review was on the impact that histone changes have on the expression of viral genes and the replication of viruses—Interactions Between Hosts and Viruses: Looking at Things Through the Lens of Epigenetics.

References

1. Riddihough G, Zahn LM. What is epigenetics? American Association for the Advancement of Science; 2010. p. 611.
2. Marcos-Villar L, Díaz-Colunga J, Sandoval J, Zamarreño N, Landeras-Bueno S, Esteller M, et al. Epigenetic control of influenza virus: role of H3K79 methylation in interferon-induced antiviral response. Sci Rep. 2018;8(1):1–13.
3. Mosnier A, Caini S, Daviaud I, Nauleau E, Bui TT, Debost E, et al. Clinical characteristics are similar across type a and B influenza virus infections. PLoS One. 2015;10(9):e0136186.
4. Hirsch A. Handbook of geographical and historical pathology. New Sydenham Society; 1883.
5. Taubenberger JK, Morens DM. The pathology of influenza virus infections. Annu Rev Pathol. 2008;3:499.
6. Bannister S, Messina NL, Novakovic B, Curtis N. The emerging role of epigenetics in the immune response to vaccination and infection: a systematic review. Epigenetics. 2020;15(6-7): 555–93.
7. Dou D, Revol R, Östbye H, Wang H, Daniels R. Influenza a virus cell entry, replication, Virion Assembly and movement. Front Immunol. 2018;9:1581.
8. Keshavarz M, Sabbaghi A, Koushki K, Miri SM, Sarshari B, Vahdat K, et al. Epigenetic reprogramming mechanisms of immunity during influenza A virus infection. Microbes Infect. 2021;23(8):104831.
9. Li S, Kong L, Yu X, Zheng Y. Host–virus interactions: from the perspectives of epigenetics. Rev Med Virol. 2014;24(4):223–41.
10. Kouzarides T. Chromatin modifications and their function. Cell. 2007;128(4):693–705.

11. Marcos-Villar L, Díaz-Colunga J, Sandoval J, Zamarreño N, Landeras-Bueno S, Esteller M, et al. Epigenetic control of influenza virus: role of H3K79 methylation in interferon-induced antiviral response. Sci Rep. 2018;8(1):1230.
12. Hughes LAE, van den Brandt PA, de Bruïne AP, Wouters KAD, Hulsmans S, Spiertz A, et al. Early life exposure to famine and colorectal cancer risk: a role for epigenetic mechanisms. PLoS One. 2009;4(11):e7951.
13. de Vogel S, Bongaerts BW, Wouters KA, Kester AD, Schouten LJ, de Goeij AF, et al. Associations of dietary methyl donor intake with MLH1 promoter hypermethylation and related molecular phenotypes in sporadic colorectal cancer. Carcinogenesis. 2008;29(9):1765–73.
14. Ling C, Rönn T. Epigenetic adaptation to regular exercise in humans. Drug Discov Today. 2014;19(7):1015–8.
15. He Y, Zhao Y, Zhang S, Chen W, Lin S, Yang Q, et al. Not polymorphism but methylation of class II transactivator gene promoter IV associated with persistent HBV infection. J Clin Virol. 2006;37(4):282–6.
16. Zheng L, Leung ETY, Wong HK, Lui G, Lee N, To K-F, et al. Unraveling methylation changes of host macrophages in Mycobacterium tuberculosis infection. Tuberculosis. 2016;98:139–48.
17. Chertow DS, Memoli MJ. Bacterial coinfection in influenza: a grand rounds review. JAMA. 2013;309(3):275–82.
18. Morens DM, Taubenberger JK, Fauci AS. Predominant role of bacterial pneumonia as a cause of death in pandemic influenza: implications for pandemic influenza preparedness. J Infect Dis. 2008;198(7):962–70.
19. Didierlaurent A, Goulding J, Patel S, Snelgrove R, Low L, Bebien M, et al. Sustained desensitization to bacterial Toll-like receptor ligands after resolutionof respiratory influenza infection. J Exp Med. 2008;205(2):323–9.
20. Mukherjee S, Vipat VC, Chakrabarti AK. Infection with influenza a viruses causes changes in promoter DNA methylation of inflammatory genes. Influenza Other Respir Viruses. 2013;7(6): 979–86.
21. Steger DJ, Lefterova MI, Ying L, Stonestrom AJ, Schupp M, Zhuo D, et al. DOT1L/KMT4 recruitment and H3K79 methylation are ubiquitously coupled with gene transcription in mammalian cells. Mol Cell Biol. 2008;28(8):2825–39.
22. Menachery VD, Schäfer A, Burnum-Johnson KE, Mitchell HD, Eisfeld AJ, Walters KB, et al. MERS-CoV and H5N1 influenza virus antagonize antigen presentation by altering the epigenetic landscape. Proc Natl Acad Sci U S A. 2018;115(5):E1012–E21.
23. Shirey KA, Perkins DJ, Lai W, Zhang W, Fernando LR, Gusovsky F, et al. Influenza "trains" the host for enhanced susceptibility to secondary bacterial infection. MBio. 2019;10:3.
24. Chen X, Liu S, Goraya MU, Maarouf M, Huang S, Chen J-L. Host immune response to influenza a virus infection. Front Immunol. 2018;9:320.
25. Ashraf U, Benoit-Pilven C, Lacroix V, Navratil V, Naffakh N. Advances in analyzing virus-induced alterations of host cell splicing. Trends Microbiol. 2019;27(3):268–81.
26. Zhang L, Wang J, Muñoz-Moreno R, Kim M, Sakthivel R, Mo W, et al. Influenza virus NS1 protein-RNA interactome reveals intron targeting. J Virol. 2018;92(24):e01634-18.
27. Heinz S, Texari L, Hayes MGB, Urbanowski M, Chang MW, Givarkes N, et al. Transcription elongation can affect genome 3D structure. Cell. 2018;174(6):1522–36.e22
28. Zhao N, Sebastiano V, Moshkina N, Mena N, Hultquist J, Jimenez-Morales D, et al. Influenza virus infection causes global RNAPII termination defects. Nat Struct Mol Biol. 2018;25(9): 885–93.
29. Bauer DLV, Tellier M, Martínez-Alonso M, Nojima T, Proudfoot NJ, Murphy S, et al. Influenza virus mounts a two-pronged attack on host RNA polymerase II transcription. Cell Rep. 2018;23 (7):2119–29.e3.
30. More S, Zhu Z, Lin K, Huang C, Pushparaj S, Liang Y, et al. Long non-coding RNA PSMB8-AS1 regulates influenza virus replication. RNA Biol. 2019;16(3):340–53.

31. Ichida Y, Matsushita K, Fujiwara Y, Hoshizaki M, Fujiwara S, Kuba K, et al. Influenza virus infection affects host epigenome structure associated with histone methylation. 日本薬理学会年会要旨集. 2018;WCP2018:PO2-11-24.
32. Marazzi I, Ho JSY, Kim J, Manicassamy B, Dewell S, Albrecht RA, et al. Suppression of the antiviral response by an influenza histone mimic. Nature. 2012;483(7390):428–33.
33. Vasileva D, Badawi A. C-reactive protein as a biomarker of severe H1N1 influenza. Inflamm Res. 2019;68(1):39–46.
34. Davey RT Jr, Lynfield R, Dwyer DE, Losso MH, Cozzi-Lepri A, Wentworth D, et al. The association between serum biomarkers and disease outcome in influenza A (H1N1) pdm09 virus infection: results of two international observational cohort studies. PLoS One. 2013;8(2): e57121.
35. Peng F, He J, Loo JFC, Yao J, Shi L, Liu C, et al. Identification of microRNAs in throat swab as the biomarkers for diagnosis of influenza. Int J Med Sci. 2016;13(1):77.
36. Hernandez-Vargas EA, Wilk E, Canini L, Toapanta FR, Binder SC, Uvarovskii A, et al. Effects of aging on influenza virus infection dynamics. J Virol. 2014;88(8):4123–31.
37. Hu J, Zhang L, Liu X. Role of post-translational modifications in influenza a virus life cycle and host innate immune response. Front Microbiol. 2020;11:517461.
38. Hatakeyama D, Shoji M, Yamayoshi S, Yoh R, Ohmi N, Takenaka S, et al. Influenza A virus nucleoprotein is acetylated by histone acetyltransferases PCAF and GCN5. J Biol Chem. 2018;293(19):7126–38.
39. Nishioka K, Daidoji T, Nakaya T. Demethylation around the transcriptional start site of the IFN-β gene induces IFN-β production and protection against influenza virus infection. Biochem Biophys Res Commun. 2019;520(2):269–76.
40. Tang B, Zhao R, Sun Y, Zhu Y, Zhong J, Zhao G, et al. Interleukin-6 expression was regulated by epigenetic mechanisms in response to influenza virus infection or dsRNA treatment. Mol Immunol. 2011;48(8):1001–8.
41. Fang J, Hao Q, Liu L, Li Y, Wu J, Huo X, et al. Epigenetic changes mediated by MicroRNA miR29 activate cyclooxygenase 2 and Lambda-1 interferon production during viral. Infection. 2012;86(2):1010–20.
42. LeMessurier KS, Rooney R, Ghoneim HE, Liu B, Li K, Smallwood HS, et al. Influenza A virus directly modulates mouse eosinophil responses. J Leukoc Biol. 2020;108(1):151–68.
43. Li W, Sun W, Liu L, Yang F, Li Y, Chen Y, et al. IL-32: a host proinflammatory factor against influenza viral replication is upregulated by aberrant epigenetic modifications during influenza a virus infection. J Immunol. 2010;185(9):5056–65.

Epigenetics of Rhinovirus

11

Md Abubakar, Eswara Rao Puppala, Bhaskar Jyoti Dutta, Krushna Ch. Maharana, Riya Thapa, S. Roshan, B. Tazneem, Abdullah Khan, and Asif Ahmad Bhat

Abstract

Epigenetics is a moderately novel field that explores the genetic and non-genetic components of transmissible phenotypic changes, which are commonly caused by ecological and metabolic factors. The epigenetic machinery of the host may employ changes like methylation of histone and acetylation, methylation of DNA and RNA, remodelling of chromatin, and non-coding RNAs to regulate gene expression. One of the most infectious and potentially lethal types of viruses is the rhinovirus. Here, we'll provide you a high-level introduction to rhinovirus epigenetics. A rhinovirus is a kind of very dangerous respiratory RNA virus. There is mounting sign that epigenetic alterations play an important role in the progress of rhinovirus diseases. A more active vaccine or safer chemotherapeutical treatments, such as epigenetic pharmaceuticals, are needed to manage with this viral pandemic and give pre- and post-exposure prophylaxes against RHINOVIRUS.

M. Abubakar
Department of Pharmacology and Toxicology, NIPER, Hajipur, India

E. R. Puppala · B. J. Dutta · K. C. Maharana
Department of Pharmacology and Toxicology, NIPER, Guwahati, India

R. Thapa · A. A. Bhat (✉)
School of Pharmacy, Suresh Gyan Vihar University, Jagatpura, Jaipur, India

S. Roshan · B. Tazneem
Deccan School of Pharmacy, Hyderabad, India

A. Khan
Faculty of Pharmacy, Quest International University, Ipoh, Perak, Malaysia

G. Gupta et al. (eds.), *Targeting Epigenetics in Inflammatory Lung Diseases*,
https://doi.org/10.1007/978-981-99-4780-5_11

Keywords

Rhinovirus · Epigenetics · Receptors · Infection

11.1 Introduction

Human rhinoviruses (HRVs) are a kind of respiratory virus that belongs to the genus *Enterovirus* and the family Picornaviridae. Human rhinoviruses (HRVs) were foremost identified in the 1950s and are responsible for the vast majority of upper respiratory tract infections. Human rhinovirus (HRV) is the leading cause of the common cold and the associated loss of work and medical expenses [1]. More than $60 billion a year is thought to be spent on treating HRV infections. Although human rhinovirus infections may happen at any period of the year and in any part of the world, they tend to peak in the early spring and late autumn. In addition to causing lung infections, HRV has been connected to the aetiology of asthma and has been shown to aggravate asthma. The demand for an HRV vaccine is increasing in tandem with the rising HRV health burden. Unfortunately, there have been several roadblocks to the creation of an efficient vaccine [2, 3] (Fig. 11.1).

Rhinoviruses are amid the furthermost common types of viruses that infect humans. About half of all occurrences of upper respiratory infection in children

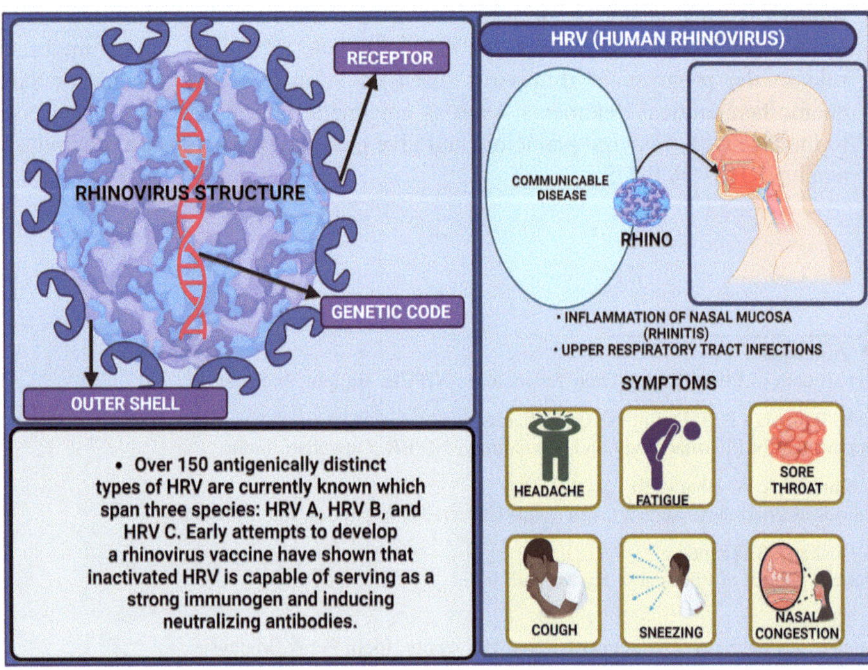

Fig. 11.1 HRV infections and disease

younger than three are caused by HRVs. This may result in as many as 8–12 sick days per year [4]. A human rhinovirus infection has a 2-day incubation period and often causes symptoms for 1–2 weeks. Nasopharyngeal infections caused by rhinoviruses are often asymptomatic in very young infants. Upper respiratory tract infections have comparable symptoms to the typical cold, including coughing, nasal congestion, sore throat, sneezing, rhinorrhea, and an overall feeling of malaise [5]. In accumulation to the common cold, infections of rhinovirus have been related to severe otitis media and rhinosinusitis, both of which are often associated with bacterial co-contamination. Human rhinoviruses (HRVs) are most at home in the somewhat colder surfaces (32–33 °C) of the respiratory tract for reproduction, but they have also been linked to lower respiratory tract infections such as pneumonia, bronchitis, bronchiolitis, and exacerbations of asthma [6]. Childhood infections with human rhinovirus (HRV) cause wheezing, which is linked to an uptick in the chance of developing asthma in later life. An HRV infection may be the underlying cause of more than half of all occurrences of asthma exacerbation. Ninety percent of hospitalised children with severe asthma attacks showed detectable HRV [7]. In addition, over half of hospitalised children and adolescents (3–18 years old) with wheeze tested positive for HRV. Although HRV does not often consist of a pathogen with high mortality, our results demonstrate that it has great latent for severe respiratory illness and the ability to generate or intensify long-lasting respiratory health concerns [8, 9]. Rhinoviruses in humans are often transmitted by direct human-to-human contact, fomite transmission, or the inhalation of infected droplets in the air. Because rhinoviruses may live for hours on inanimate surfaces, they pose a serious threat of transferring from person to person [10]. There is a correlation between a poor interferon (IFN) response, the presence of allergies or asthma, contact to air pollutants like cigarette smoke, an unhealthy diet, and stress, and an increased risk of getting and recovering from severe illnesses. Reinfection rates with certain HRV types provide evidence in HRV infection cases, immunity competent does not grow or is not maintained [11].

11.2 HRV Structure and Biology

The HRV has a 7.2 kb positive-sense single-stranded RNA (+ssRNA) that encrypts 11 proteins. The viral capsid of a HRV consists of four viral proteins (VPs): VP3, VP2, VP1, and VP4. The virus cannot multiply without the further virus-related proteins, which are accountable for reassembling the virus [12]. Antigenic variability across HRV types is due to differences in the surface expression of VP2, VP1, and VP3, whereas VP4 is critical for RNA packing throughout assembly. There is more antigenic diversity in the HRV capsid proteins because of their higher degree of heterogeneity. The research of neutralising antibodies' binding to antigens has shown a number of antigenic sites for HRV strains [13]. However, in many instances the specific sites of these markers have been lost. VP1 aids in attachment to the cell surface by connecting to several receptors there. Previous research suggested that the common HRV types bound to the intercellular adhesion molecule (ICAM)-1 receptor and that only a small fraction of HRV types bound to the low-density lipoprotein receptor (LDLR). Given the diversity of circulating HRV variations, it is plausible that other, as-yet-unidentified receptors exist despite the recent discovery of one form of cadherin-related family member (CDHR)3 as the primary receptor for HRV C species [14]. The entrance of HRV into cells has been speculated to occur by a variety of pathways, like as macropinocytosis, endocytosis of clathrin-dependent, and clathrin-independent endocytosis. After the viral genome has been translated after entry and uncoating, the virus-encoded proteases3c and 2a process it. A total of 60 copies of each protein capsid are needed to form the icosahedral shape required for virion assembly and genomic packaging, with one copy of each capsid protein present on each face [15].

11.3 Epigenetics

Scientists studying epigenetics try to determine whether or not alterations to chromatin structural/activation states may reliably alter phenotype in the absence of changes to the DNA's fundamental nucleotide sequence. Epigenomics (the study of genome epigenetics) has allowed us to get a functional understanding of the machinery of epigenetic that regulates the whole genome [16]. A large body of evidence has collected over the last several eras showing epigenetics plays a vital role in the inception and course of many common diseases, notably age-related ailments (ARD). Additionally, gene expression patterns is determined by epigenetic molecular marks produced through expansion influence of human vulnerability to numerous disorders, like viral infections [17]. Unlike mutations, which alter DNA sequences directly, epigenetic alterations only alter the structure of chromatin or nucleic acids' chemical properties. Because of this feature, alterations of epigenetic are reversible, versatile, and receptive to environmental and other experiences. In humans, revelation to fluctuating metabolic situations may have long-term impacts on cellular epigenetics [18, 19]. Epigenetic activities such as regulating gene expression aid in maintaining metabolic data consistency. Methylation of DNA and modifications of histone, in conjunction with protein modifications (prions, sirtuins, etc.) and non-coding RNA, remodel chromatin and provide entree to proteins that help in DNA transcription regulation and, by extension, protein synthesis and RNA. The epigenome represents the condition of chromatin with regard to gene activation, since it contains information on the placement and function of gene-specific activation switches in the genome [20]. The nucleosome is the fundamental unit of chromatin, which is a protein-and-DNA complex. Two of each of the four main histones make up a nucleosome (H3, H4, H2A, and H2B). Because of its distinctive chemical composition, the histone octamer regulates DNA accessibility for gene transcription and is encircled by the DNA helicase. Chromatin remodelling, which includes alterations to histones and DNA that may be reversed, controls several epigenetic events [21]. Amongst the 100 of enzymes intricate in regulation of epigenetic, histone methyltransferases (HMTs), histone deacetylases (HDACs), histone acetyltransferases (HATs), and histone kinases (HKs) are of particular relevance here. The work of enzymes is to create precise outlines that produce affinity for proteins chromatin-associated, which contributes to transitions of dynamic between active transcriptionally and chromatin silent states, which in turn contribute to cellular developmental plasticity and, in some contexts, pathological conclusions [22]. Furthermore, non-coding RNAs play a vital role in expression of genes and post-transcriptional genes by creating or regulating epigenetic procedures, including, for example, muzzling or energising genes/transcripts via diverse methods of accomplishment. There is evidence to imply that non-coding RNAs (ncRNAs) operate as messengers between cells and even control biological processes [23]. Further, lncRNAs, depending on their subcellular localisation, may attach to junctions of pre-mRNA exon/intron and affect the process of splicing, or they may imitate binding factor sites of transcription and function as a decoy. N6-adenosine methyltransferase-like 3 methylates adenine at position 6 in RNA molecules, which is only one of several examples of RNA modifications that have recently been uncovered [24]. This is especially true when m6A is present at the 5′-AGG (m6) AC-3′ consensus arrangement, where it reduces the efficiency with which mRNAs are translated. Like 5-methylcytosine, RNA of m6A may be oxidised into N6-formyladenosine (f6A), and N6-hydroximethyladenosine (hm6A) demethylating it. This may lead to a dissimilar interaction of RNA-protein and, therefore, a different pattern of gene regulation. The FTO protein, which has been linked to obesity and fat accumulation, catalyses these processes by binding to RNA. In particular, RNA methyltransferases like domain family members of NOP2/Sun and methyltransferase of DNA type 2 may methylate ribocytidines [25–27].

Enzymes called tRNA elongation factors (TEFs) play a role in both DNA and RNA translation by modifying RNA 5mC into 5-hydroxymethylcytosine (5hmC). In 7-methylguanosine (m7G), a methyl group has been added to the guanine at position 7 of riboguanosine. This modification on topped and recapitulated mRNAs is arbitrated by canonical mRNA overlaying methyltransferase to control protein synthesis from mRNA [28]. With the start of following generation sequencing technology, high-resolution epigenome maps of both healthy and sick cells can be generated, allowing for the immediate study of genomic and epigenetic modifications, as well as genome-wide association study (GWAS) and epigenome-wide association study (EWAS) [29].

11.4 Epigenetic Landscape Alteration by Viral Infection

Monitoring changes of epigenetic in pathologic situations, for example, may provide a useful window into understanding how to direct the host immune response. Viruses like the corona and influenza viruses cannot directly alter the genetic sequence of their hosts, but they may modify the epigenome [30]. Recent research has focused on the epigenetic machinery due to speculation that it plays a role in the beginning, spread, and maintenance of viral infections. New high throughput technologies have also made it probable to assess the epigenetic site on a scale that can be applied to the whole human genome [31]. The rhino may use an epigenetic approach similar to that of other virus families that have been proven to inhibit the immune system. Moreover, some reports have shown that viral interference with the regulation of the host's epigenetic network may have an effect on the host's immunological response. Using the highly infective H3N2 influenza virus as an example [32], the host's response of innate immunity is suppressed when viruses interfere with epigenetic control of gene expression. This protein may use histone mimicry, since the non-structural protein NS1 of H3N2 carboxy-terminus has been demonstrated to contain homolog systems with the histone H3 tail of the amino-terminus [33]. The viral protein of NS1 disrupts the function of the antiviral gene by transcription complex interaction, which typically H3K4 docks mark to inductee transcription. Proteins from both hepatitis C virus (HCV) and adenoviruses interfere with epigenetic functions, which in turn weakens the immune system worldwide. The restrictive histone modification is shown to physically inhibit transcription of these genes even in the presence of active transcription factors and signalling pathways, as revealed by the 2014 finding of Baric's lab [34, 35]. Downregulation of interferon (IFN)-stimulated genes (ISGs) is linked to histone 3-lysine 27 trimethylation (H3K27me3) after infection with influenza virus A/influenza/Vietnam/1203/2004. Histone methylation in H5N1 was also shown to be functionally linked to NS1's ability to suppress an immune response. Furthermore, their sequencing data revealed that methylation is specifically aimed at other regions of the genome, regions that may include crucial genes involved in viral antagonism [36]. DNA methylation was considered to be the main factor in the downregulation presentation of antigen molecules equally in diseases. They theorised that the novel viruses may exhibit characteristics similar to those of HIV-1 and herpes virus. Interferons are already critical as intermediaries of antiviral effects and as inventers immune responses of pathogen-driven, but they may become much more so if viruses acquire mechanisms of antagonistic to resist certain effectors of ISG [37]. As a matter of fact, specific epigenetic marks regulate IFN and innate immune responses during infection by modulating enzyme activity of epigenetic and generating chromatin remodelling complexes. The machinery of epigenetic serves not only to reliably regulate host responses but also to prime and remember them [38]. Fang et al. found an association between the amount of H3K9me2 and in vitro IFN expression. Heterochromatin formation and DNA methylation are both controlled by the restrictive histone mark H3K9me2. By recruiting members of the heterochromatin protein 1 family, the H3K9me2 mark blocks acetylation in a targeted manner. The levels of H3K9me2 mark in the agent section of type I interferon are inversely linked with the generation of ISGs in dendritic cells, establishing modification of histone as a major IFN response regulator [39, 40]. However, the H3K4me3 mark is often seen in vigorous promoters, and is concentrated in Toll-like receptor (TLR) promoter regions. The stimulus of dendritic cells and macrophages with lipopolysaccharide (LPS) leads to an increase in global histone acetylation and polymerase II (Pol II) binding to particular promoters. These results provide more evidence for the hypothesis that the activation of the distinctive immune retort is regulated in a targeted fashion by epigenetic mechanisms. By using ChIP-PCR, Schäfer et al. found that during H1N1 and rhino infection, the supporter regions of ISG genes had a higher concentration of histones with the dynamic mark of H3K4 than the exploitive H3K27me3 mark, open chromatin favouring and endorsing active transcription and expression of ISG [41, 42]. The promoter area of several particular ISGs subsets had increased H3K27me3 and reduced H3K4me3 in rhino-infected cells, however these ISGs were not elevated, as discovered by Menachery et al. These findings demonstrated that viruses have developed defences to avoid being attacked by the body's natural IFN response. As expected, RNA-type viruses like rhinovirus have robust relationships with RNA alterations. For instance, adenosine nucleotide modifications such as N6,2-methyladenosine (m6A) and N6,2′-O-dimethyladenosine (m6Am) have been linked to significant roles in viral replication [43]. They could alter the virus' ability to replicate, the innate immune response of host's, and some innate sensing pathways. Viral and cellular transcripts both undergo m6A RNA methylation, which controls a number of biological functions. When it comes to eukaryotic mRNAs, this modification is by far the most prevalent epitranscriptomic modification [44]. Hepatitis B (HBV) DNA virus replication concludes with the production of a pregenomic RNA intermediate, which Imam and colleagues hypothesised is regulated by m6A and associated machinery (pgRNA). These findings suggest that regulation of m6A, HBV expression of gene and reverse transcription. After inhibiting the methylase is responsible for adding the m6A alteration to the RNA, the expression levels of HBV protein rose but reverse transcription of pgRNA seemed to fall [45]. The preserved m6A accords idea in the epsilon stalk loop assembly as the site of m6A in HBV RNA. This circle is present at the 5′- and 3′-termini of the pgRNA and 3′-terminus of all HBV mRNAs. The existence of 5′ epsilon stem loop of pgRNA in the site of m6A, proves the need of m6A for effectual reverse transcription of pgRNA [46]. In addition, they found that methylation of m6A of the 3′ epsilon stem loop rendered transcripts HBV unstable, providing further evidence for a dual regulatory function for m6A in HBV RNA. The viral and cellular epitranscriptomes of m6A/m during kaposi's sarcoma-associated herpesvirus (KSHV) infection of latent and lytic, providing indication that messengers RNA of m6A and m6Am affect unique functions of cells [47]. High numbers of m6A/m changes are characteristic of KSHV transcripts, which are acquired through concealed and replication of lytic and are maintained throughout infection of diverse types of cell. Tan et al. showed that lytic replication is impaired when the N6-methyladenosine YTH RNA binding protein 2 (YTHDF2) is knocked down. The stability of viral RNA is altered by YTHDF2 interactions, which vary depending on the virus [48]. In addition, they discovered that KSHV latent infection might alter the host epitranscriptome by causing hypomethylation of the 5′ untranslated region (UTR) and hypermethylation of the 3′ UTR. This, in turn, affects epithelial-mesenchymal and oncogenic transition processes. Also, the KSHV epitranscriptome undergoes an active reprogramming when the virus undergoes lytic replication [49]. At last, Marz's group established that rhinovirus RNA always exhibited the identical 5mC methylation signature. They found that various RNAs had a tendency to be methylated in the same genomic places by comparing their 5mC contents; this led them to conclude that RNA methylation in rhinoviruses is sequence-specific or measured by RNA basic features [50].

11.5 Epigenetic Implication in Rhinovirus Infection and Therapy

There is much deeper and more nuanced knowledge of the mechanisms at work in inheritance, memory, and maturation because of the fast growth of the area of epigenetics in recent years. Scientists studying cancer, immunity, and infectious diseases are becoming more interested in studies of the human epigenome [51]. Undeniably, through the past era, research of epigenetic has revealed indication that DNA and RNA viruses have settled traits that counteract the monitoring system of the host epigenome by varying the metabolism of host's and expression of gene, so enabling viral reproduction and dispersion [52]. Changes in the host epigenome as a result of ageing have been found to have negative effects on immune cell composition and function, which in turn limits the body's capacity to generate an efficient immune response against viruses, particularly the adaptive immunological response. Rhinoviruses have been linked to epigenetic alterations, namely those that prevent presentation of antigen host or that trigger genes of interferon-response [53]. Through the analysis of DNA methylation in various blood cell types, including protected cells and other cell types, before, during, and after infection, we can gain a deeper understanding of how the aging epigenome influences disease severity and how the virus affects the aging epigenome [54]. Increased susceptibility to rhinovirus has been seen in the elderly, and it is possible that epigenetic influences on viral entry are a factor in this phenomenon. Cell surface activation of this mechanism is mediated by the virus-related protein of spike glycoprotein receptor ACE2 and the dipeptidyl peptidase-4 co-receptor (DPP4). The treatment of RHINOVIRUS infections is now limited to avoidance and supportive care since neither a vaccine nor specialised antiviral drugs are available [55]. Despite the fact that many alternative treatment routes are being investigated, further research is necessary to classify actual serums and harmless medications for treating RHINOVIRUS contaminations in order to create pre- and post-exposure therapy against the virus. Though the primary objective would be to develop vaccines against rhinovirus based on the S subtype that contains conserved epitopes and can elicit broadly neutralising antibodies or responses of virus-specific T cell, it is also significant to find and change drugs that can prevent rhinovirus from entering and replicating. Studying epigenetics may help achieve these aims by shedding light on the mechanisms behind viral chromatin change in viruses of lytic and interactions with host-virus, such as factors of genes that underwrite to defensive or detrimental host responses [56, 57]. Reducing viral replication and regulating the host immune response is thought to be aided by clinical research, epigenetic-targeted therapies licenced by the FDA, and combination therapy with antiviral treatments. New evidence reveals that epigenetic control may also affect the pharmacokinetic and pharmacodynamic properties of antivirals, which may have substantial implications for the treatment of rhinovirus infection. Multiple epigenetic mechanisms linked to rhinovirus infections have been studied by El Baba and colleagues, allowing them to identify potential treatment targets. Since HDACs regulate numerous proteins of nonstructural viral transcription involvement, duplication, and maturation, HDAC inhibitors like Vorinostat or SAHA in conjunction with antivirals may be important paraphernalia to affect with these methods [58]. Expression of ACE2 is regulated by methylation of DNA and changes of histone. Here, epigenetic enzymes including histone deacetylase 2 (HDAC2), histone acetyltransferase 1 (HAT1), DNA methyltransferase 1 (DNMT1), and lysine demethylase 5B (KDM5B) become promising therapeutic marks for regulating the immune response of host. Since viruses utilise host epigenetic machinery, there is hope that the epigenetic medications now used in cancer treatment may be harnessed for wide-ranging antiviral accomplishment and inflammatory management [59]. There is data showing that RHINOVIRUS-related mortality is mostly due to the storm of cytokine, which is characterised by an unrestrained overproduction of solvable indicators of inflammation. Decitabine is a nucleoside-based DNMT blocker often used to decrease inflammation and interferon (IFN) retort in macrophages. It is also known as 5-azadC. In particular, Decitabine has been the subject of a clinical trial for the treatment of RHINOVIRUS pneumonia-acute respiratory distress syndrome (ARDS) [60, 61]. H3K27me3 enrichment is a mechanism by which the repressive polycomb complex 2 represses transcription and may act on certain IFN-stimulated genes. PRC2 Inhibitors that are now in progressive clinical research for the treatment of cancer may be useful for RHINOVIRUS patients. Recent studies have shown that usual executioner cells and lung-inborn lymphoid cell group 2 share epigenetic pathways that sustain the enhanced innate immune response over time, a phenomenon known as trained immunity (TRIM) [62]. Cells that are exposed to a stimulus undergo metabolic, mitochondrial, and epigenetic reprogramming in response, making them more able to react to a future heterologous stimulus. The immunological dysregulation and cytokine storm caused by RHINOVIRUS are also examined, and the potential effects of glucan on these phenomena are evaluated. Their results demonstrated that -glucan-driven TRIM is in charge of defining different epigenetic modifications, indicating that it might be a potential therapeutic target for RHINOVIRUS. Recent studies suggest that vitamin supplementation and the use of natural treatments may act as epigenetic modifiers, boosting immunity and reducing inflammation in RHINOVIRUS patients [63]. Vitamin D and quercetin may be attractive for reducing rhinovirus severity because they inhibit ACE2 expression and its hypothesised effect in lowering the storm of cytokine linked with death in RHINOVIRUS patients. Research is also needed into another epigenetic strategy: RNA-based antiviral therapies. RHINOVIRUS might be prevented and treated with the use of novel techniques that utilise siRNAs, miRNAs, and LNA or GapmeRs to target, for example, the 5′URT or sections of the Spike protein [64]. Miravirsen is an antisense oligonucleotide being studied for the treatment of hepatitis C virus (HCV) that might be utilised to scavenge miRNAs important in viral replication. Results from these investigations point to the potential for optimising and using RNA-based medicines to impede rhinovirus replication and transcription. Notably, advanced bioinformatics software has allowed us to visualise and interpret the epigenomic data, revealing cellular-specific insights into an individual's genetic and epigenetic predispositions and providing an explanation for how environmental factors affect gene function by leaving permanent symbols on the genome [65]. Mapping of epigenome, along with GWAS and EWAS investigations, gives us paraphernalia in the identification of numerous communal human illnesses, suggesting that these findings may be used for separate analysis and individualised therapy: to therapeutically avoid, decrease, or converse the epigenetic modifications, to build and actualise specialised pharmacological apparatuses, and when/where to arbitrate by studying the epigenetic impacts of metabolism at every level, from genetic factor to pathways to genomes. The epigenetic enzymes answerable for the alterations also provide a novel and exciting target for therapeutic development [66, 67].

11.6 Conclusion

In terms of global health, the RHINOVIRUS is an important concern in the contemporary period. The occurrence of the virus-induced storm of cytokine that exacerbates ARDS symptoms may lead to failure of multi-organ and death. Rhinovirus infection may alter the expression of inflammatory cytokines by interacting with the host's epigenetic machinery, like interleukin (IL)-6, IFN-, IL-18, IL-1, and tumor necrosis factor (TNF)-α. By studying specific epigenetic modifiers and exploring innovative chromatin-based therapies related to various viruses, including rhinovirus, epigenomic research can pave the way for the development of antiviral drugs and reveal crucial new insights into virus-host interactions and their impact on disease severity. Previous studies isolated one or two epigenetic pathways to examine. In this chapter, we have taken a quick look at what is known so far regarding the epigenetic features of rhinovirus infection and how these studies can guide future therapies. More specifically, our findings suggest that empathetic regulation of epigenetic underlying the immune response to rhinovirus may aid in the development of novel, targeted strategies for preventing and treating the infection.

References

1. Abraham G, Colonno RJ. Many rhinovirus serotypes share the same cellular receptor. J Virol. 1984;51(2):340–5.
2. Allmaier G, et al. Monolithic anion-exchange chromatography yields rhinovirus of high purity. J Virol Methods. 2018;251:15–21.
3. Alshrari AS, et al. Bioinformatics analysis of rhinovirus capsid proteins VP1-4 sequences for cross-serotype vaccine development. J Infect Public Health. 2021;14(11):1603–11.
4. Amineva SP, et al. Comparison of rhinovirus a infection in human primary epithelial and HeLa cells. J Gen Virol. 2011;92(Pt 11):2549–57.
5. Ashraf S, et al. Biological characteristics and propagation of human rhinovirus-C in differentiated sinus epithelial cells. Virology. 2013;436(1):143–9.
6. Basnet S, Palmenberg AC, Gern JE. Rhinoviruses and their receptors. Chest. 2019;155(5): 1018–25.
7. Blaas D. Individual subunits of a rhinovirus causing common cold exhibit largely different protein-RNA contact site conformations. Commun Biol. 2020;3(1):537.

8. Bochkov YA, Gern JE. Clinical and molecular features of human rhinovirus C. Microbes Infect. 2012;14(6):485–94.
9. Choi HJ, et al. Anti-human rhinovirus activity of raoulic acid from Raoulia australis. J Med Food. 2010;13(2):326–8.
10. Croft SN, Walker EJ, Ghildyal R. RIPK1 is cleaved by 3C protease of rhinovirus A and B strains and minor and major groups. Viruses. 2021;13:12.
11. Cuevas MT, et al. Spread of different rhinovirus B genotypes in hospitalized children in Spain. Influenza Other Respir Viruses. 2013;7(5):623–8.
12. Da Costa L, et al. Structure-based drug design of potent pyrazole derivatives against rhinovirus replication. J Med Chem. 2018;61(18):8402–16.
13. da Costa Souza L, et al. Molecular and clinical characteristics related to rhinovirus infection in Brasília, *Brazil*. Braz J Microbiol. 2021;52(1):289–98.
14. da Silva ER, et al. Rhinovirus genetic diversity among immunosuppressed and immunocompetent patients presenting with a severe respiratory infection. J Clin Virol. 2013;56(1):82–3.
15. Dagher H, et al. Rhinovirus detection: comparison of real-time and conventional PCR. J Virol Methods. 2004;117(2):113–21.
16. Dolan TM, et al. Rhinovirus plaque formation in WI-38 cells with methylcellulose overlay. Appl Microbiol. 1968;16(9):1331–6.
17. Duechler M, et al. Human rhinovirus serotype 2: in vitro synthesis of an infectious RNA. Virology. 1989;168(1):159–61.
18. Dysangco AT, et al. Prolonged rhinovirus shedding in a patient with Hodgkin disease. Infect Control Hosp Epidemiol. 2017;38(4):500–1.
19. Eisa M. Isolation of a rhinovirus of bovine origin in The Sudan. Vet Rec. 1980;106(10):225–7.
20. Erkkola R, et al. Rhinovirus C is associated with severe wheezing and febrile respiratory illness in young children. Pediatr Infect Dis J. 2020;39(4):283–6.
21. Esneau C, Duff AC, Bartlett NW. Understanding rhinovirus circulation and impact on illness. Viruses. 2022;14:1.
22. Ferguson PE, et al. Human rhinovirus C in adult haematopoietic stem cell transplant recipients with respiratory illness. J Clin Virol. 2013;56(3):255–9.
23. Griggs TF, et al. Production, purification, and capsid stability of rhinovirus C types. J Virol Methods. 2015;217:18–23.
24. Gwaltney JM Jr. Rhinoviruses. Yale J Biol Med. 1975;48(1):17–45.
25. Hai le T, et al. Fatal respiratory infections associated with rhinovirus outbreak, *Vietnam*. Emerg Infect Dis. 2012;18(11):1886–8.
26. Hamre D. Rhinoviruses. Monogr Virol. 1967;1:1–85.
27. Higgins PG. Rhinovirus infections. Postgrad Med J. 1973;49(577):802–6.
28. Hrebík D, et al. ICAM-1 induced rearrangements of capsid and genome prime rhinovirus 14 for activation and uncoating. Proc Natl Acad Sci U S A. 2021;118:19.
29. Jackson DJ. The role of rhinovirus infections in the development of early childhood asthma. Curr Opin Allergy Clin Immunol. 2010;10(2):133–8.
30. Jacobs SE, et al. Human rhinoviruses. Clin Microbiol Rev. 2013;26(1):135–62.
31. Jans DA, Ghildyal R. Preface. Rhinoviruses. Methods Mol Biol. 2015;1221:v–vii.
32. Jensen LM, et al. Proteases of human rhinovirus: role in infection. Methods Mol Biol. 2015;1221:129–41.
33. Kerr SL, Mathew C, Ghildyal R. Rhinovirus and cell death. Viruses. 2021;13:4.
34. Klein KA, Jackson WT. Human rhinovirus 2 induces the autophagic pathway and replicates more efficiently in autophagic cells. J Virol. 2011;85(18):9651–4.
35. Kumar M, Blaas D. Human rhinovirus subviral a particle binds to lipid membranes over a twofold axis of icosahedral symmetry. J Virol. 2013;87(20):11309–12.
36. Lamborn IT, Su HC. Genetic determinants of host immunity against human rhinovirus infections. Hum Genet. 2020;139(6-7):949–59.
37. Lee WM, et al. Infectivity assays of human rhinovirus-A and -B serotypes. Methods Mol Biol. 2015;1221:71–81.

38. Lee WM, et al. Human rhinovirus species and season of infection determine illness severity. Am J Respir Crit Care Med. 2012;186(9):886–91.
39. Lewis-Rogers N, Seger J, Adler FR. Human rhinovirus diversity and evolution: how strange the change from major to minor. J Virol. 2017;91:7.
40. Linder JE, et al. Human rhinovirus C: age, season, and lower respiratory illness over the past 3 decades. J Allergy Clin Immunol. 2013;131(1):69–77.e1-6
41. Ling H, et al. Structural view of the 2A protease from human rhinovirus C15. Acta Crystallogr F Struct Biol Commun. 2018;74(Pt 4):255–61.
42. Loeffelholz MJ, et al. Duration of rhinovirus shedding in the upper respiratory tract in the first year of life. Pediatrics. 2014;134(6):1144–50.
43. Lu X, et al. Rhinovirus viremia in patients hospitalized with community-acquired pneumonia. J Infect Dis. 2017;216(9):1104–11.
44. Luteijn RD, van Kuppeveld FJM. Rhinoviruses usurp STING for replication. Nat Microbiol. 2022;7(5):605–6.
45. Mastromarino P, et al. Resveratrol inhibits rhinovirus replication and expression of inflammatory mediators in nasal epithelia. Antivir Res. 2015;123:15–21.
46. McIntyre CL, et al. Recombination in the evolution of human rhinovirus genomes. Arch Virol. 2013;158(7):1497–515.
47. McKinlay MA. Recent advances in the treatment of rhinovirus infections. Curr Opin Pharmacol. 2001;1(5):477–81.
48. Megremis S, et al. Rhinovirus species-specific antibodies differentially reflect clinical outcomes in health and asthma. Am J Respir Crit Care Med. 2018;198(12):1490–9.
49. Michi AN, Love ME, Proud D. Rhinovirus-induced modulation of epithelial phenotype: role in asthma. Viruses. 2020;12:11.
50. Miller EK, et al. Human rhinovirus C associated with wheezing in hospitalised children in the Middle East. J Clin Virol. 2009;46(1):85–9.
51. Miller EK, et al. Rhinovirus-associated hospitalizations in young children. J Infect Dis. 2007;195(6):773–81.
52. Monto AS, Bryan ER, Ohmit S. Rhinovirus infections in Tecumseh, Michigan: frequency of illness and number of serotypes. J Infect Dis. 1987;156(1):43–9.
53. Müller L, et al. Human rhinovirus types and association with respiratory symptoms during the first year of life. Pediatr Infect Dis J. 2015;34(8):907–9.
54. Murer L, et al. Chemical evolution of rhinovirus identifies capsid-destabilizing mutations driving low-pH-independent genome Uncoating. J Virol. 2022;96(2):e0106021.
55. Myatt TA, et al. Airborne rhinovirus detection and effect of ultraviolet irradiation on detection by a semi-nested RT-PCR assay. BMC Public Health. 2003;3:5.
56. Oliveira MA, et al. The structure of human rhinovirus 16. Structure. 1993;1(1):51–68.
57. Oluwasemowo OO, et al. Genotypes of rhinovirus detected among children in two communities of South-West Nigeria. Virus Genes. 2021;57(3):276–9.
58. Palmenberg AC. Rhinovirus C, asthma, and cell surface expression of virus receptor CDHR3. J Virol. 2017;91:7.
59. Papadopoulos NG, et al. Rhinovirus identification by BglI digestion of picornavirus RT-PCR amplicons. J Virol Methods. 1999;80(2):179–85.
60. Papadopoulos NG, et al. Mechanisms of rhinovirus-induced asthma. Paediatr Respir Rev. 2004;5(3):255–60.
61. Papi A, Johnston SL. Rhinovirus infection induces expression of its own receptor intercellular adhesion molecule 1 (ICAM-1) via increased NF-kappaB-mediated transcription. J Biol Chem. 1999;274(14):9707–20.
62. Patick AK. Rhinovirus chemotherapy. Antivir Res. 2006;71(2-3):391–6.
63. Paula NT, et al. Human rhinovirus in the lower respiratory tract infections of young children and the possible involvement of a secondary respiratory viral agent. Mem Inst Oswaldo Cruz. 2011;106(3):316–21.

64. Peltola V, et al. Rhinovirus infections in children: a retrospective and prospective hospital-based study. J Med Virol. 2009;81(10):1831–8.
65. Peltola V, et al. Clinical effects of rhinovirus infections. J Clin Virol. 2008;43(4):411–4.
66. Peltola V, et al. Rhinovirus transmission within families with children: incidence of symptomatic and asymptomatic infections. J Infect Dis. 2008;197(3):382–9.
67. Privolizzi R, et al. The application of prophylactic antibodies for rhinovirus infections. Antivir Chem Chemother. 2014;23(5):173–7.

Epigenetics of SARS-CoV2 (COVID-19) 12

Sarita Rawat, Ritu Gilhotra, Santosh Kumar Singh, Asif Ahmad Bhat, Abhijeet Ojha, Karuna Dhaundhiyal, Ishwar Singh Dhramshaktu, and Gaurav Gupta

Abstract

The study of genetic differences in gene activity without alterations to DNA sequences is referred to as "epigenetics." The focus of this extremely new field of research is on the genetic and non-genetic elements of genetic phenotypic variations. The identification of genetic risk factors for the COVID-19 pandemic is becoming increasingly crucial due to its role as the source of the COVID-19 outbreak. Since the COVID-19 pass in cells of host through the ACE2 receptor, it is suspected that the ACE2 gene serves as a gene-related risk factor for COVID-19 infections. Furin and transmembrane protease serine 2 (TMPRSS2) have a huge effect on illness severity. This chapter compiled the most recent studies showing the mutations, expression, and epigenetic modifiers that may affect a person's vulnerability to COVID-19 disease and infection prognosis.

S. Rawat
Amrapali Group of Institute, Haldwani, Nanital, Uttarakhand, India

School of Pharmacy, Suresh Gyan vihar University, Jaipur, Rajasthan, India

R. Gilhotra · S. K. Singh · A. A. Bhat · G. Gupta (✉)
School of Pharmacy, Suresh Gyan vihar University, Jaipur, Rajasthan, India

A. Ojha · K. Dhaundhiyal
Amrapali Group of Institute, Haldwani, Nanital, Uttarakhand, India

I. S. Dhramshaktu
Dr. Sushila Tiwari Medical college and Hospital, Haldwani, Nainital, Uttarakhand, India

G. Gupta et al. (eds.), *Targeting Epigenetics in Inflammatory Lung Diseases*,
https://doi.org/10.1007/978-981-99-4780-5_12

12.1 Introduction

Coronaviruses are a distinct category of human and animal diseases that include types such as the SARS-CoV and MERS-CoV. The extremely infectious coronavirus that caused the COVID-2019 epidemic in December 2019, in Wuhan, China, was also the cause of 543,902 mortality globally as of 8 July 2020, according to the WHO [1]. The family Coronaviridae in the Nidovirales order is where COVID are classified. They are tiny 65–125 nanometer in diameter and made of single-stranded ribonucleic acid (RNA) that is between 26 and 32 kb long [2]. The coronavirus family is divided into four subgroups: α, β, γ, and Δ coronaviruses. The coronavirus genus includes COVID-19. CoVs are enclosed positive-stranded ribonucleic acid viruses with a genome that is nearly 30 kb in size. Coronaviruses are single-stranded RNA viruses with an envelope that can infect humans and animals and cause respiratory, gastrointestinal, and cardiovascular disorders [3]. The viral genome's 5′ section contains the ORF1a and ORF1b coding. The 1a and 1b polyproteins interpret them. Viral and cellular proteases break these polypeptides into a group of non-structural proteins (Nsp). Coronaviruses in general contain the membrane (M), envelope (E), nucleocapsid (N), and structural proteins spike (S) [4]. Through the S1 receptor-binding domain and the S2 subunit, the spike protein, a glycoprotein, oversees virus adherence to the receptor and integration with the cell membrane. In order to create the ribonucleoprotein, the N protein binds with the viral RNA during genome replication [5]. The membrane-bound ACE2 receptor is used by the COVID-19 to bind and infect the host cells. After the virus adheres to the receptor and penetrates the cell, viral replicases aid in ribonucleic acid replication, capping, and viral particle synthesis. Since on host cells the ACE2 receptor acts as a point of entry for the virus, it is assumed that the ACE2 gene represents a hereditary risk factor for COVID-19 infection [6]. In this context, ACE2 variations that have earlier been linked to conditions including hypertension and other cardiovascular diseases offer potential possibilities for genetic factors. A person's vulnerability to COVID-19 may also be considerably influenced by variations in ACE2 expression and related epigenetic variables [4, 7]. It is known that the extremely varied DNA methylation signatures on the promoter of the angiotensin-converting enzyme 2 gene result in varied ACE2 expression in epithelial cells, which may be the main factor influencing the diverse levels of infection in different people. The term "epigenetics" relates to heritable variations in gene activity that do not need a modification in deoxyribonucleic acid sequence [8]. The biochemical modifications of histones and the methylation of cytosine in the DNA sequence are two essential mechanisms in epigenetics that play a significant role in gene regulation and differentiation. Epigenetic variations that regulate chromatin conformation have a considerable impact on genomic stability and cellular homoeostasis, and these modifications have been associated to the pathophysiology of virus infection [9, 10].

12.2 Variation in the Genes Affecting SARS-CoV-2 Entrance Genetically and Epigenetically

12.2.1 SARS-CoV-2 Entry Mechanism-Related Genes

12.2.1.1 Angiotensin-Converting Enzyme 2 (ACE2) Receptor

The ACE2 receptor located on pulmonary epithelial cells allows the highly fatal severe COVID-19 to enter the human system more easily. In lungs ACE2 levels are increased when compared to other organs, lung squamous cell carcinoma (LUSC) and lung adenocarcinoma (LUAD) exhibit much higher levels of ACE2 activity [11]. ACE2 receptors were discovered on a variety of tissues, including enterocytes of the small intestine, alveolar cells, arterial smooth muscle cells, and venous and arterial endothelial cells. Subgenomic RNAs that code for the spike protein (S), an envelope protein (E), a membrane protein (M), and a nucleocapsid protein were generated from the genomic RNA once the virus entered the host cell (N) [12]. A functional receptor for the human coronaviruses SARS and SARS-CoV-2 was generated by this protein. While the SARS-CoV-1 and SARS-CoV-2 viruses have identical attachment methods and enter epithelial cells through the same ACE2 receptor, the severity of the diseases and mortality rate vary among SARS-CoV-2 infected people [13]. It is hypothesized that the existence of furin-like protease cleavage sites in the receptor binding domain accounts for SARS-higher CoV-2's infectivity than other SARS viruses [14]. While the SARS-CoV-1 and SARS-CoV-2 viruses have identical attachment methods and enter epithelial cells through the same ACE2 receptor, the severity of the diseases and mortality rate vary among SARS-CoV-2 infected people [15].

12.2.1.2 Transmembrane Protease Serine 2 [TMPRSS2]

The TMPRSS2 gene, which is found on chromosome 21, encodes a member of the serine protease family and is essential for viral entry. TMPRSS2 plays an essential role in stimulating the S protein of SARS-CoV2 and its enzymatic degradation in order to infect target cell [16]. The TMPRSS2 protein is produced by the globally maintained TMPRSS2 gene. Since during infection, this protease cut and activated viral envelope glycoproteins, the probability of developing chronic COVID-19 was raised by TMPRSS2 polymorphisms [17]. It was postulated that because TMPRSS2 is an androgen-responsive serine protease, males may express more of it in their lungs, which aids in the pathogenesis of SARS-CoV2 and accounts for their greater fatality rate [18]. In humans, TMPRSS2 expression is modulated during development and has a linear relationship with age. Various analyses have been taken up to examine the role of these proteins in the creation of infection and severity in the case of coronavirus [19].

FURIN

FURIN is important for SARS-CoV-2 access into human airway cells and proteolytic activation., and the FURIN gene was found to be placed on chromosome 15q26.1. For SARS-CoV-2 infection and disease, the spike protein's FURIN degradation site is crucial [20]. This FURIN breakage site contains a special insertion that

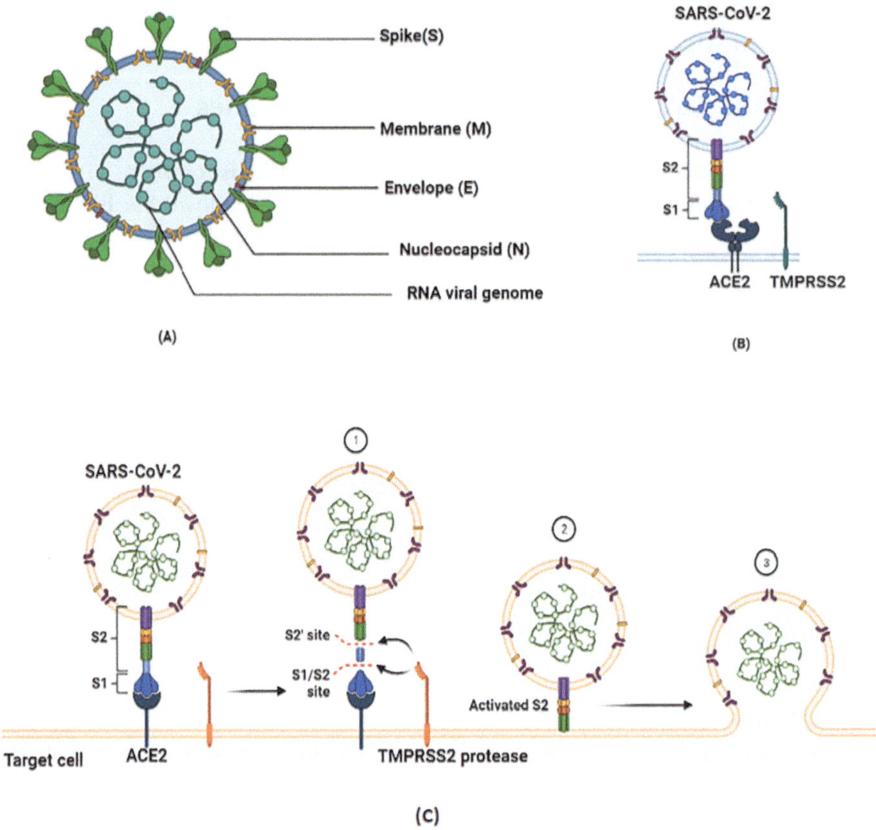

Fig. 12.1 (**a**) Human coronavirus structure. (**b**) Severe acute respiratory syndrome coronavirus-2 binding to ACE-2 receptor. (**c**) Mechanism of COVID-19 viral entry (1—Cleavage of SARS-CoV-2 S protein; 2—Activation of S2 Domain; 3—Fusion of viral and host membranes)

allows it to only cleave the spike protein of the SARS-CoV-2 coronavirus and not the closely related SARS-CoV-1 [21]. FURIN suppression has been identified as a method to reduce SARS-CoV-2 infection as a result of this information. Six inherited variants of the FURIN gene were discovered in a small cohort of Covid-19, but none of them were connected to the disease. Thirteen nonsense variants for the FURIN gene have been found that may relate to virus sensitivity, with seven of them shown to reduce the possibility of SARS-CoV-2 development in individuals [15, 22, 23] (Fig. 12.1).

12.3 Epigenetic Control of the Cytokines Storm in SARS-CoV-2

A "cytokine storm," or an increased production of proinflammatory proteins that causes increased systemic inflammation, is a common feature of serious coronavirus. If the immune system's abnormal or increased response is left uncontrolled, it might result in serious disease [24]. The severe inflammatory response called as cytokine storm, often referred as CSS, which is defined by the production of a considerable amount of proinflammatory cytokines, is one of the major suspected causes of mortality linked to coronavirus [25]. Tumor necrosis factor (TNF) and interferon (IFN) act together to create the cytokine storm that COVID-19 uses to organize the three different cell death pathways—pyroptosis, apoptosis, and necroptosis, collectively termed as PANoptosis [26]. Additionally, earlier research revealed that the DNA methylation of the gene's promoter controls the interleukin (IL)-6 expression, another significant member in the "cytokine storm" that affects the extremely serious patients diagnosed with COVID-19. It has also been discovered that oxidative stress caused by COVID-19 infection, might block the functioning of DNA methyltransferase (DNMT)1, increasing the methylation abnormalities of DNA [27, 28].

12.4 Severe Acute Respiratory Syndrome Coronavirus-2 Infection and Genetic Modifications

In order to maintain cellular homeostasis and genome integrity during viral infections, epigenetic processes are essential. By modifying the chromatin structure, epigenetic modifications function as immune system silent modulators that can affect the strength of the host response as well as the synthesis of chemical mediators of inflammation [29]. According to several studies, the results of host-pathogen interactions during COVID-19 infection are determined by a variety of epigenetic mechanisms, including non-coding RNA-mediated regulatory events, histone modifications, and DNA methylation [30].

12.4.1 DNA Methylation

One of the vital epigenetic changes that alter the chromatin structure is DNA methylation. DNA methylation is an inherited underlying genetic process in which DNMTs repair DNA at the C-5 position of the cytosine ring [31]. DNMT govern methylation of DNA. Generally, methylation of DNA is related to transcriptional gene repression, whereas DNA demethylation is linked to transcriptional gene activation [32]. In genetic alterations, CpG sites or the CpG islands are the main targets for DNA methylation [33]. CpG islands are linked to the proximal regulators of many genes in the human genome. DNA methyltransferases (MTase), which methylate CpG islands, affect chromatin condensation by forming methyl-binding proteins. CpG-island methylation in promoter areas is therefore generally linked to gene suppression [34].

12.4.2 Histone Modification

Core histone proteins are subject to significant acetylation, methylation, phosphorylation, and ubiquitination changes that either directly or indirectly affect chromatin structure [35]. The numerous possible histone alterations add an additional layer of complexity to the precise control of chromatin structure [36]. However, a potent epigenetic mechanism that can change chromatin structure is post-translational modification of histone proteins. It has been reported that H3K4me3 and H3K9me2 are the primary controllers of histone modification that could enhance IFN production in response to COVID-19 infection [37]. IL-6 is other cytokine whose production during COVID-19 may be influenced by histone modification and cis/trans-factor linkages. Patients infected with COVID-19 have enhanced levels of positive histone modification indicators in the regions of genes producing ACE2 and IL-6 [38].

12.4.3 Non-Coding RNAs

Non-coding RNAs (ribonucleic acid) are RNA molecules with important functional properties, but often lack in functional protein coding regions. Non-coding RNAs exist in a variety of forms, including transfer RNAs, ribosomal RNAs, microRNAs, and short interfering RNAs (siRNAs) [39]. By interacting with DNA, other RNAs, and proteins, non-coding RNAs have the potential to serve as epigenetic regulators. MiRNAs may have significantly inhibiting effects on cytokine storm, as non-coding miRNAs are engaged in immune system cytokine regulation [40]. As such, miRNAs are unique and developing targets for therapeutic treatment against SARS-CoV-2. Yet, miRNAs have been demonstrated to affect inflammation in the same way that lncRNAs do [41]. MiRNAs that induce inflammation include miR-1307, miR-421, miR-155, miR-106a, miR-15b, and miR-20a, while miRNAs that decrease inflammation include miR-223, miR-181, miR-146a, miR-124, miR-24, and miR-10a [42]. According to study, miR-1307 has the strongest binding for the COVID-19 genome [43]. This pathogenic miRNA has been linked to cause inflammation and has also been linked to a poor life span among patients with lung cancer. According to an in-silico study, the miR-29 family of miRNAs showed the highest binding affinity (11 locations) on the COVID-19 genome [44, 45]. A study conducted with the help of theoretical technique reported that interactions between miRNAs and seven human coronavirus RNAs, the most likely miRNA to connect to human coronavirus RNAs is miR-21, which may result in massively increased production in human lungs after infected with COVID-19 infection [46, 47].

12.5 Nutritional Epigenetics and Prognosis of the COVID-19 Illness

It was shown that SARS-CoV-2 affected people with weak immune systems and preexisting severe conditions had a higher mortality rate. In co-morbid situations, complementing nutritious meals and taking the right amounts of vitamins might result in more successful and individualized therapy [48, 49]. Diet and nutrition are a major factor in the chronic modifications in methylation of DNA sequences on chromosomes which affect age and health-related diseases. The misregulation of numerous epigenetic enzymes in cells, including DNA methyltransferases and histone acetyltransferases, by nutrients and their metabolites can lead to alterations in the transcription of crucial genes, further affecting general health and lifespan [50]. Folate is a well-known water-soluble vitamin, and studies have shown that its digestion and epigenetic modifications are closely related [51]. Additionally, it has been noted that vitamin deficiencies affect the immune system, which enhances death in people with chronic illnesses. The difference between dietary supplements and overweight has a significant impact on the intensity of COVID-19 [52]. These variables may have an impact on host epigenetic factors either directly or indirectly, and they may also deregulate gene expression. Therefore, a balanced dietary supplement is crucial for helping people with chronic conditions to control the COVID-19-related problems [53].

12.6 Conclusion

The COVID-19 pandemic is one of the most serious risks to the global health in the twenty-first century. The so-called cytokine storm carried on by the virus causes broad tissue damage and multi-organ failure as well as mortality. Epigenetic modifications affect a wide range of healthy and disease-related processes, such as cancer, epigenetic disorders, obesity, and viral infections. COVID-19 is notably more communicable than MERS-CoV and SARS-CoV-1 viruses. SARS-CoV-2 COVID-19 infected people have a lower mortality rate than MERS-CoV virus; however, co-morbid COVID-19 infected people have a greater mortality rate. Studies have shown that COVID-19 changes the host's genetic sequence in a variety of ways. There could be several causes for this broad range of medical symptoms, one of which being the processes linked with the entrance of the COVID-19. ACE2, which is mostly located on pulmonary cells, is the most important receptor for the COVID-19 spike protein. The ACE2, TMPRSS2, and FURIN genes have undergone major genetic and epigenetic changes, resulting in a major impact on COVID-19 cell pass into the host and disease severity. According to this perspective, the reversible nature of epigenetic alterations offers hope for future research that could prevent the virus from being able to infect cells. Depending on worldwide information of COVID-19 and new information regarding ways to handle medical symptoms, we concluded that the Delta variant was the most serious phase of COVID-19, with significant death globally. However, information from the Omicron phase is insufficient to assess the intensity of this variation. The formation of a successful vaccination has been one of the most critical tasks confronting COVID-19. And over 20 vaccinations have been validated thus far. The data on the effectiveness of these vaccines varies by study; nevertheless, the overall efficacy of mRNA vaccines, notably the Pfizer-BioNTech product, appears to be higher when compared to others.

References

1. A blood atlas of COVID-19 defines hallmarks of disease severity and specificity. Cell. 2022;185 (5):916–38.e58
2. Ahmad S, et al. Epigenetic underpinnings of inflammation: connecting the dots between pulmonary diseases, lung cancer and COVID-19. Semin Cancer Biol. 2022;83:384–98.
3. Anand AV, et al. Medicinal plants, phytochemicals, and herbs to combat viral pathogens including SARS-CoV-2. Molecules. 2021;26:6.
4. Arnold CG, et al. Epigenetics may characterize asymptomatic COVID-19 infection. Hum Genomics. 2022;16(1):27.
5. Aschenbrenner AC, et al. Disease severity-specific neutrophil signatures in blood transcriptomes stratify COVID-19 patients. Genome Med. 2021;13(1):7.
6. Atlante S, et al. The epigenetic implication in coronavirus infection and therapy. Clin Epigenetics. 2020;12(1):156.
7. Balnis J, et al. Blood DNA methylation and COVID-19 outcomes. Clin Epigenetics. 2021;13 (1):118.
8. Beacon TH, Delcuve GP, Davie JR. Epigenetic regulation of ACE2, the receptor of the SARS-CoV-2 virus(1). Genome. 2021;64(4):386–99.
9. Berber B, et al. Gene editing and RNAi approaches for COVID-19 diagnostics and therapeutics. Gene Ther. 2021;28(6):290–305.
10. Bernardes JP, et al. Longitudinal multi-omics analyses identify responses of megakaryocytes, erythroid cells, and plasmablasts as hallmarks of severe COVID-19. Immunity. 2020;53(6): 1296–314.e9
11. Biering SB, et al. Genome-wide bidirectional CRISPR screens identify mucins as host factors modulating SARS-CoV-2 infection. Nat Genet. 2022;54(8):1078–89.
12. Bulka CM, Enggasser AE, Fry RC. Epigenetics at the intersection of COVID-19 risk and environmental chemical exposures. Curr Environ Health Rep. 2022;9(3):477–89.
13. Calabrò L, et al. COVID and lung cancer. Curr Oncol Rep. 2021;23(11):134.
14. Cao X, et al. Accelerated biological aging in COVID-19 patients. Nat Commun. 2022;13(1): 2135.
15. Castro de Moura M, et al. Epigenome-wide association study of COVID-19 severity with respiratory failure. EBioMedicine. 2021;66:103339.
16. Cevhertas L, et al. Advances and recent developments in asthma in 2020. Allergy. 2020;75(12): 3124–46.
17. Chen F, et al. The impact of ACE2 polymorphisms on COVID-19 disease: susceptibility, severity, and therapy. Front Cell Infect Microbiol. 2021;11:753721.
18. Chi WY, et al. COVID-19 vaccine update: vaccine effectiveness, SARS-CoV-2 variants, boosters, adverse effects, and immune correlates of protection. J Biomed Sci. 2022;29(1):82.
19. Chlamydas S, Papavassiliou AG, Piperi C. Epigenetic mechanisms regulating COVID-19 infection. Epigenetics. 2021;16(3):263–70.
20. Choreño-Parra JA, et al. Clinical and immunological factors that distinguish COVID-19 from pandemic influenza A(H1N1). Front Immunol. 2021;12:593595.
21. Chotirmall SH, et al. Update in COVID-19 2020. Am J Respir Crit Care Med. 2021;203(12): 1462–71.
22. Choudhary S, et al. Role of genetic variants and gene expression in the susceptibility and severity of COVID-19. Ann Lab Med. 2021;41(2):129–38.
23. Chowdhury NU, et al. Sex and gender in asthma. Eur Respir Rev. 2021;30:162.
24. Corley MJ, et al. Genome-wide DNA methylation profiling of peripheral blood reveals an epigenetic signature associated with severe COVID-19. J Leukoc Biol. 2021;110(1):21–6.
25. Degenhardt F, Ellinghaus D, Juzenas S, Lerga-Jaso J, Wendorff M, Maya-Miles D, Uellendahl-Werth F, ElAbd H, Rühlemann MC, Arora J, Özer O, Lenning OB, Myhre R, Vadla MS, Wacker EM, Wienbrandt L, Blandino Ortiz A, de Salazar A, Garrido Chercoles A, Palom A, Ruiz A, Garcia-Fernandez AE, Blanco-Grau A, Mantovani A, Zanella A, Holten AR, Mayer A, Bandera A, Cherubini A, Protti A, Aghemo A, Gerussi A, Ramirez A, Braun A, Nebel A,

Barreira A, Lleo A, Teles A, Kildal AB, Biondi A, Caballero-Garralda A, Ganna A, Gori A, Glück A, Lind A, Tanck A, Hinney A, Carreras Nolla A, Fracanzani AL, Peschuck A, Cavallero A, Dyrhol-Riise AM, Ruello A, Julià A, Muscatello A, Pesenti A, Voza A, Rando-Segura A, Solier A, Schmidt A, Cortes B, Mateos B, Nafria-Jimenez B, Schaefer B, Jensen B, Bellinghausen C, Maj C, Ferrando C, de la Horra C, Quereda C, Skurk C, Thibeault C, Scollo C, Herr C, Spinner CD, Gassner C, Lange C, Hu C, Paccapelo C, Lehmann C, Angelini C, Cappadona C, Azuure C, COVICAT Study Group, Aachen Study (COVAS), Bianco C, Cea C, Sancho C, Hoff DAL, Galimberti D, Prati D, Haschka D, Jiménez D, Pestaña D, Toapanta D, Muñiz-Diaz E, Azzolini E, Sandoval E, Binatti E, Scarpini E, Helbig ET, Casalone E, Urrechaga E, Paraboschi EM, Pontali E, Reverter E, Calderón EJ, Navas E, Solligård E, Contro E, Arana-Arri E, Aziz F, Garcia F, García Sánchez F, Ceriotti F, Martinelli-Boneschi F, Peyvandi F, Kurth F, Blasi F, Malvestiti F, Medrano FJ, Mesonero F, Rodriguez-Frias F, Hanses F, Müller F, Hemmrich-Stanisak G, Bellani G, Grasselli G, Pezzoli G, Costantino G, Albano G, Cardamone G, Bellelli G, Citerio G, Foti G, Lamorte G, Matullo G, Baselli G, Kurihara H, Neb H, My I, Kurth I, Hernández I, Pink I, de Rojas I, Galván-Femenia I, Holter JC, Afset JE, Heyckendorf J, Kässens J, Damås JK, Rybniker J, Altmüller J, Ampuero J, Martín J, Erdmann J, Banales JM, Badia JR, Dopazo J, Schneider J, Bergan J, Barretina J, Walter J, Hernández Quero J, Goikoetxea J, Delgado J, Guerrero JM, Fazaal J, Kraft J, Schröder J, Risnes K, Banasik K, Müller KE, Gaede KI, Garcia-Etxebarria K, Tonby K, Heggelund L, Izquierdo-Sanchez L, Bettini LR, Sumoy L, Sander LE, Lippert LJ, Terranova L, Nkambule L, Knopp L, Gustad LT, Garbarino L, Santoro L, Téllez L, Roade L, Ostadreza M, Intxausti M, Kogevinas M, Riveiro-Barciela M, Berger MM, Schaefer M, Niemi MEK, Gutiérrez-Stampa MA, Carrabba M, Figuera Basso ME, Valsecchi MG, Hernandez-Tejero M, Vehreschild MJGT, Manunta M, Acosta-Herrera M, D'Angiò M, Baldini M, Cazzaniga M, Grimsrud MM, Cornberg M, Nöthen MM, Marquié M, Castoldi M, Cordioli M, Cecconi M, D'Amato M, Augustin M, Tomasi M, Boada M, Dreher M, Seilmaier MJ, Joannidis M, Wittig M, Mazzocco M, Ciccarelli M, Rodríguez-Gandía M, Bocciolone M, Miozzo M, Imaz Ayo N, Blay N, Chueca N, Montano N, Braun N, Ludwig N, Marx N, Martínez N, Norwegian SARS-CoV-2 Study group, Cornely OA, Witzke O, Palmieri O, Pa Study Group, Faverio P, Preatoni P, Bonfanti P, Omodei P, Tentorio P, Castro P, Rodrigues PM, España PP, Hoffmann P, Rosenstiel P, Schommers P, Suwalski P, de Pablo R, Ferrer R, Bals R, Gualtierotti R, Gallego-Durán R, Nieto R, Carpani R, Morilla R, Badalamenti S, Haider S, Ciesek S, May S, Bombace S, Marsal S, Pigazzini S, Klein S, Pelusi S, Wilfling S, Bosari S, Volland S, Brunak S, Raychaudhuri S, Schreiber S, Heilmann-Heimbach S, Aliberti S, Ripke S, Dudman S, Wesse T, Zheng T, STORM Study Group, The Humanitas Task Force, The Humanitas Gavazzeni Task Force, Bahmer T, Eggermann T, Illig T, Brenner T, Pumarola T, Feldt T, Folseraas T, Gonzalez Cejudo T, Landmesser U, Protzer U, Hehr U, Rimoldi V, Monzani V, Skogen V, Keitel V, Kopfnagel V, Friaza V, Andrade V, Moreno V, Albrecht W, Peter W, Poller W, Farre X, Yi X, Wang X, Khodamoradi Y, Karadeniz Z, Latiano A, Goerg S, Bacher P, Koehler P, Tran F, Zoller H, Schulte EC, Heidecker B, Ludwig KU, Fernández J, Romero-Gómez M, Albillos A, Invernizzi P, Buti M, Duga S, Bujanda L, Hov JR, Lenz TL, Asselta R, de Cid R, Valenti L, Karlsen TH, Cáceres M, Franke A. Detailed stratified GWAS analysis for severe COVID-19 in four European populations. Hum Mol Genet. 2022;31(23):3945–66. https://doi.org/10.1093/hmg/ddac158. PMCID: PMC9703941

26. Doerfler W. Adenoviral vector DNA- and SARS-CoV-2 mRNA-based Covid-19 vaccines: possible integration into the human genome—are adenoviral genes expressed in vector-based vaccines? Virus Res. 2021;302:198466.
27. Dubey H, et al. SARS-CoV-2 (COVID-19) as a possible risk factor for neurodevelopmental disorders. Front Neurosci. 2022;16:1021721.
28. Evangelou K, et al. Pulmonary infection by SARS-CoV-2 induces senescence accompanied by an inflammatory phenotype in severe COVID-19: possible implications for viral mutagenesis. Eur Respir J. 2022;60:2.
29. Fang Y, et al. Natural products as LSD1 inhibitors for cancer therapy. Acta Pharm Sin B. 2020;11(3):621–31.

30. Foolchand A, et al. A review: highlighting the links between epigenetics, COVID-19 infection, and vitamin D. Int J Mol Sci. 2022;23:20.
31. Garcia-Gallo E, et al. ISARIC-COVID-19 dataset: a prospective, standardized, global dataset of patients hospitalized with COVID-19. Sci Data. 2022;9(1):454.
32. Georg P, et al. Complement activation induces excessive T cell cytotoxicity in severe COVID-19. Cell. 2022;185(3):493–512.e25
33. Gonzalez-Perez M, et al. The BCG vaccine for COVID-19: first verdict and future directions. Front Immunol. 2021;12:632478.
34. Grant RA, et al. Circuits between infected macrophages and T cells in SARS-CoV-2 pneumonia. Nature. 2021;590(7847):635–41.
35. Hastie KM, et al. Defining variant-resistant epitopes targeted by SARS-CoV-2 antibodies: a global consortium study. Science. 2021;374(6566):472–8.
36. He X, et al. COVID-19 induces new-onset insulin resistance and lipid metabolic dysregulation via regulation of secreted metabolic factors. Signal Transduct Target Ther. 2021;6(1):427.
37. Ho JSY, et al. TOP1 inhibition therapy protects against SARS-CoV-2-induced lethal inflammation. Cell. 2021;184(10):2618–32.e17
38. Kaneko S, et al. Epigenetic mechanisms underlying COVID-19 pathogenesis. Biomedicine. 2021;9:9.
39. Kee J, et al. SARS-CoV-2 disrupts host epigenetic regulation via histone mimicry. Nature. 2022;610(7931):381–8.
40. Kgatle MM, et al. COVID-19 is a multi-organ aggressor: epigenetic and clinical Marks. Front Immunol. 2021;12:752380.
41. Knoll R, Schultze JL, Schulte-Schrepping J. Monocytes and macrophages in COVID-19. Front Immunol. 2021;12:720109.
42. Kotfis K, et al. COVID-19-the potential beneficial therapeutic effects of spironolactone during SARS-CoV-2 infection. Pharmaceuticals (Basel). 2021;14:1.
43. Krämer B, et al. Early IFN-α signatures and persistent dysfunction are distinguishing features of NK cells in severe COVID-19. Immunity. 2021;54(11):2650–69.e14
44. Li W, et al. SARS-CoV-2 RNA elements share human sequence identity and upregulate hyaluronan via NamiRNA-enhancer network. EBioMedicine. 2022;76:103861.
45. Li Z, et al. Imatinib and methazolamide ameliorate COVID-19-induced metabolic complications via elevating ACE2 enzymatic activity and inhibiting viral entry. Cell Metab. 2022;34(3):424–440.e7.
46. Liu Y, Sawalha AH, Lu Q. COVID-19 and autoimmune diseases. Curr Opin Rheumatol. 2021;33(2):155–62.
47. Losso JN. The potential of dietary bioactive compounds against SARS-CoV-2 and COVID-19-induced endothelial dysfunction. Molecules. 2022;27:5.
48. Lv T, et al. Defense of COVID-19 by human organoids. Phenomics. 2021;1(3):113–28.
49. Mantovani A, Netea MG. Trained innate immunity, epigenetics, and Covid-19. N Engl J Med. 2020;383(11):1078–80.
50. Mauvais-Jarvis F, et al. Sex and gender: modifiers of health, disease, and medicine. Lancet. 2020;396(10250):565–82.
51. Mongelli A, et al. Evidence for biological age acceleration and telomere shortening in COVID-19 survivors. Int J Mol Sci. 2021;22:11.
52. Mueller AL, McNamara MS, Sinclair DA. Why does COVID-19 disproportionately affect older people? Aging (Albany NY). 2020;12(10):9959–81.
53. Muhammad JS, Siddiqui R, Khan NA. COVID-19: does SARS-CoV-2 modulate Acanthamoeba epigenetics to enhance survival and transmission in the environment? ACS Pharmacol Transl Sci. 2021;4(2):1021–3.

Epigenetics of *Haemophilus influenzae* 13

Nitin Verma, Gagandeep Kaur, Komal Thapa, Neha Kanojia, Lata Rani, Parul Sood, and Kamal Dua

Abstract

Pathogenic bacterial strains that have adapted to solely infect humans need to go through a constant cycle of host changing, colonization, and transmission. Any population incapable of doing this might be experiencing an evolutionary dead end. An enormous amount of pressure has been placed on the creation of extraordinarily effective methods to circumvent the human innate and adaptive immunity by the ongoing selection process that occurs across several distinct cycles of transmission via various human hosts. Bacteria that have evolved to live in human surroundings as pathogens frequently exhibit the phase variation of virulence-linked genes. While homologous recombination is one possible method for this kind of random, high-frequency switching of gene expression, simple tandem DNA repeating sequences are also capable of mediating phase variation. Because of the innate instability of these DNA repeats, individuals within the bacterial population showed diversity in the expressing virulence-associated genes as a result of the regular loss or gain of repeated units in the promoter regions or the open reading frames of these genes. A population of bacteria with a diverse range of cells that have already reacted to various environmental, intra-, and inter-host constraints can be produced using a very effective contingency method called phase variation. More studies have been done on restriction modification systems with phase-variable expression. Global modifications in DNA methylation result from phase variation in these systems. Multiple

N. Verma (✉) · G. Kaur · K. Thapa · N. Kanojia · L. Rani · P. Sood
Chitkara University School of Pharmacy, Chitkara University, Baddi, India
e-mail: nitin.verma@chitkarauniversity.edu.in

K. Dua
Discipline of Pharmacy, Graduate School of Health, University of Technology Sydney, Ultimo, NSW, Australia

G. Gupta et al. (eds.), *Targeting Epigenetics in Inflammatory Lung Diseases*, https://doi.org/10.1007/978-981-99-4780-5_13

human-adapted bacterial diseases have Type III restriction-modification systems, which show how global alterations in methylation affect numerous gene expressions. Phasevarion switching aids infections in evading the human immune system by altering virulence features. The most prominent human *Haemophilus influenzae* bacterial infection is examined in this chapter with regard to the significance of phasevarions in pathogenesis and immunological resistance.

Keywords

Haemophilus influenzae · Phase varion · Immune system · DNA methylation

13.1 Introduction

Throughout the bacterial kingdom, restriction-modification (R-M) systems play a significant part in defending bacterial cells from DNA invasion. Basically, a R-M system is constituted by two of the enzymes that are restriction endonuclease and methyltransferase. Self-DNA is recognized by the methyltransferase, which also facilitates the transfer of methyl group at a particular DNA recognition region, resulting in resistance to cleavage [1, 2]. Moreover, restriction endonuclease enzyme splits into two strands of DNA that have not been methylated [3, 4]. A population of bacteria that have previously adapted to various environmental variables both within the host and between various individuals of the host can be created using the extremely efficient contingency approach known as phase variation [5]. Switching expression commonly involves genes that encode surface-expressed virulence markers and enables the emergence of a population of cells with a variety of phenotypes, among which some are better suited for survival. A variety of theories have been advanced regarding the potential of Phase variable type III R-M frameworks but none of them have received experimental support [6–8].

R-M systems have historically been assumed to offer defense against foreign DNA or bacteriophages obtained through spontaneous transformation. In this situation, phase variation might make it possible to temporarily remove this barrier, enabling the attainment of foreign DNA molecules with potential benefits [9]. However, a gene encoding a DNA methyltransferase that is localized in the cytoplasm was shown to include simple tandem DNA repeats in 2005. A regulon of genes undergoes phase variation as a result of the randomized switching of this gene [10], thus dividing the bacterial cell into two halves of alternate expression owing to a wide range of phenotypic changes. Phase variations are frequently used by bacteria as a backup plan, as shown by recent analysis of accessible genomes, which identified evidence of potential phase variations in both Types I and III restriction-modification processes in an extensive range of pathogen and non-pathogen species of bacteria [11, 12]. Heterogeneous methylation of the genomic series is one potential mechanism through which phase variable methyltransferases are employed to control genes [13]. There were no examples where the methyltransferase itself had a phase-varying expression pattern, though. DNA methylation has been found to exhibit to have an impact on gene expressing phenomenona in a number of systems in the past such as Dam methylation [14].

13.2 Bacterial DNA Methyltransferases

DNA methylation, which is accomplished by DNA methyltransferases, is one of the epigenetic regulatory mechanisms in bacteria that has received the most attention from researchers. Methyl groups are transferred from S-adenosyl-L-methionine to adenosine and cytosine bases with the help of these enzymes [15]. Both prokaryotic and eukaryotic categories undergo methylation process of adenine and cytosine bases, though cytosine methylation has been found to have a stronger effect on the regulation of gene expression in eukaryotes whereas methylation of adenine seems to have a higher regulatory role in bacteria [16]. The purpose of bacterial restriction-modification (R-M) systems is to provide defense against phage DNA, which led to the initial investigation of DNA methyltransferases of bacteria and related restriction endonucleases [17, 18].

13.2.1 Types of DNA Methyltransferases

The bulk of restriction-modification systems is connected to bacterial DNA methyltransferases. Systems for modifying restrictions serve two primary purposes: DNA cleavage caused by an endonuclease with a specified target sequence [19], and amendment at a particular region of DNA by a DNA methyltransferase that causes the inhibition of endonuclease activity to prevent the degradation of endogenous self-replicated DNA [20]. *Escherichia coli* bacteria got identified Deoxyadenosine methyltransferase (Dam) as the first orphan DNA. Adenine methylation by *E. coli* was demonstrated. Normally, both DNA strands of the *E. coli* chromosome are fully methylated, with the exception of the brief period following DNA replication [21, 22]. Because DNA replication is semi-conservative, hemimethylated DNA is produced when one DNA strand is methylated but not the other. Both DNA strands are unmethylated in dam mutants [23]. *Escherichia coli* cells without the dam gene were discovered to be alive but had a higher rate of spontaneous mutations, pointing to a function for methylation in DNA repair [24]. Furthermore, it was discovered that dam mutants lacked coordination during the initiation of chromosomal replication. Other methyltransferases separate from DNA after one catalytic cycle [25–27].

Restriction-modification processes function as a nucleic acid phase immune system to defend bacterium toward phage. Type 1 and Type 3 processes are the main processes or systems significant to phasevarions. Restriction (HsdR), Specificity (HsdS) and modification (HsdM), constituents are included in Type 1 restriction-modification processes [28]. The restriction endonuclease as well as methyltransferase, correspondingly are composed of the hsdM and hsdR genes. Every hsdS gene has 2 half-target recognition areas, every portion of which constitutes one-half of the coded HsdS proteins, particularly with specific methyl structure [29]. R2M2S is a pentamer composed of HsdM, HsdR, and HsdS, required for useful restriction activity whereas M2S is a trimer works as a standalone methyltransferase [30–32]. DNA methyltransferase and restriction endonuclease (Res) are the two main constituents of type 3 process (Mod). DNA can be methylated by Mod except for the utilization of Res. To break DNA at the target point, a tetramer complex of Mod and Res (R2M2) is required [33]. Solely the Mod subunit works as a DNA recognition constituent in Type 1 process [34].

13.2.2 Processes of Phase Variation in DNA Methyltransferases

In the bacterial restriction-modification process, phase variation has been seen to begin through two different mechanisms [35]. The larger part of Type 1 processes have been proved to phase variate by homologous recombination. This process advances swapping among different target sites of methylation instead of intervening ON/OFF changing of methylation [36]. The damage and benefit of a repeat unit in Type 1 process having direct progression repeats might effects frameshift mutation that makes the gene ON or OFF. All Type 3 systems show phase variation via the occurrence of direct sequence chains [37] (Fig. 13.1).

13.3 Phase-Variable Type III Restriction-Modification Process

H. influenzae of non-typeable strain (NTHi) is the main causative microorganism of some pulmonary-related disorders including middle ear infection [38], exacerbations of chronic obstructive pulmonary disease [39], and community-acquired pneumonia [40]. NTHi-related cancer rates have skyrocketed since the introduction of a vaccine against *H. influenzae* serotype b [41].

In a post-genomic examination of the primary gene of a free-living bacterium, ModA is among one of the earliest instances of a phasevarion discovery that occurred in a strain of *H. influenzae* Rd. [42]. From this point of view, the DNA methyltransferase was a part kind 3 process and consists of the gene's coding sequence having 5-AGCC-3 chains. Sixteen genes, comprising heat shock proteins, were differentially regulated as a result of ModA1 methyltransferase phase variation [43]. Although *H. influenzae* strain Rd. had a significant role in the development of the rationale of phasevarions, NTHi now accounts for the majority of individual sickness. Additionally, other investigations demonstrated that NTHi has a broad variety of phase variations [44]. Reports demonstrated that the innermost portion of ModA which comprises the target identification area is extremely uneven among strains having almost 21 discrete methylation target progressions (5, 22). In an investigation, one of the co-workers reported five main commonly established ModA alleles such as Mod A2, 4, 5, 9, and 10 depicting almost two-thirds of each NTHi therapeutic segregates investigated, single real-time (SMRT) steps and methylome study showed that the diverse active identification domains definitely undergoes methylation of varied chains [45].

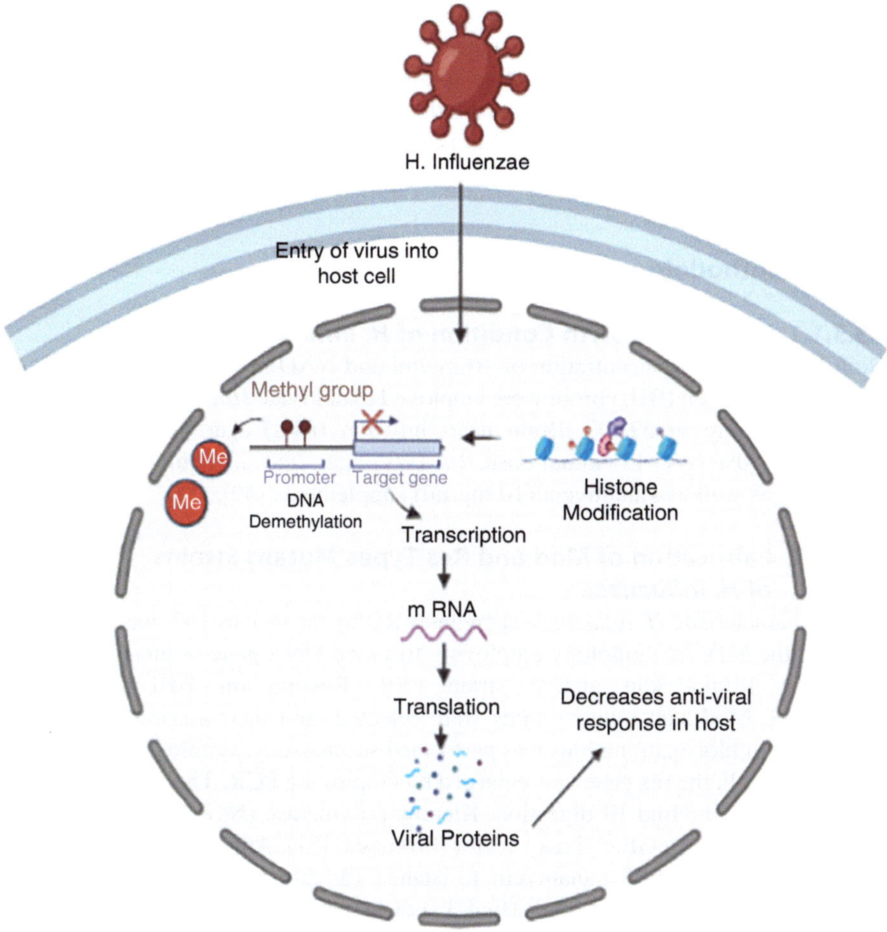

Fig. 13.1 Representation of epigenetic mechanisms of influenza H virus infection. Virus induces the modification of histones and demethylation of host DNA that results in the suppression of antiviral response in the host

Consequently, every chromosome methylates a specific DNA target-based sequence, an investigation of the expression profiles of ModA activated and inactivated strains that shows the identical allele set of allele demonstrated that every strain regulates a specific phasevarion of genes [46]. Varied resistance patterns of resistence to the drugs such as erythromycin, ampicillin, and gentamicin were demonstrated by the ModA2, ModA5, and ModA10 alleles, correspondingly [47]. High-molecular weight external membrane protein shown to exhibit phasevarion-dependent expression in ModA4 types. The specific resilience of ModA4-activated strains which showed lesser concentrations of HMW was exhibited by opsonophagocytic killing experiments utilizing HMW-specific serum, which is adaptable through ModA4 advancing exit from an adaptive immune response. ModA2 also regulated film-forming capacity and tolerance to oxidative stress, in addition to antibiotic resistance. The chinchilla OM model was found to prefer the activated over the inactive termination of modA2 in in vivo studies. Together, these findings indicate that ModA phasevarions are two-stage epigenetic alternatives with regulatory potential in immune function, pathogenicity, and niche adaptability in vivo [48].

13.3.1 Methodology

13.3.1.1 Bacterial Growth Condition of *H. influenzae*

Both Hemin in the concentration of 10 mg/ml and NAD in 2 mg/ml-supplemented brain heart infusion (BHI) broth were employed to cultivate *Haemophilus influenzae* at a temperature of 37 °C. Brain heart infusion (BHI) agar plates were made employing 10% (v/v) Levinthal base, 1% (v/v) agar, and, if required, tetracycline (5 mg/ml) as well as kanamycin (10 mg/ml) supplements [49].

13.3.1.2 Fabrication of Mod and Res Types Mutant Strains of *H. influenzae*

The uncharacteristic *H. influenzae* segregates R2866, as well as 162, were converted utilizing the MIV methodology employing trimmed DNA gene sequences obtained from the Rdmod::kan altered strain [50]. Resting on BHI plates with kanamycin, Mod::kan transformants were selected, and their assessment by PCR and Southern blot examinations was performed successfully. Employing the primers ResR and ResF, the res gene was enlarged by employing PCR. The PCR by-product was franked with Hind III digestion, Klenow polymerase (New England Biolabs), and cloning into the pGEM-Teasy vector (Promega) [51]. Among the pUC4K vector (Pharmacia), the Tn903 kanamycin resistance cassette was removed employing HincII and situated into the frank Hind 3 spot. After being linearized through NcoI digestion, the resultant plasmid, pGEMres::kan, was employed to execute the MIV methodology conversion on 162 non-typeable *H. influenzae* isolates [52].

13.3.1.3 Transformation of *H. influenzae*

After being homogenized in a 20-ml quantity of chilled sterile water having 272 millimoles sucrose (pH 7.4) and 15% glycerol, the bacteria were collected from five of the media plates of simultaneous growth on brain heart infusion (BHI) agar medium assisted with Levinthal base. Following centrifuging for 2 min at 13,000 rpm, the bacteria were mixed with 1 cc of sterile water. The process was repeated four or five times, with the cells remaining on the ice between spins. After measuring and standardizing the cell suspension optical density at 600 nm was found to be 10. The cells got one microgram of additional DNA before proceeding for incubation on ice for another 2 min. After isolating the cells by electrophoresis, they were immediately cultured in BHI broth. After incubation for 90 min at 378 °C with frequent swirling, cells were coated with tetracycline on brain heart infusion (BHI) agar. Multiple communities were selected after an overnight growth period, cultured in broth, and plasmids were formed with the use of a Qiagen Plasmid Midi Kit. Various substrates were estimated quantitatively. After selecting 12 communities to track the expansion of the mod repeat tract, a cell population sample was created to measure blunt size using the Gene Scan method, as was previously described [53]. Colony counts per milligram of DNA were used to derive the conversion efficiency. By examination of fragment size, the ratio of stimulated to unstimulated cells in the mod ON inheritor cell was examined prior to and after the transformation and was confirmed to be unaffected [51].

13.3.1.4 DNA Formation, Manipulation, and Analysis

The Sigma Proligo primers have been used in a PCR. Whole enzymes are purchased from "New England Biolabs." Primer pairs him6A and 11 were exploited to replicate the mutable region of the mod gene, whereas primer pairs him1, 3, 4, 5, as well as HI1059for and HI1052rev, were also utilized for replication of mod repeated tract and the mod/res region, respectively [54]. Primer pairs HI1054for and HI1054rev, as well as HI1054for2 and HI1054rev2, were used to amplify the area of the res gene containing well-known frameshift mutagenic alterations [55]. A Big-Dye sequencing kit and PCR procedure were used to generate the sequenced experiments (Perkin Elmer) [53].

13.3.2 Analysis of Mutant Strains of *H. influenzae*

13.3.2.1 ApoI Cleavage Assay

The Qiagen Plasmid Midi Kit can be used to generate *H. influenzae* strains with Rd. mod ON gene and mod::kan cells. According to the manufacturer's instructions, overnight ApoI digestion has been carried out on 1 μg of each plasmid. After that, the fragments were separated using TBE at 70 V for 2 h on a 2% higher-resolution agarose gel and scrutinized under UV light. Digests were performed in a manner akin to this utilizing TaqI [54].

13.3.2.2 South Western Analysis

After the membrane had been exposed to UV light to cross-link the DNA to it, it had been three times washed in "TBST" (plasmid pH Stet was retrieved employing 100 mM Tris H). The repeat tract was sequenced in order to confirm Mod ON cells. According to the manufacturer's recommendations, 5 μg of every plasmid was digested with DpnI overnight. The resulting fragments were then resolved for 2 h with TBE at 70 V on just a 1.5% elevated agarose gel. The extracted DNA fragments were deposited onto a nitrocellulose membrane through overnight capillary transfer method. Stir Hydrochloric acid of pH 8, 200 mM NaCl, 1 mM EDTA, and 0.5% Tween 20 solution was stirred gently for 5 min. After blocking in 3% BSA in TBST for 1 h, the membrane was incubated with a 1000th dilution for rabbit isolated N6-methyl-adenine monoclonal antibody at room temperature with moderate agitation. After being treated with a blocking solution that contains a goat anti-rabbit Immunoglobulin G alkaline phosphatase conjugate antibody for an additional hour, washing of the membrane was accomplished thrice for 5 min in TBST. After three further washes [55], the membrane was totally absorbed in 10 ml of Sigma FAST BCIP/NBT substrate.

13.3.2.3 5′-CGAAT-3′ as Rd. Mod Recognition Site in *H. influenzae*

In the past, studies of Rf DNA constraint systems in the *H. influenzae* strain revealed that the recognition sequence 50-CGAAT-30 was unique to type III R-M systems [56]. The mod and res genes, which are encoded by a single R-M system in the *H. influenzae* strain Rd. genomic sequence, have similarities to type III systems. A 50-AGTC-30 tetranucleotide repeat tract located within the open reading frame of the mod gene makes it phase variable in expression [57, 58]. To investigate if the Hinf-III and Mod recognition sequences are the same in strain Rd., plasmid pHStet has been produced in strain Rd. mod ON and strain Rdmod::kan cells. The generated plasmids were either in methylated or unmethylated state and then cleaved with an enzyme for whom the recognition sequence is a perfect match for the predicted methylation sequence. Protein digestion by ApoI is inhibited by the methylation of any adenine moiety in the Hinf-III region. While pHStet has many ApoI sites, only one of them coincides with a "Hinf-III" domain, suggesting that its activity could be blocked by methylation [59]. Mod's methylation of the DNA inhibited "ApoI" digestion of the plasmid, as this band was not observed in the non-methylated plasmid digested from the Rdmod::kan cells. Based on our findings, we conclude that Hinf-III and Mod Rd. share comparable site preferences like 50-CGAAT-30. Additionally, it was demonstrated that the adenine located at second position in the "Hinf-III" sequence is methylated. This is demonstrated by the fact that TaqI successfully degrades pHStet [60]. The TaqI recognition site begins at the same adenine as the first adenine in the Hinf-III sequence (50-TCGA-30). The fact that the TaqI cleavage pattern for pH Stet is the same when it is taken from "Rd mod ON and mod::kan cells" demonstrates that this base is not methylated by Mod. This lends credence to previous studies that demonstrated DNA methylation via the "Hinf-III" enzyme isolated from *H. pylori*, i.e., S-[methyl-3 H] adenosyl methionine was utilized along TaqI to cleave influenza strain Rf [61].

13.3.2.4 Distinction of Mod Recognition Site in Various Strains of *H. influenzae*

Aligning of the four mod gene sequences isolated from four different *H. influenzae* strains are accomplished. Genomes of influenza have been sequenced and made available, along with the discovery of the fact that mod gene sequence was segmented into 3 domains, as is distinctive characteristic of mod genes of type III R-M framework [62]. The amino and carboxy terminus of the peptide chain exhibited around 90–96% and 29–31% sequence homology, respectively, among the genome sequenced strains, but the "core domain" of the protein reveals just 29–30% similarity. This is reliable with previous studies signifying that Mod proteins display significant conservation only in the N- and C-terminal points of the protein [63], while the central region shows much lower preservation. It has been suggested that this core region of the protein functions in the recognition and binding of target sequences as well as the binding of the donor of methyl [64]. The conserved areas are thought to be involved in Mod and Res' protein–protein interactions. The fact that the antisera were able to bind to every piece of the fragmented DNA showed that the plasmid's numerous sites were being methylated by endogenous methylases. One of the DNA fragments was found to exhibit differential antiserum binding to plasmid recovered from R2866 mod ON and mod::kan cells, displaying that this DNA segment includes several Mod methylation sites that cause this detectable variation in antiserum binding between same genus strains. The dissimilarity in antisera binding between mod mutant and wild-type strains proves that mod is an efficient methyltransferase in this strain [65].

13.3.2.5 Differences in the Mod Gene Sequence Classify 15 Distinct Clusters of *H. influenzae*

H. influenzae-derived mod gene is analyzed phylogenetically to observe its heterogenic sequencing that exists in genome-sequenced isolates of *H. influenzae*. Moreover, experimental data support the proportionality of recognition site specificity and mod gene sequences. Non-typeable *H. influenzae* (NTHi) causes otitis media and respiratory problems, while encapsulated strains are linked to more severe illnesses such as meningitis and pneumonia [66]. Significant variability was seen in the mod repeat tract. The *H. influenzae* survey acknowledged 15 mod sequence groupings, wherein the variable region sequence was >95% identical amino acids within a group. The differential region of proteins presented only a 29–38% sequence similarity between groups. There is a connection between the type of mod sequence and the capsular serotype. With the exception of the ATCC 9833 strain, all members of a given capsular serotype have the same mod sequence type. Every single one of the capsular-typed strains has its own one-of-a-kind mod sequence type [67]. Both capsular types e and f have a unique mod sequence type that cannot be found in any other strain whether it is a capsular or non-typeable one. There was no evidence of the mod or res genes, suggesting that these strains either never had them or never picked up the R-M system. The absence of these genes was validated by a Southern blot analysis using a probe specific to the 30 variable regions of the mod. It has also been established that the length of a mod repeat is related to the sequence family from which it was derived. A repeat tract in the mod gene has only been found in three of these 15 sequence variations. Isolates 6 and 7 are completely NTHi, while Isolates 8 include both capsular types e and f. Phase variability at the identified mod gene is not anticipated from them. Previous studies have linked the occurrence of mod phase fluctuations to the length of the repeat section in strain Rd. The fact that different mod alleles have different functional effects [68] may explain the variations occurring in different phases of mod gene expression and the regulating the pace at which this develops. In our analysis, nil correlation was found to exist among mod sequence groups or its length and phenotypic characteristics of disease. Given the great degree of preserving sequence of mod gene among strains of *H. influenzae* and *Neisseria*, the homology in the DNA repeated strands involved for phase variation, and the genetic similarity via horizontal gene transfer, the naming convention has been proposed [69]. Genes are labeled with the letter modA followed by a number that identifies their place in the mod grouping generated on the basis of the allele DNA recognition marker. The ability to recognize specific DNA sequences is crucial [70].

13.3.2.6 Multiple Strains Carry res Gene Mutations That Render It Inactive

Res sequence existing in between Rd. and NTHi genes is pretty much conserved than the mod gene sequencing of *H. influenzae* strains. The protein length predicted by the res ORF differed widely between strains 86-028NP (722 aa) and Rd. (930 aa). Res gene of Rd. Strain produces a protein that is composed of 930 amino acids, while strains R2866 and R2846 produce proteins that are 929 aa in length [71]. Only in strain 86-028NP, a single base pair deletion resulted in a frameshift mutation involving 207 amino acids at the carboxy terminus of the res gene, bringing it closer to the full-length genetic sequence. The subunit Res present in type III R-M networks shares so much resemblance in sequence to that of RNA and DNA helicases superfamily II [72]. An ATP-binding site is an example of a conserved motif (TGxGKT). Metal binding in this region of the protein is essential for the DNA-cleaving activity of a number of restriction enzymes. It was found that the res gene was deleted at location-1 in mod group 2 strains as well. For whatever reason, the res genes of four strains in mod group 2 were missing a specific number of nucleotide base pairs. This alteration resulted in a frameshift that severed further amino acids from the peptide's carboxy terminus, shortening it to 495 in total [73].

13.3.2.7 Experiments on Type III Restriction Activity by Transformation with Methylation or Unmethylated Plasmids Created by Mod

Mod plasmids were tested for their ability to convert *H. influenzae* strain Rd. 162 with and without methylation obtained from ON: mod or Kan: mod cells, respectively, to determine the efficacy of the type III Restriction Mononucleases framework. Since the mod is constitutively generated in strain 162, it was not essential to analyze the methylated derivative of plasmids derived from strain Rd. using the Apo protein-I inhibitory assay and by scanning the mod repetition tract prior to transformation. When a plasmid that had not been methylated was put into a homologous strain, transformation efficiency plummeted (mod ON cells of strain Rd. 162). There were just twice as many successful plasmid conversions using the Mod modified plasmids as there were using the non-methylated plasmids. Type III R-M system function alone accounts for the observed variance in transformation efficiency [74], in contrast to the other R-M systems which are unaffected in the wild type or mod::kan strain combinations that are mostly isogenic in nature. Changes in methylation systems apart from those involved in type III recombination account for the substantial variation in transformation efficiency between these two strains. Evaluating the ability to convert unmethylated plasmids is one way to determine the contribution of the Mod-Res system toward the restriction blockade [75]. Although the transformation efficiency is drastically lower than that of Mod-methylated plasmids, only 43% of the plasmids are successful in transforming the cells. These results suggest that type III R-M systems contribute only marginally to the cellular defense mechanism against foreign DNA.

13.4 Future Perspectives of Epigenetics

Both oral infections and the host may employ epigenetic pathways to modulate the immune response. Because infectious diseases are so intricate and multifaceted, it is reasonable to think that epigenetic events might have a role in either the onset or the development of the disease. Research in this area has the potential to advance our understanding of the pathogenesis of influenzae bacterial infection because epigenetic pathways have been shown to affect our understanding of inflammatory illnesses.

References

1. Boyer HW. DNA restriction and modification mechanisms in bacteria. Annu Rev Microbiol. 1971;25(1):153–76.
2. Iida S, Meyer J, Bächi B, Stålhammar-Carlemalm M, Schrickel S, Bickle TA, Arber W. DNA restriction—modification genes of phage P1 and plasmid p15B: Structure and in vitro transcription. J Mol Biol. 1983;165(1):1–8.
3. Phillips ZN, Husna AU, Jennings MP, Seib KL, Atack JM. Phasevarions of bacterial pathogens–phase-variable epigenetic regulators evolving from restriction–modification systems. Microbiology. 2019;165(9):917–28.
4. De Bolle X, Bayliss CD, Field D, Van De Ven T, Saunders NJ, Hood DW, Moxon ER. The length of a tetranucleotide repeat tract in Haemophilus influenzae determines the phase variation rate of a gene with homology to type III DNA methyltransferases. Mol Microbiol. 2000;35(1): 211–22.
5. Ahmad S, Ahmad M, Khan S, Ahmad F, Nawaz S, Khan FU. An overview on phase variation, mechanisms and roles in bacterial adaptation. J Pak Med Assoc. 2017;67(2):285–91.
6. Weiser JN, Williams A, Moxon ER. Phase-variable lipopolysaccharide structures enhance the invasive capacity of Haemophilus influenzae. Infect Immun. 1990;58(10):3455–7.
7. van Ham SM, van Alphen L, Mooi FR, van Putten JP. Phase variation of *H. influenzae* fimbriae: transcriptional control of two divergent genes through a variable combined promoter region. Cell. 1993;73(6):1187–96.
8. Srikhanta YN, Dowideit SJ, Edwards JL, Falsetta ML, Wu HJ, Harrison OB, Fox KL, Seib KL, Maguire TL, Wang AH, Maiden MC. Phasevarions mediate random switching of gene expression in pathogenic Neisseria. PLoS Pathog. 2009;5(4):e1000400.
9. John Roche R, High NJ, Moxon ER. Phase variation of Haemophilus influenzae lipopolysaccharide: characterization of lipopolysaccharide from individual colonies. FEMS Microbiol Lett. 1994;120(3):279–83.
10. Srikhanta YN, Maguire TL, Stacey KJ, Grimmond SM, Jennings MP. The phasevarion: a genetic system controlling coordinated, random switching of expression of multiple genes. Proc Natl Acad Sci. 2005;102(15):5547–51.
11. Yuan R, Hamilton DL. Type I and type III restriction-modification enzymes. In: DNA methylation. New York: Springer; 1984. p. 11–37.
12. Van Der Woude MW, Bäumler AJ. Phase and antigenic variation in bacteria. Clin Microbiol Rev. 2004;17(3):581–611.

13. Maskell DJ, Szabo MJ, Butler PD, Williams AE, Moxon ER. Phase variation of lipopolysaccharide in Haemophilus influenzae. Res Microbiol. 1991;142(6):719–24.
14. Razin A, Cedar H. DNA methylation and gene expression. Microbiol Rev. 1991;55(3):451–8.
15. Jeltsch A. Beyond Watson and Crick: DNA methylation and molecular enzymology of DNA methyltransferases. Chembiochem. 2002;3(4):274–93.
16. Adhikari S, Curtis PD. DNA methyltransferases and epigenetic regulation in bacteria. FEMS Microbiol Rev. 2016;40(5):575–91.
17. Murphy J, Mahony J, Ainsworth S, Nauta A, van Sinderen D. Bacteriophage orphan DNA methyltransferases: insights from their bacterial origin, function, and occurrence. Appl Environ Microbiol. 2013;79(24):7547–55.
18. Nye TM, Fernandez NL, Simmons LA. A positive perspective on DNA methylation: regulatory functions of DNA methylation outside of host defense in Gram-positive bacteria. Crit Rev Biochem Mol Biol. 2020;55(6):576–91.
19. Dryden DT. Bacterial DNA methyltransferases. S-Adenosylmethionine-dependent methyltransferases: structures and functions. Singapore: World Scientific Publishing; 1999. p. 283–340.
20. Abdelraheem E, Thair B, Varela RF, Jockmann E, Popadić D, Hailes HC, Ward JM, Iribarren AM, Lewkowicz ES, Andexer JN, Hagedoorn PL. Methyltransferases: functions and applications. Chembiochem. 2022;12:e202200212.
21. Marinus MG, Morris NR. Isolation of deoxyribonucleic acid methylase mutants of Escherichia coli K-12. J Bacteriol. 1973;114(3):1143–50.
22. Brooks JE, Blumenthal RM, Gingeras TR. The isolation and characterization of the Esherichia Coli DNA adenine methylase (dam) gene. Nucleic Acids Res. 1983;11(3):837–51.
23. Smith HO, Tomb JF, Dougherty BA, Fleischmann RD, Venter JC. Frequency and distribution of DNA uptake signal sequences in the Haemophilus influenzae Rd genome. Science. 1995;269 (5223):538–40.
24. Marinus MG, Morris NR. Biological function for 6-methyladenine residues in the DNA of Escherichia coli K12. J Mol Biol. 1974;85(2):309–22.
25. Boye E, Løbner-Olesen A, Skarstad K. Timing of chromosomal replication in Escherichia coli. Biochim Biophys Acta. 1988;951(2–3):359–64.
26. Gonzalez D, Kozdon JB, McAdams HH, Shapiro L, Collier J. The functions of DNA methylation by CcrM in Caulobacter crescentus: a global approach. Nucleic Acids Res. 2014;42(6): 3720–35.
27. Gonzalez D, Collier J. DNA methylation by CcrM activates the transcription of two genes required for the division of C aulobacter crescentus. Mol Microbiol. 2013;88(1):203–18.
28. DebRoy S, Shropshire WC, Tran CN, Hao H, Gohel M, Galloway-Peña J, Hanson B, Flores AR, Shelburne SA. Characterization of the type I restriction modification system broadly conserved among group A streptococci. Msphere. 2021;6(6):e00799–21.
29. Oliver MB, Swords WE. Comparative analysis of streptococcus pneumoniae type i restriction-modification loci: variation in hsds gene target recognition domains. Pathogens. 2020;9(9):712.
30. Loenen WA, Dryden DT, Raleigh EA, Wilson GG. Type I restriction enzymes and their relatives. Nucleic Acids Res. 2014;42(1):20–44.
31. Dryden DT, Murray NE, Rao DN. Nucleoside triphosphate-dependent restriction enzymes. Nucleic Acids Res. 2001;29(18):3728–41.
32. Kennaway CK, Taylor JE, Song CF, Potrzebowski W, Nicholson W, White JH, Swiderska A, Obarska-Kosinska A, Callow P, Cooper LP, Roberts GA. Structure and operation of the DNA-translocating type I DNA restriction enzymes. Genes Dev. 2012;26(1):92–104.
33. Raghavendra NK, Bheemanaik S, Rao DN. Mechanistic insights into type III restriction enzymes. Front Biosci (Landmark Ed). 2012;17(3):1094–107.
34. Rao DN, Dryden DT, Bheemanaik S. Type III restriction-modification enzymes: a historical perspective. Nucleic Acids Res. 2014;42(1):45–55.

35. Atack JM, Yang Y, Seib KL, Zhou Y, Jennings MP. A survey of Type III restriction-modification systems reveals numerous, novel epigenetic regulators controlling phase-variable regulons; phasevarions. Nucleic Acids Res. 2018;46(7):3532–42.
36. Jayaraman R. Phase variation and adaptation in bacteria: A'Red Queen's Race'. Curr Sci. 2011;25:1163–71.
37. Fox KL, Srikhanta YN, Jennings MP. Phase variable type III restriction-modification systems of host-adapted bacterial pathogens. Mol Microbiol. 2007;65(6):1375–9.
38. Murphy TF, Faden H, Bakaletz LO, Kyd JM, Forsgren A, Campos J, Virji M, Pelton SI. Nontypeable Haemophilus influenzae as a pathogen in children. Pediatr Infect Dis J. 2009;28(1):43–8.
39. Sethi S, Murphy TF. Infection in the pathogenesis and course of chronic obstructive pulmonary disease. N Engl J Med. 2008;359(22):2355–65.
40. Johnson RH. Community-acquired pneumonia: etiology, diagnosis, and treatment. Clin Ther. 1988;10(5):568–73.
41. Bakaletz LO, Novotny LA. Nontypeable haemophilus influenzae (nthi). Trends Microbiol. 2018;26(8):727–8.
42. Fleischmann RD, Adams MD, White O, Clayton RA, Kirkness EF, Kerlavage AR, Bult CJ, Tomb JF, Dougherty BA, Merrick JM, McKenney K. Whole-genome random sequencing and assembly of Haemophilus influenzae Rd. Science. 1995;269(5223):496–512.
43. Srikhanta YN, Fox KL, Jennings MP. The phasevarion: phase variation of type III DNA methyltransferases controls coordinated switching in multiple genes. Nat Rev Microbiol. 2010;8(3):196–206.
44. Hood DW, Deadman ME, Jennings MP, Bisercic M, Fleischmann RD, Venter JC, Moxon ER. DNA repeats identify novel virulence genes in Haemophilus influenzae. Proc Natl Acad Sci. 1996;93(20):11121–5.
45. Atack JM, Srikhanta YN, Fox KL, Jurcisek JA, Brockman KL, Clark TA, Boitano M, Power PM, Jen FE, McEwan AG, Grimmond SM. A biphasic epigenetic switch controls immunoevasion, virulence and niche adaptation in non-typeable Haemophilus influenzae. Nat Commun. 2015;6(1):1–2.
46. Wong S, Akerley BJ. Identification and analysis of essential genes in Haemophilus influenzae. Methods Mol Biol. 2008;416:27–44.
47. Williams JD, Moosdeen F. Antibiotic resistance in Haemophilus influenzae: epidemiology, mechanisms, and therapeutic possibilities. Rev Infect Dis. 1986;8(Supplement_5):S555–61.
48. Brockman KL, Branstool MT, Atack JM, Robledo-Avila F, Partida-Sanchez S, Jennings MP, Bakaletz LO. The ModA2 phasevarion of nontypeable Haemophilus influenzae regulates resistance to oxidative stress and killing by human neutrophils. Sci Rep. 2017;7(1):1–1.
49. Starner TD, Zhang N, Kim G, Apicella MA, McCray PB Jr. Haemophilus influenzae forms biofilms on airway epithelia: implications in cystic fibrosis. Am J Respir Crit Care Med. 2006;174(2):213–20.
50. Hallet B. Playing Dr Jekyll and Mr Hyde: combined mechanisms of phase variation in bacteria. Curr Opin Microbiol. 2001;4(5):570–81.
51. Sambrook J, Fritsch EF, Maniatis T. Molecular cloning: a laboratory manual. Cold Springer Harbor: Cold Springer Harbor Laboratory Press; 1989.
52. Herriott RM, Meyer EM, Vogt M. Defined nongrowth media for stage II development of competence in Haemophilus influenzae. J Bacteriol. 1970;101(2):517–24.
53. Erwin AL, Allen S, Ho DK, Bonthius PJ, Jarisch J, Nelson KL, Tsao DL, Unrath WC, Watson ME Jr, Gibson BW, Apicella MA. Role of lgtC in resistance of nontypeable Haemophilus influenzae strain R2866 to human serum. Infect Immun. 2006;74(11):6226–35.
54. Fox KL, Dowideit SJ, Erwin AL, Srikhanta YN, Smith AL, Jennings MP. Haemophilus influenzae phasevarions have evolved from type III DNA restriction systems into epigenetic regulators of gene expression. Nucleic Acids Res. 2007;35(15):5242–52.
55. Sikkema DJ, Murphy TF. Molecular analysis of the P2 porin protein of nontypeable Haemophilus influenzae. Infect Immun. 1992;60(12):5204–11.

56. Piekarowicz A, Bickle TA, Shepherd JC, Ineichen K. The DNA sequence recognised by the HinfIII restriction endonuclease. J Mol Biol. 1981;146(1):167–72.
57. Fox KL, Yildirim HH, Deadman ME, Schweda EK, Moxon ER, Hood DW. Novel lipopolysaccharide biosynthetic genes containing tetranucleotide repeats in Haemophilus influenzae, identification of a gene for adding O-acetyl groups. Mol Microbiol. 2005;58(1): 207–16.
58. Musser JM, Barenkamp SJ, Granoff DM, Selander RK. Genetic relationships of serologically nontypable and serotype b strains of Haemophilus influenzae. Infect Immun. 1986;52(1): 183–91.
59. Lancashire JF, Terry TD, Blackall PJ, Jennings MP. Plasmid-encoded Tet B tetracycline resistance in Haemophilus parasuis. Antimicrob Agents Chemother. 2005;49(5):1927–31.
60. Low DA, Weyand NJ, Mahan MJ. Roles of DNA adenine methylation in regulating bacterial gene expression and virulence. Infect Immun. 2001;69(12):7197–204.
61. Musser JM, Kroll JS, Granoff DM, Moxon ER, Brodeur BR, Campos J, Dabernat H, Frederiksen W, Hamel J, Hammond G, Høiby EA. Global genetic structure and molecular epidemiology of encapsulated Haemophilus influenzae. Rev Infect Dis. 1990;12(1):75–111.
62. Hümbelin M, Suri B, Rao DN, Hornby DP, Eberle H, Pripfl T, Kenel S, Bickle TA. Type III DNA restriction and modification systems EcoP1 and EcoP15: Nucleotide sequence of the EcoP1 operon, the EcoP15mod gene and some EcoP1mod mutants. J Mol Biol. 1988;200(1): 23–9.
63. Bickle TA, Krüger DH. Biology of DNA restriction. Microbiol Rev. 1993;57(2):434–50.
64. Rao DN, Page MG, Bickle TA. Cloning, over-expression and the catalytic properties of the EcoP15 modification methylase from Escherichia coli. J Mol Biol. 1989;209(4):599–606.
65. Glatman LI, Kravets AN. Sistemy restriktsii--modifikatsii DNK [DNA restriction-modification systems]. Antibiot Khimioter. 1989;34(12):932–8.
66. Turk DC. The pathogenicity of Haemophilus influenzae. J Med Microbiol. 1984;18(1):1–6.
67. St Geme JW 3rd. Nontypeable Haemophilus influenzae disease: epidemiology, pathogenesis, and prospects for prevention. Infect Agents Dis. 1993;2(1):1–6.
68. Meats E, Feil EJ, Stringer S, Cody AJ, Goldstein R, Kroll JS, Popovic T, Spratt BG. Characterization of encapsulated and noncapsulated Haemophilus influenzae and determination of phylogenetic relationships by multilocus sequence typing. J Clin Microbiol. 2003;41 (4):1623–36.
69. Cody AJ, Field D, Feil EJ, Stringer S, Deadman ME, Tsolaki AG, Gratz B, Bouchet V, Goldstein R, Hood DW, Moxon ER. High rates of recombination in otitis media isolates of non-typeable Haemophilus influenzae. Infect Genet Evol. 2003;3(1):57–66.
70. Kroll JS, Wilks KE, Farrant JL, Langford PR. Natural genetic exchange between Haemophilus and Neisseria: intergeneric transfer of chromosomal genes between major human pathogens. Proc Natl Acad Sci. 1998;95(21):12381–5.
71. Gorbalenya AE, Koonin EV. Endonuclease (R) subunits of type-I and type-III restriction-modification enzymes contain a helicase-like domain. FEBS Lett. 1991;291(2):277–81.
72. Bist P, Sistla S, Krishnamurthy V, Acharya A, Chandrakala B, Rao DN. S-adenosyl-L-methionine is required for DNA cleavage by type III restriction enzymes. J Mol Biol. 2001;310(1): 93–109.
73. Saha S, Ahmad I, Reddy YV, Krishnamurthy V, Rao DN. Functional Analysis of Conserved Motifs in Type III. Restriction-Modification Enzymes.
74. Seib KL, Srikhanta YN, Atack JM, Jennings MP. Epigenetic regulation of virulence and immunoevasion by phase-variable restriction-modification systems in bacterial pathogens. Annu Rev Microbiol. 2020;74:655–71.
75. Janscak P, Sandmeier U, Szczelkun MD, Bickle TA. Subunit assembly and mode of DNA cleavage of the type III restriction endonucleases EcoP1I and EcoP15I. J Mol Biol. 2001;306 (3):417–31.

Targeting Epigenetics in Pulmonary Arterial Hypertension

14

K. M. Taufiqur Rahman, Tanim Islam, Md Fahmid Islam,
Roberto G. Carbone, Nicholas C. Butzin, and Md Khadem Ali

Abstract

Pulmonary arterial hypertension (PAH) is a fatal and enigmatic disease of the pulmonary circulatory system for which there is currently no cure. The intricate pathogenesis of PAH poses a barrier to identifying novel therapeutic targets, resulting in high morbidity and mortality rates. To identify potential drugs or biomarkers from the bench to the bedside, there is a need to expand the current knowledge of the pathogenesis of PAH. Alternative therapies and treatment strategies may slow down the progression of this disease, but a complete cure requires uncovering the underlying novel mechanisms. Emerging epigenetics-based studies are paving the way for understanding the pathophysiology of several complicated disorders, such as cancer, peripheral hypertension, and asthma. Thus, epigenetic studies may help to comprehend the multifaceted nature

K. M. T. Rahman
Department of Biology and Microbiology, South Dakota State University, Brookings, SD, USA

T. Islam
Department of Veterinary Pathobiology, Oklahoma State University, Stillwater, OK, USA

M. F. Islam
Department of Pediatrics, College of Medicine, University of Saskatchewan, Saskatoon, Saskatoon, SK, Canada

R. G. Carbone
Department of Internal Medicine, University of Genoa, Genoa, Italy

N. C. Butzin
Department of Biology and Microbiology, South Dakota State University, Brookings, SD, USA

Department of Chemistry and Biochemistry, South Dakota State University, Brookings, SD, USA

M. K. Ali (✉)
Pre-Professional Health Academic Program, California State University, East Bay, Hayward, CA, USA

G. Gupta et al. (eds.), *Targeting Epigenetics in Inflammatory Lung Diseases*,
https://doi.org/10.1007/978-981-99-4780-5_14

of PAH. This chapter compiles the current knowledge and emerging therapeutic and biomarker potential of epigenetic factors, such as DNA methylation, histone modifications, and noncoding RNAs, in PAH. As no animal models can fully recapitulate human PAH features, we highlight the emerging microfluidic lab-on-a-chip (LoC) technology as an excellent model for the disease and testing therapeutics.

Keywords

Pulmonary arterial hypertension · Epigenetics · DNA methylation · Noncoding RNAs · Microfluidics · Lab-on-a-chip

Abbreviations

ABCA1	ATP binding cassette 1
APOA4	Apolipoprotein A4
BER	Base excision repair
BMPR2	Bone morphogenetic protein type II receptor
BRDP	Bromodomain-containing protein
CAP	Concentrated ambient particles
ChIP	Chromatin immunoprecipitation
COPD	Chronic obstructive pulmonary disease
DNMT	DNA methyltransferase
ELISA	Enzyme-linked immunosorbent assay
HDAC	Histone deacetyl transferases
HME	Histone modifying enzymes
IL	Interleukins
LoC	Lab-on-a-chip
MBD	Methyl CpG binding
MCT	Monocrotaline
OoC	Organ-on-a-chip
PAC	Pulmonary arterial cells
PAEC	Pulmonary endothelial cells
PAH	Pulmonary arterial hypertension
PARP	Poly (ADP-ribose) polymerases
PASMC	Pulmonary arterial smooth muscle cell
PDGF-BB	Platelet-derived growth factor B
PH	Pulmonary hypertension
PPI	Protein–protein interactions
PVOD	Pulmonary veno-occlusive disease
RAAS	Renin–angiotensin–aldosterone system
RISC	RNA-induced silencing complex
RSRS	Renal sodium retention system
RVLM	Rostral ventrolateral medulla

SAM	S-adenyl methionine
SMC	Smooth muscle cell
SNS	Sympathetic nervous system
TET	Ten-eleven translocation
TGF	Transforming growth factor
TLR	Toll-like receptors
TNF	Tumor necrosis factor
UHRF	Ubiquitin-like, containing PHD and RING finger domain
VEGF	Vascular endothelial growth factors

14.1 Introduction

Pulmonary arterial hypertension (PAH) is a chronic cardiovascular condition characterized by vascular remodeling that ultimately results in right heart failure and death [1, 2]. Current therapeutics have primarily concentrated on addressing issues with the pulmonary vasculature and abnormal signaling pathways such as those involving prostacyclin, endothelin 1, and nitric oxide. The goal of these therapies is to alleviate the burden on the right side of the heart by reducing afterload [3, 4], yet a universally effective drug is under investigation. The leading causes of PAH are multifactorial and intricate, such as endothelial cell dysfunction [5], smooth muscle cell hyperproliferation [6], vascular inflammation, and immune dysregulation [7], germline mutations [8], and many others. Existing combination therapies can slow down the progression of the disease, but remain incurable [9]. Thus, there is an unmet need to understand this disease mechanism to discover effective drugs for therapeutic purposes.

Pathobiological research, diagnostic, and treatment strategies for PAH have advanced significantly, although the most precise mechanism is still enigmatic [10]. The involvement of aberrant cell signaling, genetics (mutations), and environmental variables, such as hypoxia, viral infections, and anorectic agents in the pathogenesis of PAH has been the subject of much exploration. But less emphasis has been paid to the relationship between epigenetics and PAH [11]. Epigenetics may be a potential pharmacological target since compelling data indicates that epigenetic mechanisms play a critical role in the pathogenesis of PAH [11–14]. Therefore, targeting epigenetic processes may offer a promising avenue for pharmacological interventions. However, the field of epigenetics investigates alterations in gene expression that can be inherited without changes to the DNA sequence [15]. The three primary mechanisms of epigenetics are DNA methylation, histone modification, and microRNA (miRNA) regulation. These processes play significant roles in modulating gene expression and regulating cellular activity [16]. Some miRNAs including miR-34a [17], miR-17–92 [18], and miR-212-5p [19] were identified as potential therapeutic targets in PAH. Despite the crucial role of epigenetics in the pathogenesis of PAH, the role of histone modification and DNA methylation in this condition has received relatively little research attention [11]. Moreover, a new class of RNAs called circular RNAs (CircRNAs) is emerging as diagnostic and therapeutic agents in various diseases including PAH [20]. Thus, targeting epigenetics in PAH could open the door to addressing many unanswered questions. However, understanding disease mechanisms and evaluating therapeutic effectiveness rely heavily on the fidelity of animal models. Unfortunately, no animal models have been identified that can replicate fully the pathophysiology of PAH in humans [21]. Modern scientific advancements have produced a rapid and resilient microfluidics-based technology that has the potential to connect traditional cell cultures, animal models, and human subjects, thereby bridging the gap between them [22]. One of the most promising advancements in this field is the lab-on-a-chip (LoC) technology, combined with tissue engineering and organ-on-a-chip (OoC) technology [23]. Together, these technologies can accurately emulate human pathobiology and evaluate the efficacy of drugs [24, 25]. In concert with a state-of-the-art microscope imaging system, this might be a revolutionary approach for unraveling the mystery of a putative PAH mechanism in humans. This chapter aims to provide an overview of the functions and mechanisms of epigenetic regulators in the development of PAH, and their potential therapeutic targets. Furthermore, this chapter will discuss the potential of microfluidics-based lab-on-a-chip (LoC) technologies for PAH research.

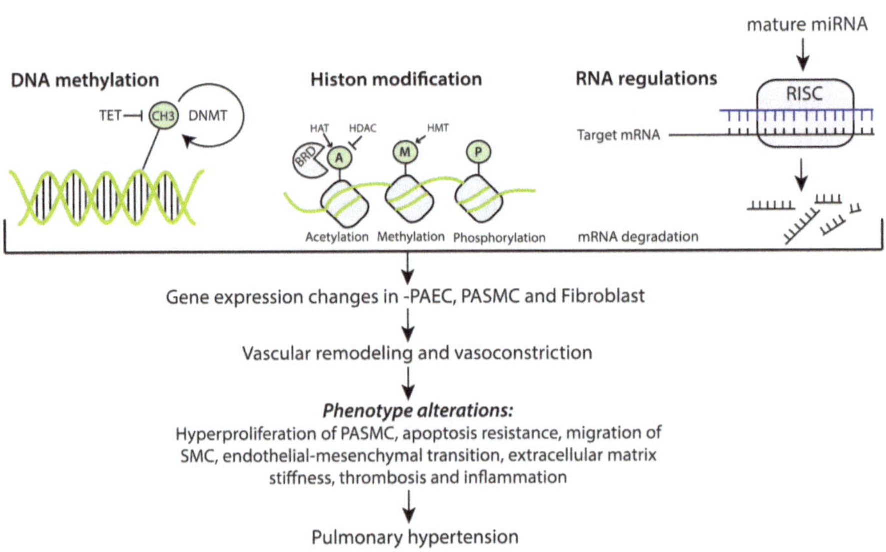

Fig. 14.1 The role of epigenetic mechanisms in the pathogenesis of PH

14.2 Targeting Epigenetics in PAH

Epigenetics refers to the investigation of heritable changes in phenotype that do not involve changes in DNA sequences. There are three primary epigenetic mechanisms: histone modifications (such as acetylation and methylation), noncoding RNA (including microRNA and long noncoding RNA), and DNA methylation, all of which can activate or deactivate genes [26, 27]. A brief overview of these three mechanisms will be discussed later in this chapter.

We depicted the role of epigenetic mechanisms in the pathogenesis of PH in Fig. 14.1. DNMT (DNA methyltransferase) is used to catalyze and maintain DNA methylation. The TET (ten-eleven translocation) proteins can block DNMT. Histones found in the nucleosome's N-terminal tails can be acetylated, deacetylated, methylated, and other posttranslational changes. The epigenetic reader BRD4 (bromodomain-containing protein 4) can identify the acetylated lysine residues on histone tails. Mature miRNA is incorporated into the RISC (RNA-induced silencing complex) to mediate target mRNA destruction or translation inhibition. These three epigenetic variables have a significant impact on PAH and pulmonary vascular remodeling (dysfunction of key vascular cells such as pulmonary artery endothelial cells (PAECs), smooth muscle cells (PASMCs), and fibroblasts). PAECs play a critical role in regulating thrombosis and inflammation. PASMCs contribute to increase in vascular resistance in pulmonary hypertension [27, 28]. Besides, vascular extracellular matrix (ECM) remodeling and homeostasis are regulated by fibroblasts [29].

14.2.1 DNA Methylation

In 1948, the major epigenetic DNA methylation was first discovered by Rollin Hotchkiss while preparing calf thymus. He found a modified version of the nucleotide cytosine and stated that it existed naturally in DNA [30]. This finding led to the hypothesis that this fragment of the DNA was 5-methylcytosine (5 mC) [30]. In 1980s, numerous studies showed that DNA methylation, which involves a methyl group addition to the DNA molecule at the fifth carbon of the cytosine ring in a CpG dinucleotide sequence, is a critical mechanism for gene regulation and a major epigenetic factor controlling gene activities in eukaryotes [31]. The process of DNA methylation is catalyzed by a family of enzyme known as DNA methyltransferases (DNMTs), during which a methyl group from S-adenyl methionine (SAM) is transferred to fifth carbon of cytosine [32]. The enzymes DNMT3a and DNMT3b are known as de novo DNMT as these can transfer the methyl group into a naked DNA molecule while the DNMT1 functions by copying the methylation pattern of the DNA during replication [33]. There are three classes of enzymes known as writers, erasers, and readers, which are involved in establishment, recognition, and removal of DNA methylation, respectively. The addition of methyl groups to cytosine residues is catalyzed by writers, whereas the modifications and removal events are done by erasers, and lastly, readers are responsible for recognizing and binding these methyl groups to induce subsequent gene expressions [33]. The three DNMT members (DNMT1, DNMT3a, and DNMT3b) catalyze the methylation event. These enzymes have unique expression patterns even though they are structurally similar [34]. The native DNA methylation pattern is conserved by DNMT1, often referred to as the maintenance DNMT. During DNA replication, DNMT1 binds to the replication fork and localizes to the newly synthesized hemimethylated DNA strand, where it mimics the original methylation pattern [35, 36]. DNMT3a and DNMT3b are unique when compared to DNMT1, as these when overexpressed can methylate both the synthetic and native DNA [37], and thus they are also known as de novo DNMT due to their capability of methylating naked DNA. The key property that differentiates DNMT3a from DNMT3b is the pattern of their gene expression [33]. DNA demethylation can be categorized into either passive or active forms. During cell division, inhibition of the DNMT1 leads to low levels of methylation overall, as the newly synthesized cytosine residues remain unmethylated [33]. However, active demethylation can occur in both dividing and non-dividing cells, where the 5mC is regressed back to a naked cytosine by enzymatic reactions [38, 39]. During demethylation, a series of enzymatically catalyzed chemical reactions occur, with subsequent oxidation or deamination of the 5mC to a product, which is further identified by the base excision repair (BER) pathway [40]. There are some other proposed mechanisms of active demethylation, for example, one of the suggested mechanisms is driven by the TET enzymes, in which these enzymes catalyze the addition of a hydroxyl group to the methyl group of 5 mC forming 5 hmC [41, 42], which is reverted back to a naked cytosine by two other separate processes such as iterative oxidation of 5 hmC or 5 hmC deamination by AID/APOBE complex (activation-induced cytidine deaminase/apolipoprotein B mRNA-editing enzyme) [43].

There are three different protein families known as the MBD protein, the UHRF (ubiquitin-like, containing PHD and RING finger domain) proteins and the zinc-finger proteins that can recognize DNA methylation. Some of the proteins of the MBD family are the MBD1, MBD2, MBD3, MBD4, and MeCP2 [44]. There is a methyl CpG binding (MBD) domain conserved in the MBD proteins. This domain has a high affinity for single methylated CpG sites [45]. MBD3 and MBD4 exhibit atypical characteristics among the MBD proteins. MBD3 lacks the ability to directly bind to DNA due to a mutation in the MBD domain [46], while MBD4 binds when it recognizes a mismatched guanine residue with thymine, uracil, or 5-fluorouracil [47–49]. Conversely, MeCP2 plays a double function by acting as both a transcriptional repressor and participating in the maintenance of DNA methylation [33]. The UHRF family has multi-domain proteins such as UHRF1 and UHRF2 which uses a SET- and RING-connected DNA binding domain to bind the methylated cytosine residues [50]. During DNA replication, this protein family primarily binds to the DNMT1 and maintains DNA methylation by projection it to a hemimethylated DNA [51, 52]. The final set of methyl binding proteins uses a zinc-finger domain to bind the methylated DNA. This family consists of Kaiso, ZBTB4, and ZBTB38 [53, 54]. These domains are unique as both Kaiso and ZBTB4 have binding affinities for a sequence motif that lacks methylcytosine [55], where Kaiso can also bind to two successive methylated CpG sites [56].

DNA methylation plays a crucial role in gene regulation and transcriptional silencing [57]. In particular, when DNA is hypermethylated in regions of the genome that contain a high density of CpG sites (known as CpG islands), this is often associated with transcriptional repression [58]. In a study of PH, where the pulmonary arterial smooth muscle cells (PASMCs) were isolated from fawn hooded rats, a diminish in the SOD2 gene expression was observed. These observations were also seen in the plexiform lesions of PH [14, 16]. It was later demonstrated that when CpG islands are selectively hypermethylated in both the promoter and the intronic regions of the SOD2 gene, the expression is reduced by approximately 50% [14]. Another study claimed that epigenetic regulation was influenced by the hypermethylation of the GNLY genes, specifically in the case of pulmonary veno-occlusive disease (PVOD) [59]. Some of the key biological pathways, where DNA methylation is involved in epigenetic regulations are the sympathetic nervous system (SNS), renin–angiotensin–aldosterone system (RAAS), and renal sodium retention system (RSRS) [60]. The SNS system plays a key role in the pathophysiology and pathogenesis of hypertension [61], whereas the RAAS contributes to the occurrence of the disease [62].

A vast amount of data suggests that DNA methylation has an integral contribution in pulmonary vascular remodeling in PAH. In a study, where some of the genes such as the ATP binding cassette 1 (ABCA1), adiponectin (ADIPOQ), and apolipoprotein A4 (APOA4) were detected in PAH patients, the authors saw that ABCA1 was majorly hypermethylated, indicating that the metabolism of cholesterol might be related with PAH [63]. Some of the key genetic risk factors of heritable PAH (HPAH) is the heterozygous loss of function (LOF) mutation in the *BMPR2* gene [64]. A study performed bisulfite sequencing to determine the DNA methylation profile of the *BMPR2* gene and found that this gene was hypermethylated in the HPAH patients, whereas the wild type allele was more hypermethylated when compared to the aberrant allele [64]. In another study, a group investigated and demonstrated that when rodents with PAH were administered with monocrotaline or exposed to hypoxia, the global DNA methylation increased in their lungs [65]. Their data shows the upregulation of the enzyme DNMT3b in both rodent models and PAH patients. Also, the authors showed that knockout DNMT3b$^{-/-}$ rats, suffered from severe pulmonary vascular remodeling and regular inhibition of DNMT3b caused proliferation of PASMCs as a response to deprivation of the platelet-derived growth factor-BB. Overall, their study reveals that DNMT3b mediates the pathogenesis of PAH coupling epigenetic regulations with vascular remodeling [65].

Pulmonary hypertension shares common risk factors with hypertension, such as age, obesity, alcohol consumption, drugs, etc. Evidence suggests that variations DNA methylation is associated with these factors [66, 67]. A study demonstrated that alcohol consumption enhances the DNA methylation of the ADD1 gene promoter, thus increasing the hypertension susceptibility [68]. Furthermore, a study claimed that differential variability in methylation can influence obesity. After analyzing genome wide methylation profiles of CpG sites in both obese and lean individuals, the authors found a significant number of differential methylated CpG in obese cases [66]. Age is also a common risk factor in both PH and hypertension, and genome wide methylation studies of CpG sites showed that approximately 28.8% of the CpG sites were associated with age and around 7.9% CpG sites were capable to predict age, making this association between DNA methylation strong and ubiquitous [67]. In addition, a cross-over trial, with human exposure to concentrated ambient particles (CAP) showed that these particles diminished methylation of toll-like receptors (TLR4) and Alu [69].

14.2.2 Histone Modification

Histones are protein that is basic in nature and contains high abundance of arginine and lysine residues, located in nuclei of eukaryotes [70]. These proteins play essential roles in protecting DNA from becoming tangled and in process being damaged [71]. There are five known histone proteins, among which H1 and H5 are the linker histones, while H2, H3, and H4 are core histones [70]. Nucleosomes are higher order structures that wrap 146 base pairs of DNA around core histone proteins [16]. Histones are key proteins that maintain the chromatin structure and influence long-term gene regulations [72]. Even though the chromatin exhibits a compact structure, histone modifications result in the chromatin losing its compactness [73]. Histone acetylation, ubiquitylation, and histone methylation process take place at the N-terminal where acetylation and methylation processes are carried out by acetyltransferases and methyltransferases, respectively [73]. The discovery of histone H1 and histone H3 phosphorylation was made in regard to the condensation of the chromosomes during meiosis [73].

Histone modifying enzymes (HME) carry out the posttranslational modifications of histone tails and thus can regulate gene expression. These enzymes (Table 14.1) can be categorized in two ways, as writers (induce modifications) and as erasers (revert modifications). Apart from the common modifications stated above, some of

Table 14.1 Some of the known histone modifying enzymes

Process of modification	Writer	Eraser
Methylation	Histone methyltransferase	Histone demethylase
Ubiquitination	Ubiquitin ligase	Deubiquitinating enzyme
Phosphorylation	Protein kinase	Protein phosphatase
Acetylation	Histone acetyltransferase	Histone deacetylase
ADP ribosylation	Poly ADP-ribose polymerase	(ADP-ribosyl) hydrolases ARH-1& ARH3
Proline isomerization	Fpr4	Fpr4
Sumoylation	E3 sumo ligase	Sumo specific protease
O-GlcNAcylation	O-GlcNAc transferase	O-GlcNAcase

the other histone modifications including SUMOylation, ADP ribosylation, O-GlcNAcylation, and proline isomerization [74–77].

In epigenetic mechanisms, it is integral for the HMEs to collaborate with other cofactors which involve numerous protein–protein interactions (PPI). Furthermore, these histone modifying enzymes can form functional complexes by interacting with non-histone proteins [78]. The activity of these enzymes is crucial, as any imbalance in their activity along with the change in the compactness of the chromatin is associated with several diseases including the vascular remodeling in pulmonary arterial hypertension [79]. Some of these enzymes can also be used as potential drug targets against PAH. For example, SAHA, a broad spectrum histone deacetylase (HDAC) inhibitor drug, has demonstrated the ability to decrease the migration of monocyte induced by fibroblasts, suggesting that such HDAC inhibiting drugs reduce vascular inflammation [13].

Amongst the five histone proteins, H3 maintains the chromatin structure in eukaryotes. Modifications of the N-terminal H3 tail by adding acetyl or methyl groups to arginine or lysine residues can be done post-translations. Also, phosphorylation of the serine or threonine residues is possible [72]. Gene expression can affect the methylation pattern of lysine-9, as this residue is hypermethylated when the gene expression decreases while it gets monomethylated when the gene gets activated [65]. The acetylation of H3 lysine residues is carried out by the enzyme histone acetyltransferase. Such acetylation events may associate with hypertension as they induce the glial cell line-derived neurotrophic factor [80]. This factor is key for the survival of dopaminergic neurons and can be followed up with melatonin treatment [80]. The neurons expressing melatonin can give input to the neurons of the rostral ventrolateral medulla (RVLM), which can regulate the sympathetic outflow to blood vessels. In humans, a dysfunctional RVLM can act as a mechanism to promote hypertension [81]. Post acetylation, the histone proteins are more accessible to the RNA polymerase due to losing structural compactness [82]. Thus, this enables the adjacent genes to get transcribed. Furthermore, bromodomain-containing proteins (BRDPs) can bind with these acetylated histones and recruit chromatin-modifying factors and induce transcription [83].

There are three families of enzymes that regulate histone modifications, out of which the histone deacetyl transferases (HDACs) family is the most researched in regard to PAH [84]. This enzyme family contains a highly conserved domain and maintains the equilibrium of lysine acetylation by removing acetyl groups in histones. One of the very first studies on HDACs and their association with PAH reported that when PAH was induced by hypoxia in bovine models, the phenotype of the pulmonary adventitial fibroblasts showed abnormality in the activity of class I HDACs. The authors also reported that these fibroblasts had an increased levels of proteins due to an elevation in the activity of the class I HDAC [85]. In another study, observation of the human idiopathic PAH lung homogenates showed that out of six screened HDACs, the expression levels of HDAC1 and HDAC5 were elevated [13]. Boucherat and colleagues demonstrated that in PAH patients, the HDAC6 gene is upregulated in the lungs, in the pulmonary endothelial cells, distal pulmonary arteries, and PASMCs [86]. The MEF2 transcription factor family is known to be linked with class II HDACs and plays a crucial part in cardiovascular development, wherein HDACs maintain transcriptional inactivity of MEF2s [87]. In PAH, the nucleus of pulmonary arterial endothelial cells experiences an abundant accumulation of HDAC4 and HDAC5, which causes a suppression of MEF2's transcriptional activity [88].

14.2.3 Crosstalk of DNA Methylation with Histone Modification

It is difficult to find concrete data on how epigenetic mechanisms influence pulmonary hypertension (PH), but several hypotheses claim that they have substantial roles in initiation, advancement, and establishment stages of PH [89]. Usually, in case of eukaryotes, histone proteins are bound with DNA to help package the molecule into small nuclear pockets. The N-terminal of the amino acids in the histones is chemically modified by process like acetylation, methylation, phosphorylation, and ubiquitination [33]. The DNMTs usually target the CpG sites and methylate the DNA to repress gene expression [33]. The histone modifications that lead to tighter packaging of DNA induce gene repression. DNMT1 and DNMT3a enzymes are associated with the histone methyltransferase SUV39H1 that inhibits gene expressions [90], while DNMT1 and DNMT3b interact with histone deacetylases to condense the DNA more tightly represses transcription [91]. In areas with active transcription, DNA methylation is removed by the TET enzyme, and inhibition of DNMT binding to the unmethylated CpG sites is done by the histone tails containing $H3K4me^3$ [33]. DNA methylation and histone modifications are crucial epigenetic mechanisms involved in the development of PAH. The subsequent section provides a detailed explanation of their roles in this context.

14.2.4 DNA Methylation and Histone Modifications as Therapeutic Targets of PAH

PAH becomes severe as it progresses, eventually causing right ventricular failure and death. The therapies available at present are not effective enough to reverse the disease and therefore result in high mortality and morbidity [92]. Hence, it is an utmost challenge and requirement to develop newer therapeutics for treating PAH. There are several therapeutic targets to treat PAH, such as drugs targeting estrogen signaling, drugs against growth factors, or even drugs targeting properties like epigenetic modifications and DNA damage [92]. PAH can influence DNA damage as it is associated with oxidative stress and inflammation [93]. Recently, the drug Olaparib which is approved for ovarian cancer has been in clinical trials to treat PAH [92]. This study is sponsored by Laval University, and their drug is under a preclinical study targeting to inhibit DNA damage and poly (ADP-ribose) polymerases (PARP) where they have claimed that PARP1 inhibition is cardio protective and can also reverse PAH disease conditions in an animal model [94]. It has been reported that PARP1 expression and activation levels are elevated in

PAH-PASMCs and PAH distal PAs [95]. This particular study also showed that in PH rats induced with both hypoxia and MCT, the PARP1 inhibiting drug veliparib was successful in reversing the PAH conditions [95]. Another drug that is being clinically studied is Apabetalone, which targets epigenetic modifications. This drug targets the inhibition of bromodomain-containing protein 4 (BRD4), which plays a significant role in the pathogenesis of PAH. Their results indicate that BRD4 inhibition can also reverse the disease conditions in animal models [96]. According to a study conducted in 2015, patients with PAH have more expression of the BRD4 gene in their distal PAs and PASMCs [97]. They also claimed that such overexpression of the BRD4 gene results in multiplication of the PASMC [97].

14.2.5 MicroRNAs (miRNAs)

MicroRNAs (miRNAs), firstly discovered in 1990s, are type of small regulatory molecules consisting of 18 to 24 nucleotides [98, 99]. The involvement of miRNAs in development and diseases is being perceived firmly over time due to their gene regulatory roles, conservation among species, and clear understanding of their mechanism of functions based on sequence complementarity [100–102]. The development of high-throughput technologies, computational methods, and experimental techniques enhanced the discovery process of miRNAs [98]. According to an estimation, there is around 1000 miRNAs in human genome with targeting potency of one-third of the entire human genes [103, 104]. MiRNAs can be generated singly or as a cluster from any part of the human genome through transcriptional machinery, irrespective of coding or noncoding regions [98]. A single miRNA can have multiple targets while sometimes multiple miRNAs can work together to regulate the same target, indicating their gene regulatory network attribute [98]. MiRNAs are generated and processed in a multi-step fashion, taking together several molecular players [105]. After originating from nucleus, miRNAs come to the cytoplasm and exert their functions [105]. MiRNAs are known to play a role in the development and progression of various human diseases, such as different types of cancers, neuronal development disorders, cardiovascular diseases, and skin diseases. Particularly, miRNAs have been widely studied in pulmonary hypertension [106–108].

MicroRNAs are associated with PAH [109, 110]. Aberrant miRNA expressions contribute to the abnormal remodeling of vascular cells, including adventitial fibroblast (AdvFB) migration, PASMC proliferation, and PAEC dysfunction in PAH [111]. Incomprehension of miRNAs role in complex cellular networks blocks the way to developing new drugs targeting pathological pathways [112]. The miRNA, miR-34a is a promising candidate for regulating the development of PAH. Reducing of its expression leads to elevated proliferation of PASMC, while overexpression produces the opposite effect [17]. Silencing of the miR-17–92 in smooth muscle cells (SMCs) induces PAH in mice [18]. Recently, miR-212-5p, an anti-proliferative miRNA, was suggested as a potential therapeutic target. Here, miR-212-5p was shown to be consistently and selectively elevated in both animal and human models of PAH [19]. Another study reported that miR-30a could be a promising miRNA for therapeutic targets [113]. Overexpression of miR-483 can suppress the PAH gene, resulting in the development of PAH. This finding indicates that miR-483 may serve as a potential disease biomarker and a target for therapeutic interventions [114]. In a recent review, it was found that miR-29, miR-124, miR-140, and miR-204 are crucial miRNAs that share a common expression pattern in both human and animal models [112]. However, several studies have previously documented the list of miRNAs that are dysregulated in PAH [106, 112, 115–117].

14.2.5.1 MiRNAs as a Therapeutic Target for PAH

In PAH, miRNA-based therapy is still in its infancy due to several limitations, such as miRNA mimics/antagonist, animal models, and instability of RNA molecules [117, 118]. However, miRNAs have a shorter length with known and conserved sequences and can target multiple genes within a network. Also, miRNAs have potential preclinical evidence against human diseases. So miRNA could be a promising therapeutic target molecule [106, 112]. Currently, four clinical trials related to miRNA and PAH studies have been discovered while searching https://clinicaltrials.gov/ (searched on November 9, 2022). The study (NCT00806312, completed) evaluated the expression and implications of miRNA profiles and inflammation markers in PAH patients. The study (NCT04489251, recruiting) looked into the miRNA and TGF pathways in pediatric PAH patients. The results of the other two studies were either withheld or unknown.

14.2.6 Long Noncoding RNAs (lncRNAs)

In the past, some DNA was regarded as "junk DNA" because it was ubiquitously transcribed but did not encode proteins; its functions were also unknown [119]. With the advent of the human genome project, however, the activities of these proteins were uncovered. This essential protein is referred to as "non-coding RNAs" (ncRNAs), which play a key role in protein-coding gene expression as well as several complex biological processes and disease progressions [120]. However, ncRNAs have been classified into two heterogenous subclasses: small ncRNAs with 15–200 nucleotides (nt) and long ncRNAs (lncRNAs) >200 nt long. There are thousands of ncRNA have been identified in the eukaryotic genome [121–123]. This section will give an overview of lncRNAs, what role they play in PAH, and how they might be used to find potential therapeutic targets. In the 1990s, the first lncRNAs were discovered using deep-sequencing and microarray technology. The importance of understanding the role of lncRNAs in epigenetic silencing and the evolution of multicellular organisms readily became increasingly apparent [124]. LncRNAs vary in size, structure (which resembles a flexible scaffold), and functions [125]. Their synthesis takes place in the nucleus via similar processes to mRNA transcription. LncRNAs are transcribed mostly by RNA polymerase II, but also by other polymerases [126]. Interestingly, lncRNAs can act in either the nucleus or the cytoplasm; hence, they are also classed as nuclear or cytoplasmic lncRNAs [127].

Based on recent classification, lncRNAs include long intergenic noncoding RNAs (lincRNAs) [128], natural antisense transcripts (NATs) [129], intronic lncRNAs, bidirectional transcripts [130, 131], sense lncRNA, enhancer RNA (eRNA) [132], and circular RNA (circRNAs) [133]. Nearly all recognized lncRNAs are similar to mRNA, but they have fewer exons, are less abundant, are nuclear-localized, and are not evolutionarily conserved [134–136]. They are capable to interact with DNA, RNA, and proteins [127]. Recent research has demonstrated that lncRNAs play a wide range of functions in cellular processes, including regulation of mRNA stability/turnover, chromatic architecture, and interference with posttranslational modification in the cytoplasm [137]; some of them can also translate into short polypeptides [138, 139]. A recent study also indicated that lncRNAs have a role in the development of various human diseases, including neurological, cancerous, and cardiovascular disorder [140].

14.2.6.1 LncRNAs in PAH

The pathogenesis of PAH is significantly influenced by lncRNAs, which exhibit differential expression patterns in pulmonary vascular cells such as PAECs and PASMCs. This differential expression leads to vascular remodeling and the subsequent development of PAH [141]. PASMC dysfunction is primarily attributed to various signaling pathways, including but not limited to PDGF, TGF-β, Wnt, hedgehog, estrogen, Notch, PI3K/AKT/mTOR, and MAPK signaling, as well as apoptotic pathways, hypoxia, and the influence of noncoding RNAs (lncRNAs and miRNAs) [142]. The platelet-derived growth factor B (PDGF-BB), a pro-inflammatory cytokine, was shown to be elevated in the lungs of PAH patients [143, 144]. Intriguingly, PDGF has a strong link with PASMCs proliferation [145]. However, lncRNA H19, a highly evolutionarily conserved maternally transcribed lncRNA, has been observed to be overexpressed in cancer patients cells when stimulated by TNF- α, IL-1, IL-6, TGF-1, and PDGF-BB [146, 147]. Several studies suggest that H19 plays a crucial role in the origin of PAH by increasing the proliferation of PASMCs, whereas decreasing H19 expression inhibits the development of PAH [148]. How precisely PDGF modulates H19 is currently unknown. Therefore, additional research is required to establish H19 as an efficacious PAH therapeutic target [141].

PAH's pathogenesis also involves endothelial dysfunction, where lncRNAs play a crucial role in the dysregulation process. In particular, dysregulation of lncRNAs has been implicated in the pathogenesis of chronic thromboembolic pulmonary hypertension (CTEPH), a subtype of PAH that is caused by chronic pulmonary embolism. Analysis of microarrays has identified 185 differentially expressed lncRNAs in CTEPH tissue when compared to healthy controls. Among these lncRNAs, NR_001284, NR_036693, NR_033766, and NR_027783 were significantly altered [149]. The MIR22 host gene (MIR22HG), MIR210HG, H19, MALAT1, and MEG9 lncRNAs were differentially expressed in endothelial cells in response to hypoxia, based on NGS studies [150]. Another lncRNA (TYKRIL) can stimulate the growth of PASMC. An ex vivo study using precision-cut lung slices (PCLS) demonstrated that TYKRIL knockdown can reverse pulmonary vascular remodeling. These findings suggest that TYKRIL may serve as a promising therapeutic candidate [151]. Smooth muscle enriched long noncoding RNA (SMILR) was also identified as a promising treatment target, as it is substantially expressed in PAH patients [152]. However, understanding the role of lncRNAs in the pathogenesis of PAH and CTEPH may lead to the identification of new diagnostic and therapeutic targets for these conditions. However, further research is needed to fully elucidate the molecular mechanisms underlying the dysregulation of lncRNAs in these diseases.

14.2.6.2 Therapeutic Application of lncRNAs in PAH

LncRNAs have a functional correlation with many human diseases, in particular PAH. Dysregulation of lncRNAs induces cell proliferation, leading to disease formation. As a result, this could be used as a promising biomarker or as a tool for personalized disease treatment. A study reported that, in the Chinese population, lncRNA MALAT1 (rs619586 A $>$ G) has been found to be associated with a reduced risk of developing PAH, indicating its potential as a biomarker [153]. Studies have reported that BC-819 (BioCancell Therapeutics Inc.), which is a double-stranded DNA plasmid containing diphtheria toxin under the H19 gene promoter regulation, can decrease the size of bladder tumors in mice [154].

A phase II clinical study involving patients with non-muscle-invasive bladder cancer reported that combining BC-819 with Bacillus Calmette-Guerin (BCG) led to an improvement in patient condition [141]. However, there are some potential drawbacks to studying lncRNAs as a promising drug target identification for cardiovascular diseases, specifically PAH. Although much research has been implemented in search of potential lncRNAs, therapeutic development is still in the early phases. RNA-based drug stability, an effective mode of delivery, establishing the secondary structure of lncRNAs, regulating off-target effects, and altering patient-specific dosages are typical challenges [155].

14.2.7 Circular RNAs (circRNAs)

The earliest evidence of any form of circular RNA (circRNA) molecule was reported when Diener found genome of viroids causing disease in potato plants in 1971 [156, 157]. The viroids, consisting of single-stranded and closed RNA molecules, were further confirmed by several studies in the 1970s. Interestingly, various types of circular RNAs were reported so far across different life forms (from plant viruses to humans) [158]. Nigro et al. confirmed a type of circular RNA transcript in human cells for the very first time while studying a tumor suppressor gene [159]. As experimental and computational challenges existed during the earlier studies, circRNAs coming from transcription process were thought to be splicing artifacts in most of the cases [160]. But the advancement of next-generation sequencing technologies and bioinformatics tools in the recent times made it possible to identify large numbers of circRNAs in different species [20, 158, 160]. Salzman et al. were the first to apply high-throughput RNA sequencing technology (RNA-seq) to detect circRNAs in the human body in 2012 [133]. These current approaches of circRNA discovery are further equipped with advanced experimental techniques for validation, functional characterization, and mechanistic study of circRNAs [20].

CircRNAs, unlike their linear counterparts, have covalently closed loop and single-stranded structure generated from a kind of non-canonical splicing called back splicing [20, 158, 160]. Majority of these circRNAs are widely and abundantly expressed in eukaryotes including humans and other mammals [161, 162]. This circularized chemical structure also provide higher stability, causing greater abundance compared to the linear mRNAs [163]. CircRNAs are mainly endogenous and show high level of conservation and specificity in tissue types and development stages [164–168]. Circular RNAs are classified into three primary categories based on the biochemical processes involved in their formation: exonic circRNAs (EcircRNAs), intronic circRNAs (CiRNAs), and exon-intron circRNAs (EIciRNAs). The major biogenesis mechanisms include intron pairing-driven circularization, lariat-driven circularization, RBP/trans-factor-driven circularization, cis-regulation based on variable circularization [133, 162, 167–175]. Though EcircRNAs are found mostly in cytoplasm, EIciRNAs and CiRNAs are predominantly reside in nucleus [20, 158, 173, 176]. Several groups also studied circRNAs degradation mechanisms in the recent times [176–180]. CircRNAs act as a novel class of genomic regulators by playing significant roles in various biological and physiological processes [20, 158, 160]. The most predominant function of circRNAs is to regulate target gene expression by competing with a type of noncoding RNA called micro RNAs (miRNAs) [165, 181]. Moreover, they can interact with RNA binding proteins (RBP) or other proteins; work as translational regulators or protein scaffolds, regulate transcription, post transcription or alternate splicing; participate in translation of peptides or proteins; and involve in N6-methyladenosine (m6A) modification of RNAs [20, 160, 182, 183].

14.2.7.1 CircRNAs in PAH

The dysregulated expression of circRNAs has been linked to the pathogenesis, progression, and development of numerous human diseases, including cancer, diabetes, neurodegenerative disorders, and cardiovascular diseases [20, 157, 183]. Specifically, circRNAs showed significant level of involvement in different forms of pulmonary hypertension (PH), e.g., pulmonary arterial hypertension (PAH), idiopathic pulmonary arterial hypertension (IPAH), hypoxia-induced pulmonary hypertension (HPH), and chronic thromboembolic pulmonary hypertension (CTEPH). CircRNA profiling studies have been conducted on clinical samples such as blood and lungs of PH patients, as well as PH rodent and in vitro models, resulting in identification of some differentially expressed circRNAs related to PH [184–191]. Pulmonary vascular cells, e.g., PASMCs, are impacted by regulatory effects of circRNAs which eventually lead to pulmonary vascular remodeling [187–198]. Furthermore, circRNAs have been shown to cause functional abnormalities in PAECs and result in right ventricular (RV) remodeling in PH [20, 158, 160]. CircRNAs are found to be overexpressed in PH condition in most of the instances leading to mechanistic effects at molecular level [20, 158, 160]. One example of a differentially expressed circRNA related to PH is circ-CALM4, which has been found to be upregulated in the lungs of HPH mice. It regulates PASMC pyroptosis induced by hypoxia through circ-Calm4/miR-1243p/programmed cell death protein 6 signaling and also regulates the proliferative properties of PASMCs through the miR-337-3p/Myo10 signaling axis [192, 196]. Another circRNA, hsa_circ_0016070, has been found to be overexpressed in the lungs of COPD-PAH patients and enhances PASMC proliferation through miR-942/CCND1 [197]. In contrast, hsa_circNFXL1_009 is downregulated in blood samples of COPD-PAH patients and hypoxic PASMCs and has been shown to play regulatory roles in the proliferation, migration, apoptosis, and potassium channel activation of PASMCs [199]. Additionally, hsa_circ_0046159 is upregulated in the blood of CTEPH patients and exerts regulatory roles through the predicted mechanism of the hsa_circ_0046159/miR-1226-3p/ATP2A2 axis. Circ-GSAP has been demonstrated to be linked with the development and unfavorable outcomes of IPAH and is downregulated in PBMCs and lung tissues of IPAH patients, as well as in lung tissues of monocrotaline-PAH rats, SU5416/hypoxia-induced PAH rats, and hypoxic PMECs [195].

CircRNAs are considered promising biomarkers in different disease conditions due to their stability and abundance [20]. Circ-006848 and circ-GSAP are examples of circRNAs that have been identified as potential diagnostic biomarkers for PH. For instance, Circ-006848 is overexpressed in PAH patients serum and has been shown to be a useful tool in predicting the diagnosis of right ventricular hypertrophy (RVH) in these patients [200]. RVH is a common complication of PAH and is associated with a poor prognosis. Therefore, identifying biomarkers that can accurately predict RVH in PAH patients is important for early diagnosis and management of the disease. On the other hand, circ-GSAP is downregulated in the PBMCs of IPAH patients, and this has been found to correlate with the development of IPAH as well as poor clinical outcomes associated with IPAH [195]. IPAH is a severe form of PAH, identifying diagnostic and prognostic biomarkers for IPAH is crucial for early diagnosis, appropriate treatment, and improved outcomes. Current progress in the field of RNA therapeutics and the relevant attractive properties (stability, conservation, specificity, and abundance) make circRNAs a very promising group of therapeutically potential molecules [20, 158, 160]. Though there is still no evidence of any clinical trials regarding circRNA therapeutics, this promise is being further strengthened by different layers of studies (in vitro, in vivo, and clinical) [20].

14.2.8 Microfluidic Based LoC Technology

Uncovering disease mechanisms is key to developing diagnostic tools and new therapeutic targets. Rapid and resilient technology is needed to simplify complex disease phenomena. Microfluidics technology has emerged together with tissue engineering, microfabrication, and Lab-on-a-Chip (LoC) methods to build ultra-precise models to study various aspects of molecular biology, cell, and synthetic biology research [201]. This emerging technology reduces chemical consumption, minimizes cost, and provides rapid, reliable, and high-throughput screening [202, 203]. This technology can potentially fill the gaps between conventional cell cultures, animal models, or human subjects in various fields, particularly in biomedical research and drug development [22]. Furthermore, recently developed organ-on-a-chip (OoC) systems combine biology with microtechnology [23] that replicates the organ's physiological environment, allowing researchers to study organ-level responses to drugs, toxins, or diseases without the need for animal models [24, 25] The advent of OoC technology facilitates the building of on-a-chip devices: heart-on-a-chip [204], lung-on-a-chip [205, 206], animal-on-a-chip [207], and body-on-a-chip [208]. These on-a-chip systems mimic biological function and allow for precision sciences. For example, Michas and colleagues recently engineered a microfabricated living cardiac pump on a chip mimicking the human ventricular chamber [209]. This new generation of sophisticated systems leverages the benefit of studying a replicated human organ in a tissue-chip model with structurally, biologically, and biomechanically similar functions [209].

14.2.8.1 Promising LoC Study with PAH

Advancement of modern science integrated micro total analysis system (μTAS), microfluidic LoC, and bio-microelectromechanical systems to develop new tools to investigate complex disease mechanisms. Several research groups focused on designing microfluidics devices to study pulmonary hypertension for potential therapeutic intervention. One such device, constructed by Lee and colleagues, is a microfluidic diagnostic tool integrated with an electrochemical assay system to rapidly identify pulmonary hypertension-associated biomarkers including fibrinogen, adiponectin, low-density lipoprotein, and 8-isoprostane [210]. To understand the vascular phenomenon, such as endothelial cell morphology and proliferation, Bogorad et al., described a 3D printed tissue-engineered microvessel model [211]. A microfluidic cell stretch device showed stretch stress that facilitates cell proliferation in PAH [212]. One elegant study described the fabrication process to create a near-ideal miniature vascular structure mimicking human's healthy and stenotic blood vessels using 3D printing system [213]. This method can overcome the limitation of conventional 2D wafer-based soft lithography or other biofabrication processes [214–216]. However, an advanced ultra-precise 3D printing fabrication process and a conceptual "PAH-chip" design is described later in this section.

Some other promising microfluidics devices have recently been developed to study PAH. Al-hilal and colleagues recently reported PAH-on-a-chip, simulating five human pulmonary artery layers: luminal, intimal, medial, adventitial, and perivascular. PAH-on-a-chip resembles human PAH pathophysiology enabling the growth of different types of pulmonary arterial cells (PAC) including endothelial, smooth muscle, and adventitial cells without altering their phenotypes. Disease-affected PAC cells generally migrate to their designated layers, this phenomenon exactly showed in an experiment with "PAH-on-a-chip" [217], which validates the device's accuracy and ability to recapitulate human pathophysiology. However, a critical phenomenon was observed in PAH patients where women are affected more by PAH but survive longer than men, known as "sex-paradox" [218–220]. The key mechanism of this sex disparity is unknown; however, it is interesting that both sexes follow the same treatment plan but vary in response. The reason could be the unavailability of screening tools and animal models. Moreover, the gold standard in vitro assays is insufficient to recapitulate the intricate cellular interaction [221]. In one study, male and female cells were separately loaded into a PAH-on-a-chip and demonstrated that the chip accurately mimics human PAH. So, this chip might be useful for detecting sex hormones and therapeutic impacts on PAH progression [217, 222]. This PAH-chip has also demonstrated promising capabilities in testing the effectiveness of the anti-PAH drug including fasudil (a Rho kinase inhibitor). The results were shown consistency with the in vivo studies using both Sugen-5416-plus-hypoxia (SuHx) and monocrotaline (MCT) rat model of PAH [223] as well as in patients with PAH [224]. This indicates that the chip could serve as a valuable tool to investigate the effectiveness of other anti-PAH drugs such as ASK-1 inhibitor [225], GS-44421731, or TGF-β trap therapeutics [226].

Another PAH-chip was developed to study pulmonary endothelial and smooth muscle cells by monitoring functional and transcriptomic changes [21]. The study was performed using a combination of BMPR2 (bone morphogenetic protein type II receptor) knockdown and hypoxia (2% O_2). BMPR2 is a type of rare genetic mutation that can be initiated by various factors (e.g., hypoxia, inflammation) [227] and can increase the susceptibility to PAH [228]. PAH-chip data showed that combination treatment significantly increased the proliferation of pulmonary artery smooth muscle cells (PASMCs) compared to separate treatments, similar to the effect observed in static cell culture [229]. A receptor tyrosine kinase inhibitor (10 μM of imatinib mesylate) was used to inhibit the proliferation. This demonstrates PAH-chip can measure therapeutic effectiveness. This sophisticated PAH-chip has the potential to trigger PAH diseases and evaluating their response to various possible treatments, leveraging studies on pulmonary vascular remodeling and drug target analysis [21]. By providing a platform to investigate cellular interactions and responses to potential therapies, this advanced chip technology can significantly improve our understanding of PAH and potentially lead to more effective treatment strategies in the future.

14.2.8.2 Microfluidic Chromatin Immunoprecipitation (ChIP) Assays for Epigenetic Study

Epigenetics refers to altering gene expression without changing the DNA sequence. Epigenetic factors play a key role in regulating cellular processes like stem cell pluripotency/differentiation and tumorigenesis [15]. Epigenetic regulators such as DNA methylation, histone modifications, and noncoding RNAs (micro or miRNA) are the potential PAH drug targets [230]. Epigenetic mechanisms can be effectively investigated using chromatin immunoprecipitation (ChIP), which require millions of cells, making it challenging to conduct large-scale studies due to the need for many human or animal models [15]. Presently, the common approach is to use animal models to assess a drug's efficacy or probe the possible disease mechanism. Various rodent models, including genetically modified mice and those induced by chronic hypoxia, Sugen-5416-plus-hypoxia (SuHx), and monocrotaline (MCT), are frequently employed in PAH research [217, 231]. However, these animal or cellular models have limitations in fully replicating the pathophysiology of human PAH, hindering their ability to demonstrate disease severity, histopathology, and sex-related differences in response to therapy [222, 231–234]. A new approach has emerged, shifting the paradigm in PAH research. Microfluidic ChIP assays offer the potential to study PAH without the reliance on animal models [235]. Recent studies have shown that microfluidics device enable quick, sensitive, and high-throughput testing with a significantly reduced number of cells (only 2000 cells are required) [236]. Additionally, combining sonication and immunoprecipitation (IP) in the same device has improved the accuracy of the assay. This method enables the testing of just 100 cross-linked cells for ChIP or 500 pg of genomic DNA for methylated DNA IP in only one hour [237]. In PAH research, miRNAs have been identified as promising therapeutic target molecules (as discussed earlier). However, the challenge lies in the divergence of miRNA expression patterns between human PAH and animal models [112]. To overcome all the barriers, future miRNA based therapies and interventions might rely on organ-chip systems that better mimic the human PAH phenotype.

14.2.8.3 Quantification of Endothelial Cells of PAH by LoC Method

Early diagnosis of PAH reduces disease progression and identifies effective treatments. Developing early diagnostic tools will save lives and increase the quality of life. Various molecular and cellular biomarkers have been studied in PAH patients to gain insights into disease progression and potential therapeutic interventions. Among these biomarkers, cellular biomarkers, such as endothelial progenitor cells (EPC) and circulating endothelial cells (CEC), are potential candidates to study the progression of PAH disease and therapeutic intervention as they contribute to vascular repair and remodeling [238]. There are some challenges to study of these cells such as the endothelial dysfunction and EPC number are inversely correlated [239, 240]; how the EPC promotes vascular repairment is still unclear [241]. Therefore, cell isolation and accurate quantification are necessary to explore PAH disease mechanisms and therapeutic targets. Standard techniques, such as enzyme-linked immunosorbent assay (ELISA) and polymerase chain reaction (PCR), make it relatively simple to measure molecular biomarkers; nevertheless, it is difficult and time-consuming to isolate and quantify cellular biomarkers (EPCs and CECs) using standard methods such as flow cytometry, magnetic bead-based techniques, and colony-forming assays [242, 243]. Furthermore, the quantification of EPC in PAH patients is highly inconsistent [244, 245], showing either an increase [244, 246], decrease [247–249], or no-change [248, 250] compared to the healthy controls. To overcome this problem, disposable microfluidics chips, "EPC capture chip" [242] and "CEC capture chip" [251], were designed to trap and count the EPCs and CECs with amazingly small amounts of whole blood (200–400 μL) without sample pre-preprocessing. The reliability of these chips was validated using flow cytometry, resulting in a strong correlation (EPCs, $R = 0.83$; CECs, $R = 0.89$). These innovative capture chips could serve as a robust method for measuring, monitoring, and screening samples from PAH patients, as well as those with other cardiovascular

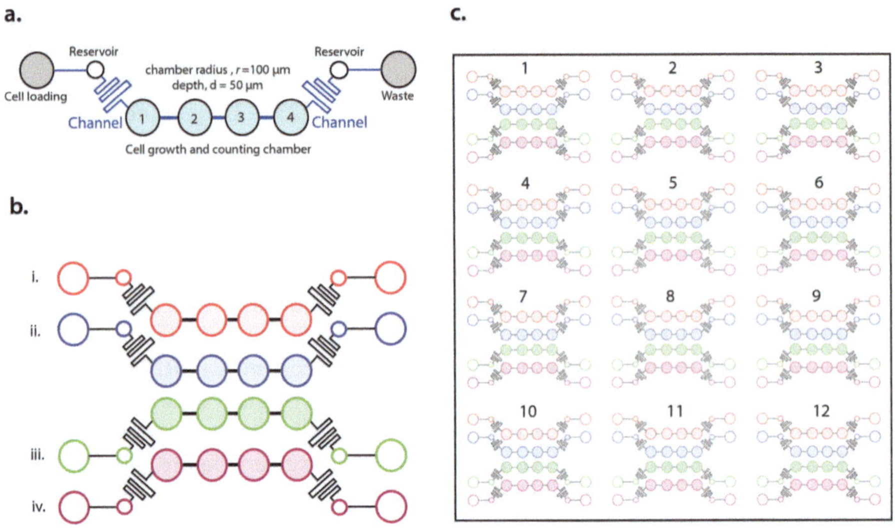

Fig. 14.2 A conceptual PAH-chip model. (**a**) A prototype design displaying the various components of a chip, such as a cell loading chamber, reservoirs, cell growth and counting chambers, channels, and waste chamber. (**b**) Multiple chips can be combined into one for concurrent sample loading. (**c**) Multiple replicate analyses are possible on a single chip; here showed 12 chips

and neurodegenerative diseases [252] and cancer [253]. The development of accurate and reliable microfluidic chips represents a significant advancement in PAH research and has the potential to revolutionize disease diagnosis, monitoring, and the discovery of targeted treatments. These microfluidic devices offer new possibilities for understanding disease mechanisms, identifying biomarkers, and eventually improving patient outcomes. As technology continues to evolve, we can expect even more sophisticated and precise microfluidic solutions that will continue to drive progress in PAH and other areas of medicine.

14.2.8.4 Conceptual Model Design for a PAH-Chip

A straightforward conceptual "PAH-chip" model (Fig. 14.2a) is presented to describe how a microfluidic chip is designed and also briefly explains the printing and fabrication process using cutting-edge 3D printing technology. This model "PAH-chip" contains four chambers, each of which has the same size (radius, $r = 100$ µm and depth, $d = 50$ µm). However, the dimensions of other components (such as the channel, reservoirs, cell loading zone, and outlet to release waste) are not shown, as this is simply a conceptual model. Based on the experimental model and other requirements, design customization is feasible. These chips can be made for simultaneous loading and screening. For example, Fig. 14.2b design includes four chips under platform (or one chip) that can be simultaneously used for four different sample/cell types. The "PAH-chips" device allows the growth of three major types of PAC cells simultaneously in one chamber. These PAC cells are endothelial cells, smooth muscle cells, and adventitial cells [217].

One advantage of using the "PAH-chips" device is that it allows researchers to study the interactions between these three types of cells in a controlled environment that closely mimics the in vivo conditions of the pulmonary arterial system. Additionally, the device allows for one chamber to be used as a control, which enables researchers to compare the behavior of cells in the experimental chamber to that of cells in the control chamber. However, multiple replicates analysis is also possible using the model design in Fig. 14.2c, which reduces time and provides consistent and rapid experiment. By using 'PAH-chips', researchers can better understand the complex interactions between different types of PAC cells and identify promising therapeutic targets for pulmonary arterial diseases. Notably, conventional and standardized approaches (such as flow cytometry and colony-forming assay) are employed to isolate and quantify cells, which is challenging and time-consuming [242, 243]. Advanced microfluidics-based approaches can solve this limitation, as discussed earlier in this section.

Making a "PAH-chip" starts with the design of a 3D model. The "PAH-chip" prototype can be designed using an open-source CAD (computer-aided design) program (e.g., AutoCAD, Fusion 360). An ideal microfluidics device comprises of an inlet (cell loading), reservoir (flow turbulence control), chambers, and an outlet (waste) to expel by-products. Here, the design is similar to Fig. 14.2a. Next, the designed CAD model (e.g., stereolithography file format) needs to be transferred to a slicer software (e.g., DeScribe, Nanoscribe, Germany) in order to make the 3D printing compatible file format (e.g., GWL, Nanoscribe, Germany). Recently, an ultra-high precision two-photon polymerization (2PP)-based 3D printer (e.g., Nanoscribe, Germany) offers a powerful platform to print micro- to meso-level structures. However, after printing, polydimethylsiloxane (PDMS) can be used to construct a PDMS-based microchip. PDMS (a prepolymer and cross-linker at a 10:1 (w/w) ratio) is a widely used elastomer in the field of microfluidics and lab-on-a-chip devices. However, the ratio could be changed based on necessity. These devices are produced based on experimental specifications, testing, and they often require modification (multiple generations of a device) for the desired outcome. The general microfabrication process has already been established by Jeff Hasty's lab [254] and other pioneered microfluidics research groups in pulmonary hypertension [21, 212, 217, 255].

In the last 20 years, most microfluidic devices were designed using photolithography, which was initially developed for the semiconductor industry [254]. In recent years, the 3D printing of microfluidic devices and molds of microfluidic devices have gained popularity because of the preciseness and complexity of structures that can be printed [256]. Microfluidic devices for bacteria are often challenging to produce due to the small size of bacteria. In addition, devices that monitor multiple-sized cells (e.g., bacteria and plant cells) can be challenging to produce accurately using photolithography. However, 3D printing provides the accuracy and speed to make these types of structures at a relatively low cost. For example, the Butzin lab's 3D microfab facility [257] has recently built several new microfluidic devices to monitor bacteria, bacteria–plant interactions, and accurately and quickly count cells or particles precisely and reliably regardless of size and shape (data not shown).

Recently, microfluidics coupled with an advanced microscope system (e.g., Nikon Eclipse Ti2 inverted fluorescence) has shown great success. Live cell imaging, single cell analysis, and cell tracking have already been implemented in antibiotic resistance studies [258–260] and other prominent research including single cell immunology [261], cancer cell trapping [262], and host–microbe interaction at single cell level [263]. This integrated knowledge might help to create a better microfluidic system and data analysis platform to probe disease mechanisms and potential drug target identification.

14.2.8.5 Current Limitation of Microfluidic-Based LoC Study in PAH

Although some microfluidic studies were employed to explore pulmonary hypertension, several notable gaps were highlighted in each study. An organ-chip model can mimic human pathophysiology, albeit there is not much work showing the application of therapeutics using chip methods. Some microfluidic devices were built using conventional 2D wafer-based methods; however, they could not accurately recreate the in vivo 3D model [242]. Sato and Costa's research team examined shear stress to comprehend the dynamics and nature of hypertension and thrombosis. They proposed further customization of their device may aid in clearly understanding the PAH mechanism [212, 213]. Numerous microfluidics research concentrated on certain aspects of cardiovascular function, including angiogenesis [264, 265], transport properties [266–268], shear stress [269], cancer metastasis [270], and hemodynamics [271]. Based on the current knowledge, an integrated organ-on-a-chip model with cutting-edge tissue engineering, 3D printing, and microfabrication technologies may facilitate the exploration of the complex processes behind pulmonary hypertension and the identification of effective therapeutic targets.

14.3 Conclusion

Epigenetic studies in concert with PAH have steadily increased in recent times. Clinical and experimental data suggested that the plasticity of epigenetic changes in PAH could be a plausible candidate for future research and therapeutic development. Indeed, microfluidic LoC technology has the potential to offer a novel study design mimicking human pathobiology to resolve the current constraints in PAH studies. Epigenetics, in conjunction with microfluidic LoC technology, is perhaps the most useful research in unraveling the mystery of PAH pathogenesis and identifying novel drugs to treat this chronic progressive cardiopulmonary disease.

Acknowledgments We thank the members of Dr. Butzin's lab for the critical review of this manuscript and for sharing their experiences. This work is supported by the National Science Foundation Award Numbers 1922542 and 1849206, and by a USDA National Institute of Food and Agriculture Hatch project grant number SD00H653-18/ project accession no. 1015687.

References

1. Thenappan T, et al. Pulmonary arterial hypertension: pathogenesis and clinical management. BMJ. 2018;360:j5492.
2. Tonelli AR, et al. Causes and circumstances of death in pulmonary arterial hypertension. Am J Respir Crit Care Med. 2013;188(3):365–9.
3. Humbert M, et al. Advances in therapeutic interventions for patients with pulmonary arterial hypertension. Circulation. 2014;130(24):2189–208.
4. Humbert M, Ghofrani H-A. The molecular targets of approved treatments for pulmonary arterial hypertension. Thorax. 2016;71(1):73–83.
5. Giaid A, et al. Expression of endothelin-1 in the lungs of patients with pulmonary hypertension. N Engl J Med. 1993;328(24):1732–9.
6. Morrell NW, et al. Altered growth responses of pulmonary artery smooth muscle cells from patients with primary pulmonary hypertension to transforming growth factor-β1 and bone morphogenetic proteins. Circulation. 2001;104(7):790–5.
7. Perros F, et al. Pulmonary lymphoid neogenesis in idiopathic pulmonary arterial hypertension. Am J Respir Crit Care Med. 2012;185(3):311–21.
8. Ma L, et al. A novel channelopathy in pulmonary arterial hypertension. N Engl J Med. 2013;369:351–61.
9. Lau EM, et al. Epidemiology and treatment of pulmonary arterial hypertension. Nat Rev Cardiol. 2017;14(10):603–14.
10. Luna R, et al. Insights on the epigenetic mechanisms underlying pulmonary arterial hypertension. Braz J Med Biol Res. 2018;51:e7437.
11. Kim J-D, et al. Epigenetic modulation as a therapeutic approach for pulmonary arterial hypertension. Exp Mol Med. 2015;47(7):e175.
12. Tuder RM, et al. Relevant issues in the pathology and pathobiology of pulmonary hypertension. J Am Coll Cardiol. 2013;62(25S):D4–D12.
13. Zhao L, et al. Histone deacetylation inhibition in pulmonary hypertension: therapeutic potential of valproic acid and suberoylanilide hydroxamic acid. Circulation. 2012;126(4):455–67.
14. Archer SL, et al. Epigenetic attenuation of mitochondrial superoxide dismutase 2 in pulmonary arterial hypertension: a basis for excessive cell proliferation and a new therapeutic target. Circulation. 2010;121(24):2661–71.
15. Ho L, et al. Epigenetic mechanisms as emerging therapeutic targets and microfluidic chips application in pulmonary arterial hypertension. Biomedicine. 2022;10(1):170.
16. Kim GH, et al. Epigenetic mechanisms of pulmonary hypertension. Pulm Circ. 2011;1(3): 347–56.
17. Wang P, et al. miRNA-34a promotes proliferation of human pulmonary artery smooth muscle cells by targeting PDGFRA. Cell Prolif. 2016;49(4):484–93.
18. Chen T, et al. Loss of microRNA-17∼ 92 in smooth muscle cells attenuates experimental pulmonary hypertension via induction of PDZ and LIM domain 5. Am J Respir Crit Care Med. 2015;191(6):678–92.
19. Chen T, et al. MicroRNA-212-5p, an anti-proliferative miRNA, attenuates hypoxia and sugen/ hypoxia-induced pulmonary hypertension in rodents. Mol Ther Nucleic Acids. 2022;29:204–16.
20. Ali MK, et al. The role of circular RNAs in pulmonary hypertension. Eur Respir J. 2022;
21. Wojciak-Stothard B et al. A microfluidic chip for pulmonary arterial hypertension. 2021.
22. Low L, Tagle D. Tissue chips–innovative tools for drug development and disease modeling. Lab Chip. 2017;17(18):3026–36.
23. Leung CM, et al. A guide to the organ-on-a-chip. Nat Rev Methods Primers. 2022;2(1):1–29.
24. Bhatia SN, Ingber DE. Microfluidic organs-on-chips. Nat Biotechnol. 2014;32(8):760–72.
25. Ingber DE. Human organs-on-chips for disease modelling, drug development and personalized medicine. Nat Rev Genet. 2022;23(8):467–91.

26. Cheng X, Wang Y, Du L. Epigenetic modulation in the initiation and progression of pulmonary hypertension. Hypertension. 2019;74(4):733–9.
27. Bisserier M, et al. Targeting epigenetic mechanisms as an emerging therapeutic strategy in pulmonary hypertension disease. Vasc Biol. 2020;2(1):R17–34.
28. Shimoda LA, Laurie SS. Vascular remodeling in pulmonary hypertension. J Mol Med. 2013;91(3):297–309.
29. Wang A, et al. Substrate stiffness and stretch regulate profibrotic mechanosignaling in pulmonary arterial adventitial fibroblasts. Cell. 2021;10(5):1000.
30. Hotchkiss RD. The quantitative separation of purines, pyrimidines, an d nucleosides by paper chromatography. J Biol Chem. 1948;175(1):315–32.
31. Holliday R, Pugh JE. DNA modification mechanisms and gene activity during development: developmental clocks may depend on the enzymic modification of specific bases in repeated DNA sequences. Science. 1975;187(4173):226–32.
32. Bestor TH, Verdine GL. DNA methyltransferases. Curr Opin Cell Biol. 1994;6(3):380–9.
33. Moore LD, Le T, Fan G. DNA methylation and its basic function. Neuropsychopharmacology. 2013;38(1):23–38.
34. Xie S, et al. Cloning, expression and chromosome locations of the human DNMT3 gene family. Gene. 1999;236(1):87–95.
35. Hermann A, Goyal R, Jeltsch A. The Dnmt1 DNA-(cytosine-C5)-methyltransferase methylates DNA processively with high preference for hemimethylated target sites. J Biol Chem. 2004;279(46):48350–9.
36. Li E, Bestor TH, Jaenisch R. Targeted mutation of the DNA methyltransferase gene results in embryonic lethality. Cell. 1992;69(6):915–26.
37. Okano M, et al. DNA methyltransferases Dnmt3a and Dnmt3b are essential for de novo methylation and mammalian development. Cell. 1999;99(3):247–57.
38. Mayer W, et al. Demethylation of the zygotic paternal genome. Nature. 2000;403(6769): 501–2.
39. Zhang F, et al. Active tissue-specific DNA demethylation conferred by somatic cell nuclei in stable heterokaryons. Proc Natl Acad Sci. 2007;104(11):4395–400.
40. Bhutani N, Burns DM, Blau HM. DNA demethylation dynamics. Cell. 2011;146(6):866–72.
41. Tahiliani M, et al. Conversion of 5-methylcytosine to 5-hydroxymethylcytosine in mammalian DNA by MLL partner TET1. Science. 2009;324(5929):930–5.
42. Ito S, et al. Role of Tet proteins in 5mC to 5hmC conversion, ES-cell self-renewal and inner cell mass specification. Nature. 2010;466(7310):1129–33.
43. Ito S, et al. Tet proteins can convert 5-methylcytosine to 5-formylcytosine and 5-carboxylcytosine. Science. 2011;333(6047):1300–3.
44. Meehan RR, et al. Identification of a mammalian protein that binds specifically to DNA containing methylated CpGs. Cell. 1989;58(3):499–507.
45. Nan X, Meehan RR, Bird A. Dissection of the methyl-CpG binding domain from the chromosomal protein MeCP2. Nucleic Acids Res. 1993;21(21):4886–92.
46. Hendrich B, Bird A. Identification and characterization of a family of mammalian methyl CpG-binding proteins. Genet Res. 1998;72(1):59–72.
47. Bellacosa A, et al. MED1, a novel human methyl-CpG-binding endonuclease, interacts with DNA mismatch repair protein MLH1. Proc Natl Acad Sci. 1999;96(7):3969–74.
48. Petronzelli F, et al. Biphasic kinetics of the human DNA repair protein MED1 (MBD4), a mismatch-specific DNA N-glycosylase. J Biol Chem. 2000;275(42):32422–9.
49. Wong E, et al. Mbd4 inactivation increases C→ T transition mutations and promotes gastrointestinal tumor formation. Proc Natl Acad Sci. 2002;99(23):14937–42.
50. Hashimoto H, et al. The SRA domain of UHRF1 flips 5-methylcytosine out of the DNA helix. Nature. 2008;455(7214):826–9.
51. Sharif J, et al. The SRA protein Np95 mediates epigenetic inheritance by recruiting Dnmt1 to methylated DNA. Nature. 2007;450(7171):908–12.

52. Achour M, et al. The interaction of the SRA domain of ICBP90 with a novel domain of DNMT1 is involved in the regulation of VEGF gene expression. Oncogene. 2008;27(15): 2187–97.
53. Prokhortchouk A, et al. The p120 catenin partner Kaiso is a DNA methylation-dependent transcriptional repressor. Genes Dev. 2001;15(13):1613–8.
54. Filion GJ, et al. A family of human zinc finger proteins that bind methylated DNA and repress transcription. Mol Cell Biol. 2006;26(1):169–81.
55. Sasai N, Nakao M, Defossez P-A. Sequence-specific recognition of methylated DNA by human zinc-finger proteins. Nucleic Acids Res. 2010;38(15):5015–22.
56. Daniel JM, et al. The p120 ctn-binding partner Kaiso is a bi-modal DNA-binding protein that recognizes both a sequence-specific consensus and methylated CpG dinucleotides. Nucleic Acids Res. 2002;30(13):2911–9.
57. Ehrlich M. DNA hypomethylation in cancer cells. Epigenomics. 2009;1(2):239–59.
58. Gao F, et al. Global analysis of DNA methylation in hepatocellular carcinoma by a liquid hybridization capture-based bisulfite sequencing approach. Clin Epigenetics. 2015;7(1):1–11.
59. Perros F, et al. Cytotoxic cells and granulysin in pulmonary arterial hypertension and pulmonary veno-occlusive disease. Am J Respir Crit Care Med. 2013;187(2):189–96.
60. Han L, et al. DNA methylation and hypertension: emerging evidence and challenges. Brief Funct Genomics. 2016;15(6):460–9.
61. DeLalio LJ, Sved AF, Stocker SD. Sympathetic nervous system contributions to hypertension: updates and therapeutic relevance. Can J Cardiol. 2020;36(5):712–20.
62. Ji L, et al. Association between polymorphisms in the renin-angiotensin-aldosterone system genes and essential hypertension in the Han Chinese population. PLoS One. 2013;8(8): e72701.
63. Hautefort A, et al. Pulmonary endothelial cell DNA methylation signature in pulmonary arterial hypertension. Oncotarget. 2017;8(32):52995.
64. Liu D, et al. Hypermethylation of BMPR2 promoter occurs in patients with heritable pulmonary arterial hypertension and inhibits BMPR2 expression. Am J Respir Crit Care Med. 2017;196(7):925–8.
65. Yan Y, et al. DNA methyltransferase 3B deficiency unveils a new pathological mechanism of pulmonary hypertension. Sci Adv. 2020;6(50):eaba2470.
66. Xu X, et al. A genome-wide methylation study on obesity: differential variability and differential methylation. Epigenetics. 2013;8(5):522–33.
67. Smith JA, et al. Epigenomic indicators of age in African Americans. Hereditary Genet Curr Res. 2014;3:3.
68. Han L, et al. The interactions between alcohol consumption and DNA methylation of the ADD1 gene promoter modulate essential hypertension susceptibility in a population-based, case–control study. Hypertens Res. 2015;38(4):284–90.
69. Bellavia A, et al. DNA hypomethylation, ambient particulate matter, and increased blood pressure: findings from controlled human exposure experiments. J Am Heart Assoc. 2013;2 (3):e000212.
70. Lehninger AL, Nelson DL, Cox MM. Lehninger principles of biochemistry. Macmillan; 2005.
71. Redon C, et al. Histone H2a variants H2AX and H2AZ. Curr Opin Genet Dev. 2002;12(2): 162–9.
72. Millis RM. Epigenetics and hypertension. Curr Hypertens Rep. 2011;13(1):21–8.
73. Gupta S, Yel L. Molecular biology and genetic engineering. Middleton's Allergy: Principles and Practice. 2013;10:162–18.
74. Sakabe K, Wang Z, Hart GW. β-N-acetylglucosamine (O-GlcNAc) is part of the histone code. Proc Natl Acad Sci. 2010;107(46):19915–20.
75. Nathan D, et al. Histone sumoylation is a negative regulator in Saccharomyces cerevisiae and shows dynamic interplay with positive-acting histone modifications. Genes Dev. 2006;20(8): 966–76.

76. Wang Y, et al. Human PAD4 regulates histone arginine methylation levels via demethylimination. Science. 2004;306(5694):279–83.
77. Nelson CJ, Santos-Rosa H, Kouzarides T. Proline isomerization of histone H3 regulates lysine methylation and gene expression. Cell. 2006;126(5):905–16.
78. Meng F, et al. Discovery and development of small molecules targeting epigenetic enzymes with computational methods. In: Epi-informatics. Elsevier; 2016. p. 75–112.
79. Gerthoffer W. Epigenetic targets for oligonucleotide therapies of pulmonary arterial hypertension. Int J Mol Sci. 2020;21(23):9222.
80. Irmak M, Sizlan A. Essential hypertension seems to result from melatonin-induced epigenetic modifications in area postrema. Med Hypotheses. 2006;66(5):1000–7.
81. Morimoto S, et al. Sympathetic activation and contribution of genetic factors in hypertension with neurovascular compression of the rostral ventrolateral medulla. J Hypertens. 1999;17 (11):1577–82.
82. Cohen I, et al. Histone modifiers in cancer: friends or foes? Genes Cancer. 2011;2(6):631–47.
83. Bannister AJ, Kouzarides T. Regulation of chromatin by histone modifications. Cell Res. 2011;21(3):381–95.
84. Chelladurai P, et al. Targeting histone acetylation in pulmonary hypertension and right ventricular hypertrophy. Br J Pharmacol. 2021;178(1):54–71.
85. Li M, et al. Emergence of fibroblasts with a proinflammatory epigenetically altered phenotype in severe hypoxic pulmonary hypertension. J Immunol. 2011;187(5):2711–22.
86. Boucherat O, et al. HDAC6: a novel histone deacetylase implicated in pulmonary arterial hypertension. Sci Rep. 2017;7(1):1–14.
87. Desjardins CA, Naya FJ. The function of the MEF2 family of transcription factors in cardiac development, cardiogenomics, and direct reprogramming. J Cardiovasc Dev Dis. 2016;3(3): 26.
88. Kim J, et al. Restoration of impaired endothelial myocyte enhancer factor 2 function rescues pulmonary arterial hypertension. Circulation. 2015;131(2):190–9.
89. Chelladurai P, Seeger W, Pullamsetti SS. Epigenetic mechanisms in pulmonary arterial hypertension: the need for global perspectives. Eur Respir Rev. 2016;25(140):135–40.
90. Fuks F, et al. The DNA methyltransferases associate with HP1 and the SUV39H1 histone methyltransferase. Nucleic Acids Res. 2003;31(9):2305–12.
91. Fuks F, et al. DNA methyltransferase Dnmt1 associates with histone deacetylase activity. Nat Genet. 2000;24(1):88–91.
92. Ali MK, Ichimura K, Spiekerkoetter E. Promising therapeutic approaches in pulmonary arterial hypertension. Curr Opin Pharmacol. 2021;59:127–39.
93. Federici C, et al. Increased mutagen sensitivity and DNA damage in pulmonary arterial hypertension. Am J Respir Crit Care Med. 2015;192(2):219–28.
94. Provencher S. Olaparib for PAH: a multicenter clinical trial. 2018.
95. Meloche J, et al. Role for DNA damage signaling in pulmonary arterial hypertension. Circulation. 2014;129(7):786–97.
96. Provencher S. Apabetalone for pulmonary arterial hypertension: a pilot study. 2018.
97. Meloche J, et al. Bromodomain-containing protein 4: the epigenetic origin of pulmonary arterial hypertension. Circ Res. 2015;117(6):525–35.
98. Sun BK, Tsao H. Small RNAs in development and disease. J Am Acad Dermatol. 2008;59(5): 725–37.
99. Arasu P, Wightman B, Ruvkun G. Temporal regulation of lin-14 by the antagonistic action of two other heterochronic genes, lin-4 and lin-28. Genes Dev. 1991;5(10):1825–33.
100. Lee RC, Feinbaum RL, Ambros V. The C. elegans heterochronic gene lin-4 encodes small RNAs with antisense complementarity to lin-14. Cell. 1993;75(5):843–54.
101. Wightman B, Ha I, Ruvkun G. Posttranscriptional regulation of the heterochronic gene lin-14 by lin-4 mediates temporal pattern formation in C. elegans. Cell. 1993;75(5):855–62.
102. Reinhart BJ, et al. The 21-nucleotide let-7 RNA regulates developmental timing in Caenorhabditis elegans. Nature. 2000;403(6772):901–6.

103. Cummins J, Velculescu V. Implications of micro-RNA profiling for cancer diagnosis. Oncogene. 2006;25(46):6220–7.
104. Lewis BP, Burge CB, Bartel DP. Conserved seed pairing, often flanked by adenosines, indicates that thousands of human genes are microRNA targets. Cell. 2005;120(1):15–20.
105. Zeng Y. Principles of micro-RNA production and maturation. Oncogene. 2006;25(46): 6156–62.
106. Zhou G, Chen T, Raj JU. MicroRNAs in pulmonary arterial hypertension. Am J Respir Cell Mol Biol. 2015;52(2):139–51.
107. Lee A, et al. Therapeutic implications of microRNAs in pulmonary arterial hypertension. BMB Rep. 2014;47(6):311.
108. Negi V, Chan SY. Discerning functional hierarchies of microRNAs in pulmonary hypertension. JCI insight. 2017;2(5):e91327.
109. Chen J-Q, et al. The role of microRNAs in the pathogenesis of autoimmune diseases. Autoimmun Rev. 2016;15(12):1171–80.
110. Meloche J, et al. Therapeutic potential of microRNA modulation in pulmonary arterial hypertension. Curr Vasc Pharmacol. 2015;13(3):331–40.
111. Boucherat O, Potus F, Bonnet S. microRNA and pulmonary hypertension. Adv Exp Med Biol. 2015;888:237–52.
112. Santos-Ferreira CA, et al. Micro-RNA analysis in pulmonary arterial hypertension: current knowledge and challenges. Basic Transl Sci. 2020;5(11):1149–62.
113. Ma W, et al. Inhibition of microRNA-30a alleviates vascular remodeling in pulmonary arterial hypertension. Mol Ther Nucleic Acids. 2021;26:678–93.
114. Zhang J, et al. Micro RNA-483 amelioration of experimental pulmonary hypertension. EMBO Mol Med. 2020;12(5):e11303.
115. Xu J, et al. MicroRNAs in pulmonary hypertension, from pathogenesis to diagnosis and treatment. Biomol Ther. 2022;12(4):496.
116. Wang Y, et al. Epigenetic regulation and its therapeutic potential in pulmonary hypertension. Front Pharmacol. 2018;9:241.
117. Carregal-Romero S, et al. MicroRNA nanotherapeutics for lung targeting. Insights into pulmonary hypertension. Int J Mol Sci. 2020;21(9):3253.
118. Haussecker D. Current issues of RNAi therapeutics delivery and development. J Control Release. 2014;195:49–54.
119. Ohno S, Much S. Junk. in DNA in our genome. Brookhaven Symp Biol. 1972.
120. Beermann J, et al. Non-coding RNAs in development and disease: background, mechanisms, and therapeutic approaches. Physiol Rev. 2016;
121. Castel SE, Martienssen RA. RNA interference in the nucleus: roles for small RNAs in transcription, epigenetics and beyond. Nat Rev Genet. 2013;14(2):100–12.
122. Engreitz JM, et al. Local regulation of gene expression by lncRNA promoters, transcription and splicing. Nature. 2016;539(7629):452–5.
123. Batista PJ, Chang HY. Long noncoding RNAs: cellular address codes in development and disease. Cell. 2013;152(6):1298–307.
124. Brown CJ, et al. The human XIST gene: analysis of a 17 kb inactive X-specific RNA that contains conserved repeats and is highly localized within the nucleus. Cell. 1992;71(3): 527–42.
125. Derrien T, et al. The GENCODE v7 catalog of human long noncoding RNAs: analysis of their gene structure, evolution, and expression. Genome Res. 2012;22(9):1775–89.
126. Statello L, et al. Gene regulation by long non-coding RNAs and its biological functions. Nat Rev Mol Cell Biol. 2021;22(2):96–118.
127. Lesizza P, et al. Noncoding RNAs in cardiovascular disease, In Nucleic acid nanotheranostics. 2019, Elsevier. p. 43–87.
128. Ulitsky I, et al. Conserved function of lincRNAs in vertebrate embryonic development despite rapid sequence evolution. Cell. 2011;147(7):1537–50.

129. Katayama S, et al. Antisense transcription in the mammalian transcriptome. Science. 2005;309 (5740):1564–6.
130. Rearick D, et al. Critical association of ncRNA with introns. Nucleic Acids Res. 2011;39(6): 2357–66.
131. Buratowski S. Transcription. Gene expression--where to start? Science (New York, NY). 2008;322(5909):1804–5.
132. Wang D, et al. Reprogramming transcription by distinct classes of enhancers functionally defined by eRNA. Nature. 2011;474(7351):390–4.
133. Salzman J, et al. Circular RNAs are the predominant transcript isoform from hundreds of human genes in diverse cell types. PLoS One. 2012;7(2):e30733.
134. Guo C-J, et al. Distinct processing of lncRNAs contributes to non-conserved functions in stem cells. Cell. 2020;181(3):621–36. e22
135. Quinn JJ, et al. Rapid evolutionary turnover underlies conserved lncRNA–genome interactions. Genes Dev. 2016;30(2):191–207.
136. Hezroni H, et al. Principles of long noncoding RNA evolution derived from direct comparison of transcriptomes in 17 species. Cell Rep. 2015;11(7):1110–22.
137. Yao R-W, Wang Y, Chen L-L. Cellular functions of long noncoding RNAs. Nat Cell Biol. 2019;21(5):542–51.
138. Anderson DM, et al. A micropeptide encoded by a putative long noncoding RNA regulates muscle performance. Cell. 2015;160(4):595–606.
139. Nelson BR, et al. A peptide encoded by a transcript annotated as long noncoding RNA enhances SERCA activity in muscle. Science. 2016;351(6270):271–5.
140. Wapinski O, Chang HY. Long noncoding RNAs and human disease. Trends Cell Biol. 2011;21(6):354–61.
141. Han Y, et al. Role of long non-coding RNAs in pulmonary arterial hypertension. Cell. 2021;10 (8):1892.
142. Zahid KR, et al. Pathobiology of pulmonary artery hypertension: role of long non-coding RNAs. Cardiovasc Res. 2020;116(12):1937–47.
143. Rieg AD, et al. PDGF-BB regulates the pulmonary vascular tone: impact of prostaglandins, calcium, MAPK-and PI3K/AKT/mTOR signalling and actin polymerisation in pulmonary veins of Guinea pigs. Respir Res. 2018;19(1):1–18.
144. Nogueira-Ferreira R, et al. Exploring the monocrotaline animal model for the study of pulmonary arterial hypertension: a network approach. Pulm Pharmacol Ther. 2015;35:8–16.
145. Wang Z, et al. Long non-coding RNA MEG3 mediates high glucose-induced endothelial cell dysfunction. Int J Clin Exp Pathol. 2018;11(3):1088.
146. Stuhlmüller B, et al. Detection of oncofetal h19 RNA in rheumatoid arthritis synovial tissue. Am J Pathol. 2003;163(3):901–11.
147. Wang S-H, et al. Upregulation of H19 indicates a poor prognosis in gallbladder carcinoma and promotes epithelial-mesenchymal transition. Am J Cancer Res. 2016;6(1):15.
148. Su H, et al. LncRNA H19 promotes the proliferation of pulmonary artery smooth muscle cells through AT1R via sponging let-7b in monocrotaline-induced pulmonary arterial hypertension. Respir Res. 2018;19(1):1–18.
149. Gu S, et al. Aberrant expression of long noncoding RNAs in chronic thromboembolic pulmonary hypertension. Mol Med Rep. 2015;11(4):2631–43.
150. Voellenkle C, et al. Implication of long noncoding RNAs in the endothelial cell response to hypoxia revealed by RNA-sequencing. Sci Rep. 2016;6:24141.
151. Zehendner CM, et al. Long noncoding RNA TYKRIL plays a role in pulmonary hypertension via the p53-mediated regulation of PDGFRβ. Am J Respir Crit Care Med. 2020;202(10): 1445–57.
152. Lei S, et al. LncRNA-SMILR modulates RhoA/ROCK signaling by targeting miR-141 to regulate vascular remodeling in pulmonary arterial hypertension. Am J Phys Heart Circ Phys. 2020;319(2):H377–91.

153. Zhuo Y, et al. Functional polymorphism of lncRNA MALAT1 contributes to pulmonary arterial hypertension susceptibility in Chinese people. Clin Chem Lab Med. 2017;55(1): 38–46.
154. Smaldone MC, Davies BJ. BC-819, a plasmid comprising the H19 gene regulatory sequences and diphtheria toxin A, for the potential targeted therapy of cancers. Curr Opin Mol Ther. 2010;12(5):607–16.
155. Gomes CP, et al. The function and therapeutic potential of long non-coding RNAs in cardiovascular development and disease. Mol Ther Nucleic Acids. 2017;8:494–507.
156. Diener T. Potato spindle tuber virus: a plant virus with properties of a free nucleic acid: III. Subcellular location of PSTV-RNA and the question of whether virions exist in extracts or in situ. Virology. 1971;43(1):75–89.
157. Diener T. Potato spindle tuber "virus": IV. A replicating, low molecular weight RNA. Virology. 1971;45(2):411–28.
158. Wang Q, et al. Circular RNAs in pulmonary hypertension: emerging biological concepts and potential mechanism. Animal Model Exp Med. 2022;5(1):38–47.
159. Nigro JM, et al. Scrambled exons. Cell. 1991;64(3):607–13.
160. Wang J, et al. Circular RNAs: a rising star in respiratory diseases. Respir Res. 2019;20(1): 1–10.
161. Kristensen LS, et al. The biogenesis, biology and characterization of circular RNAs. Nat Rev Genet. 2019;20(11):675–91.
162. Jeck WR, et al. Circular RNAs are abundant, conserved, and associated with ALU repeats. RNA. 2013;19(2):141–57.
163. Jeck WR, Sharpless NE. Detecting and characterizing circular RNAs. Nat Biotechnol. 2014;32 (5):453–61.
164. Salzman J, et al. Cell-type specific features of circular RNA expression. PLoS Genet. 2013;9 (9):e1003777.
165. Memczak S, et al. Circular RNAs are a large class of animal RNAs with regulatory potency. Nature. 2013;495(7441):333–8.
166. Dong R, et al. Increased complexity of circRNA expression during species evolution. RNA Biol. 2017;14(8):1064–74.
167. Werfel S, et al. Characterization of circular RNAs in human, mouse and rat hearts. J Mol Cell Cardiol. 2016;98:103–7.
168. Chen W, Schuman E. Circular RNAs in brain and other tissues: a functional enigma. Trends Neurosci. 2016;39(9):597–604.
169. Li X, Yang L, Chen L-L. The biogenesis, functions, and challenges of circular RNAs. Mol Cell. 2018;71(3):428–42.
170. Starke S, et al. Exon circularization requires canonical splice signals. Cell Rep. 2015;10(1): 103–11.
171. Zhang Y, et al. Circular intronic long noncoding RNAs. Mol Cell. 2013;51(6):792–806.
172. Li Z, et al. Exon-intron circular RNAs regulate transcription in the nucleus. Nat Struct Mol Biol. 2015;22(3):256–64.
173. Ashwal-Fluss R, et al. circRNA biogenesis competes with pre-mRNA splicing. Mol Cell. 2014;56(1):55–66.
174. Conn SJ, et al. The RNA binding protein quaking regulates formation of circRNAs. Cell. 2015;160(6):1125–34.
175. Zhang X-O, et al. Complementary sequence-mediated exon circularization. Cell. 2014;159(1): 134–47.
176. Gao Y, et al. Comprehensive identification of internal structure and alternative splicing events in circular RNAs. Nat Commun. 2016;7(1):1–13.
177. Lee Y, et al. Molecular mechanisms driving mRNA degradation by m6A modification. Trends Genet. 2020;36(3):177–88.
178. Liu C-X, et al. Structure and degradation of circular RNAs regulate PKR activation in innate immunity. Cell. 2019;177(4):865–80. e21

179. Hansen TB, et al. miRNA-dependent gene silencing involving Ago2-mediated cleavage of a circular antisense RNA. EMBO J. 2011;30(21):4414–22.
180. Lasda E, Parker R. Circular RNAs co-precipitate with extracellular vesicles: a possible mechanism for circRNA clearance. PLoS One. 2016;11(2):e0148407.
181. Hansen TB, et al. Natural RNA circles function as efficient microRNA sponges. Nature. 2013;495(7441):384–8.
182. Yu C-Y, Kuo H-C. The emerging roles and functions of circular RNAs and their generation. J Biomed Sci. 2019;26(1):1–12.
183. Liu C, et al. A narrative review of circular RNAs as potential biomarkers and therapeutic targets for cardiovascular diseases. Ann Transl Med. 2021;9(7):578.
184. Xu S-L, et al. Regulation of circular RNAs act as ceRNA in a hypoxic pulmonary hypertension rat model. Genomics. 2021;113(1):11–9.
185. Miao R, et al. Microarray expression profile of circular RNAs in chronic thromboembolic pulmonary hypertension. Medicine. 2017;96(27):e7354.
186. Wang J, et al. Characteristics of circular RNA expression in lung tissues from mice with hypoxia-induced pulmonary hypertension. Int J Mol Med. 2018;42(3):1353–66.
187. Huang CX, et al. Hsa_circ_0016070/micro-340-5p Axis accelerates pulmonary arterial hypertension progression by upregulating TWIST1 transcription via TCF4/β-catenin complex. J Am Heart Assoc. 2022;11(14):e024147.
188. Jing X, et al. Circular RNA Sirtuin1 represses pulmonary artery smooth muscle cell proliferation, migration and autophagy to ameliorate pulmonary hypertension via targeting microRNA-145-5p/protein kinase-B3 axis. Bioengineered. 2022;13(4):8759–71.
189. Diao W, et al. Evaluating the effect of Circ-Sirt1 on the expression of SIRT1 and its role in pathology of pulmonary hypertension. Cell Transplant. 2022;31:09636897221081479.
190. Huang Y, et al. Expression and clinical significance of circular RNA hsa_circ_0003416 in pediatric pulmonary arterial hypertension associated with congenital heart disease. J Clin Lab Anal. 2022;36(4):e24273.
191. Li S-S, et al. hsa_circWDR37_016 regulates hypoxia-induced proliferation of pulmonary arterial smooth muscle cells. Cardiovasc Ther. 2022;2022:7292034.
192. Jiang Y, et al. Circular RNA Calm4 regulates hypoxia-induced pulmonary arterial smooth muscle cells pyroptosis via the Circ-Calm4/miR-124-3p/PDCD6 axis. Arterioscler Thromb Vasc Biol. 2021;41(5):1675–93.
193. Ma C, et al. circRNA CDR1as promotes pulmonary artery smooth muscle cell calcification by upregulating CAMK2D and CNN3 via sponging miR-7-5p. Mol Ther Nucleic Acids. 2020;22: 530–41.
194. Wang Y, et al. Hsa_circ_0002062 promotes the proliferation of pulmonary artery smooth muscle cells by regulating the Hsa-miR-942-5p/CDK6 signaling pathway. Front Genet. 2021;12:1201.
195. Yuan P, et al. Impact of circGSAP in peripheral blood mononuclear cells on idiopathic pulmonary arterial hypertension. Am J Respir Crit Care Med. 2021;203(12):1579–83.
196. Zhang J, et al. Circ-calm4 serves as an miR-337-3p sponge to regulate Myo10 (Myosin 10) and promote pulmonary artery smooth muscle proliferation. Hypertension. 2020;75(3): 668–79.
197. Zhou S, et al. Circular RNA hsa_circ_0016070 is associated with pulmonary arterial hypertension by promoting PASMC proliferation. Mol The Nucleic Acids. 2019;18:275–84.
198. Guo J, et al. CircATP2B4 promotes hypoxia-induced proliferation and migration of pulmonary arterial smooth muscle cells via the miR-223/ATR axis. Life Sci. 2020;262:118420.
199. Jin X, et al. hsa_circNFXL1_009 modulates apoptosis, proliferation, migration, and potassium channel activation in pulmonary hypertension. Mol Ther Nucleic Acids. 2021;23:1007–19.
200. Guo HM, Liu ZP. Up-regulation of circRNA_0068481 promotes right ventricular hypertrophy in PAH patients via regulating miR-646/miR-570/miR-885. J Cell Mol Med. 2021;25(8): 3735–43.

201. Duncombe TA, Tentori AM, Herr AE. Microfluidics: reframing biological enquiry. Nat Rev Mol Cell Biol. 2015;16(9):554–67.
202. Nguyen T, et al. Point-of-care devices for pathogen detections: the three most important factors to realise towards commercialization. TrAC Trends Anal Chem. 2020;131:116004.
203. Nguyen T, et al. From lab on a chip to point of care devices: the role of open source microcontrollers. Micromachines. 2018;9(8):403.
204. Jayne RK, et al. Direct laser writing for cardiac tissue engineering: a microfluidic heart on a chip with integrated transducers. Lab Chip. 2021;21(9):1724–37.
205. Zhang B, et al. Advances in organ-on-a-chip engineering. Nat Rev Mater. 2018;3(8):257–78.
206. Huh D, et al. Reconstituting organ-level lung functions on a chip. Science. 2010;328(5986): 1662–8.
207. Sin A, Baxter GT, Shuler ML. Animal on a chip: a microscale cell culture analog device for evaluating toxicological and pharmacological profiles. In: Microfluidics and BioMEMS. 2001. SPIE.
208. Sin A, et al. The design and fabrication of three-chamber microscale cell culture analog devices with integrated dissolved oxygen sensors. Biotechnol Prog. 2004;20(1):338–45.
209. Michas C, et al. Engineering a living cardiac pump on a chip using high-precision fabrication. Sci Adv. 2022;8(16):eabm3791.
210. Lee G, et al. Single microfluidic electrochemical sensor system for simultaneous multi-pulmonary hypertension biomarker analyses. Sci Rep. 2017;7(1):1–8.
211. Bogorad MI, et al. Tissue-engineered 3D microvessel and capillary network models for the study of vascular phenomena. Microcirculation. 2017;24(5):e12360.
212. Sato K, Nitta M, Ogawa A. A microfluidic cell stretch device to investigate the effects of stretching stress on artery smooth muscle cell proliferation in pulmonary arterial hypertension. Inventions. 2018;4(1):1.
213. Costa PF, et al. Mimicking arterial thrombosis in a 3D-printed microfluidic in vitro vascular model based on computed tomography angiography data. Lab Chip. 2017;17(16):2785–92.
214. Westein E, et al. Atherosclerotic geometries exacerbate pathological thrombus formation poststenosis in a von Willebrand factor-dependent manner. Proc Natl Acad Sci. 2013;110 (4):1357–62.
215. Nesbitt WS, et al. A shear gradient–dependent platelet aggregation mechanism drives thrombus formation. Nat Med. 2009;15(6):665–73.
216. Jain A, et al. Assessment of whole blood thrombosis in a microfluidic device lined by fixed human endothelium. Biomed Microdevices. 2016;18(4):1–7.
217. Al-Hilal TA, et al. Pulmonary-arterial-hypertension (PAH)-on-a-chip: fabrication, validation and application. Lab Chip. 2020;20(18):3334–45.
218. Humbert M, et al. Pulmonary arterial hypertension in France: results from a national registry. Am J Respir Crit Care Med. 2006;173(9):1023–30.
219. Shapiro S, et al. Sex differences in the diagnosis, treatment, and outcome of patients with pulmonary arterial hypertension enrolled in the registry to evaluate early and long-term pulmonary arterial hypertension disease management. Chest. 2012;141(2):363–73.
220. Chung L, et al. Survival and predictors of mortality in systemic sclerosis-associated pulmonary arterial hypertension: outcomes from the pulmonary hypertension assessment and recognition of outcomes in scleroderma registry. Wiley Online Library; 2014.
221. de Souza Carvalho C, Daum N, Lehr C-M. Carrier interactions with the biological barriers of the lung: advanced in vitro models and challenges for pulmonary drug delivery. Adv Drug Deliv Rev. 2014;75:129–40.
222. Sarkar T, et al. A protocol for fabrication and on-chip cell culture to recreate PAH-afflicted pulmonary artery on a microfluidic device. Micromachines. 2022;13(9):1483.
223. Mouchaers K, et al. Fasudil reduces monocrotaline-induced pulmonary arterial hypertension: comparison with bosentan and sildenafil. Eur Respir J. 2010;36(4):800–7.

224. Ruan H, et al. The acute effects of 30 mg vs 60 mg of intravenous Fasudil on patients with congenital heart defects and severe pulmonary arterial hypertension. Congenit Heart Dis. 2019;14(4):645–50.
225. Budas GR, et al. ASK1 inhibition halts disease progression in preclinical models of pulmonary arterial hypertension. Am J Respir Crit Care Med. 2018;197(3):373–85.
226. Roman BL, St. Hilaire C. Catching a disease: a molecular trap as a therapy for pulmonary arterial hypertension. American Thoracic Society; 2016. p. 1047–9.
227. Yuan JX-J, Rubin LJ. Pathogenesis of pulmonary arterial hypertension: the need for multiple hits. Am Heart Assoc. 2005:534–8.
228. Southgate L, et al. Molecular genetic framework underlying pulmonary arterial hypertension. Nat Rev Cardiol. 2020;17(2):85–95.
229. Burton VJ, et al. Bone morphogenetic protein receptor II regulates pulmonary artery endothelial cell barrier function. Blood, The Journal of the American Society of Hematology. 2011;117(1):333–41.
230. Barroso M, et al. Inhibition of cellular methyltransferases promotes endothelial cell activation by suppressing glutathione peroxidase 1 protein expression. J Biol Chem. 2014;289(22): 15350–62.
231. Zaiman A, et al. One hundred years of research in the pathogenesis of pulmonary hypertension. Am J Respir Cell Mol Biol. 2005;33(5):425–31.
232. Heath D. The rat is a poor animal model for the study of human pulmonary hypertension. Cardioscience. 1992;3(1):1–6.
233. Robbins IM. Advancing therapy for pulmonary arterial hypertension: can animal models help? American Thoracic Society; 2004. p. 5–6.
234. Schermuly RT, et al. Mechanisms of disease: pulmonary arterial hypertension. Nat Rev Cardiol. 2011;8(8):443–55.
235. Lu C, Verbridge SS. Microfluidic methods for molecular biology, vol. 1. Springer; 2016.
236. Wu AR, et al. Automated microfluidic chromatin immunoprecipitation from 2,000 cells. Lab Chip. 2009;9(10):1365–70.
237. Cao Z, Lu C. A microfluidic device with integrated sonication and immunoprecipitation for sensitive epigenetic assays. Anal Chem. 2016;88(3):1965–72.
238. Pezzuto B, et al. Circulating biomarkers in pulmonary arterial hypertension: update and future direction. J Heart Lung Transplant. 2015;34(3):282–305.
239. Asahara T, et al. Isolation of putative progenitor endothelial cells for angiogenesis. Science. 1997;275(5302):964–6.
240. Urbich C, Dimmeler S. Endothelial progenitor cells: characterization and role in vascular biology. Circ Res. 2004;95(4):343–53.
241. Yan F, et al. Paracrine mechanisms of endothelial progenitor cells in vascular repair. Acta Histochem. 2022;124(1):151833.
242. Hansmann G, et al. Design and validation of an endothelial progenitor cell capture chip and its application in patients with pulmonary arterial hypertension. J Mol Med. 2011;89(10):971–83.
243. Kraan J, et al. Clinical value of circulating endothelial cell detection in oncology. Drug Discov Today. 2012;17(13–14):710–7.
244. Asosingh K, et al. Letter by Asosingh et al regarding article, "circulating endothelial progenitor cells in patients with eisenmenger syndrome and idiopathic pulmonary arterial hypertension". Circulation. 2009;119(9):e230.
245. Diller G-P, et al. Response to letter regarding article,"circulating endothelial progenitor cells in patients with Eisenmenger syndrome and idiopathic pulmonary arterial hypertension". Circulation. 2009;119(9):e231.
246. Toshner M, et al. Evidence of dysfunction of endothelial progenitors in pulmonary arterial hypertension. Am J Respir Crit Care Med. 2009;180(8):780–7.
247. Diller G-P, et al. Circulating endothelial progenitor cells in patients with Eisenmenger syndrome and idiopathic pulmonary arterial hypertension. Circulation. 2008;117(23): 3020–30.

248. JunHui Z, et al. Reduced number and activity of circulating endothelial progenitor cells in patients with idiopathic pulmonary arterial hypertension. Respir Med. 2008;102(7):1073–9.
249. Fadini GP, et al. Depletion of endothelial progenitor cells may link pulmonary fibrosis and pulmonary hypertension. Am J Respir Crit Care Med. 2007;176(7):724–5.
250. Smadja DM, et al. Circulating endothelial cells: a new candidate biomarker of irreversible pulmonary hypertension secondary to congenital heart disease. Circulation. 2009;119(3): 374–81.
251. Sallmon H, et al. Circulating endothelial cell quantification by microfluidics Chip in pulmonary arterial hypertension. Am J Respir Cell Mol Biol. 2017;56(5):680–2.
252. Lee S, et al. Reduced circulating angiogenic cells in Alzheimer disease. Neurology. 2009;72 (21):1858–63.
253. Dome B, et al. Circulating endothelial cells, bone marrow-derived endothelial progenitor cells and proangiogenic hematopoietic cells in cancer: from biology to therapy. Crit Rev Oncol Hematol. 2009;69(2):108–24.
254. Ferry MS, Razinkov IA, Hasty J. Microfluidics for synthetic biology: from design to execution. In: Methods in enzymology. Elsevier; 2011. p. 295–372.
255. Dani K. An integrated microfluidic platform for vascular studies and more efficient preclinical development. Johns Hopkins University; 2020.
256. Prabhakar P, et al. 3D-printed microfluidics and potential biomedical applications. Front Nanotechnol. 2021;3:609355.
257. 3D-Microfab. https://www.sdstate.edu/3d-microfabrication-shared-facility.
258. Deter HS, Hossain T, Butzin NC. Antibiotic tolerance is associated with a broad and complex transcriptional response in E. coli. Sci Rep. 2021;11(1):1–15.
259. Hossain T, et al. Antibiotic tolerance, persistence, and resistance of the evolved minimal cell, Mycoplasma mycoides JCVI-Syn3B. Iscience. 2021;24(5):102391.
260. Hossain T, Singh A, Butzin NC. Escherichia coli cells are primed for survival before lethal antibiotic stress. Microbiol Spectr. 2023:e01219-23.
261. Jammes FC, Maerkl SJ. How single-cell immunology is benefiting from microfluidic technologies. Microsyst Nanoeng. 2020;6(1):1–14.
262. Hisey CL, et al. A versatile cancer cell trapping and 1D migration assay in a microfluidic device. Biomicrofluidics. 2019;13(4):044105.
263. Delincé MJ, et al. A microfluidic cell-trapping device for single-cell tracking of host–microbe interactions. Lab Chip. 2016;16(17):3276–85.
264. Chung S, et al. Cell migration into scaffolds under co-culture conditions in a microfluidic platform. Lab Chip. 2009;9(2):269–75.
265. Song JW, Munn LL. Fluid forces control endothelial sprouting. Proc Natl Acad Sci. 2011;108 (37):15342–7.
266. Young EW, et al. Technique for real-time measurements of endothelial permeability in a microfluidic membrane chip using laser-induced fluorescence detection. Anal Chem. 2010;82 (3):808–16.
267. Douville NJ, et al. Fabrication of two-layered channel system with embedded electrodes to measure resistance across epithelial and endothelial barriers. Anal Chem. 2010;82(6): 2505–11.
268. Shao J, et al. A microfluidic chip for permeability assays of endothelial monolayer. Biomed Microdevices. 2010;12(1):81–8.
269. Reinitz A, et al. Human brain microvascular endothelial cells resist elongation due to shear stress. Microvasc Res. 2015;99:8–18.
270. Song JW, et al. Microfluidic endothelium for studying the intravascular adhesion of metastatic breast cancer cells. PLoS One. 2009;4(6):e5756.
271. Shevkoplyas SS, et al. Direct measurement of the impact of impaired erythrocyte deformability on microvascular network perfusion in a microfluidic device. Lab Chip. 2006;6(7):914–20.

Future Prospects and Challenges

15

Olorunfemi R. Molehin, Adeniyi S. Ohunayo, and Frank A. Ogundolie

Abstract

The abnormality is the state of health that can be attached to various things. This abnormality could be caused by exposure to infectious diseases, reoccurring underlying health conditions, metabolic disorders, comorbidities, and genetic mutations. There are many factors attached to the general well-being of the human body, some of which include immunity and genetic variations. However, epigenetic studies have opened newer ways of determining disease conditions and severity in humans. Epigenetics can be scientifically described as the study which particularly elaborates on the effect of human behavior and immediate environment on individuals regarding their susceptibilities to infections. Epigenetic changes observed in human has been associated with the ever-changing human environments, age, and developmental growth. The future perspective of epigenetics is to link the susceptibility of human to infections, cancers, and metabolic disorders. In this chapter, future prospects and challenges of epigenetics studies will be critically reviewed while providing information on their effect on gene expression.

O. R. Molehin (✉)
Department of Science Laboratory Technology, Faculty of Science, Ekiti State University, Ado-Ekiti, Nigeria
e-mail: olorunfemi.molehin@eksu.edu.ng

A. S. Ohunayo
Department of Biochemistry, Faculty of Science, Ekiti State University, Ado-Ekiti, Nigeria
e-mail: adeniyi.ohunayo@eksu.edu.ng

F. A. Ogundolie
Department of Biotechnology, Baze University, Abuja, Nigeria
e-mail: frank.ogundolie@bazeuniversity.edu.ng

G. Gupta et al. (eds.), *Targeting Epigenetics in Inflammatory Lung Diseases*,
https://doi.org/10.1007/978-981-99-4780-5_15

Keywords

Epigenetics · Behavior · Gene · Infections · Cancer · and Diseases

15.1 Introduction

Over two decades, massive efforts have been made to confer effective treatment to several inflammatory pulmonary diseases including pulmonary embolism, PH, ARDS, IPF, lung tuberculosis, COPD, and asthma [1–5]. As a result of developmental technologies in different fields of –omics, substantial advances have been made to delineate the biology of immune system and function of structural cells associated with pulmonary diseases [2]. Till now, scientists are, however, struggling to develop a perfectly effective and fundamentally novel therapy for many of these lung diseases; as some of those developed have not only failed in some cases but are not able to reverse the established airway damage [2]. For instance, 5–10% of severe persistent asthmatics are insensitive to an established therapy, inhalation of corticosteroids contributing largely to fatal asthma episodes, thereby requiring an alternative approach such as antibody-based anti-inflammatory therapy [2]. The overwhelming production rate of ROS which incapacitate the neutralizing effort of defense mechanisms of endogenous antioxidant during inflammation also serve has bottleneck to establishment of effective therapy to these mortality that have affected innumerable individuals worldwide [4, 6].

Heritable imprints and alterations in the genetic makeup, expression, and functions that do not involve deviations in the DNA base sequence, epigenetics; remain a major challenging factor in conquering inflammatory lung disease and its comorbidities [1, 3–7]. Epigenetic mechanisms that involve changes in phenotypic traits or regulation of gene expression independent of DNA sequence alterations, play intermediate role between genetic and environmental factors associated with most diseases. These changes in phenotype may have many potential causes, like environmental exposure, aging, and disease. Considering the role partaken in the development and progression of complex comorbidity, epigenetics has become an increasingly important area of research toward ILDs, such as asthma, PE, IPF, ARDS, PH, and COPD [1–5]. Interestingly, scientists in recent times proposedly adopt the hypothesis of understanding the underlying mechanisms of epigenetic in developing novel therapy that promises vascular remodeling, long-time suppression, and reversal of pathological airways in inflammatory lung disease [2, 4]. Epigenetic therapies are therefore being developed as a new and alternative treatment option for inflammatory lung diseases. These therapies aim to target specific epigenetic mechanisms in order to change the expression of genes that lead to the illness. For example, drugs that target enzymes that add or remove methyl groups from DNA are being developed as a therapy for asthma [2]. However, intensive research is still required to improve in exploring such promising therapy, as several factors such as the dynamism of epigenetic markers, remain as bottleneck.

Overall, the future of epigenetics studies in ILD is promising, as it may provide new insights into the causes behind these diseases and lead to the development of more effective treatments. However, more studies are needed to completely comprehend the multifaceted interactions between epigenetics, genetics, and the environment in the development of these diseases.

15.2 Epigenetics and General Mechanisms

The word epigenetics originated from the Greek words; *epi* which means "over" or "above," and *genetics* which stands for the science of genetic transfer [5, 8]. Simply put, epigenetics is the study of phenotypic traits inheritable from the mother cell, which resulted from changes in the chromosomes without any alterations in the sequence of DNA molecules [9]. Modifications to phenotypic features or gene expression that are not caused by changes to the DNA sequence are known as epigenetic processes. Therefore, any disease that triggers adaptive changes in cellular response without any alteration in the genetic makeup (gene expression and function) must have been induced by some kind of epigenetic marks. In the core aspect of progression and development of ILD, epigenetic marks have been stated to show the mediating role between environmental and genetic factors including physical such as sunlight and irradiation; chemical such as smokes, silica, industrial agents, and asbestos; and biological such as microbial pathogen and toxin [5, 8, 10–12]. Meanwhile, regulation of genes and proteins expression and functions through epigenetic mechanisms involves a set of highly conserved processes associated with changes in normal cellular development and adaptation, and organ homeostasis [2].

Among the most important epigenetic processes in inflammatory lung disease is the role of methylation, which is a process that can turn off genes. Alterations in methylation patterns have been demonstrated in certain genes that may help with the growth and improvement of ILDs. For example, research has identified specific methylation patterns implicated in immune response and inflammation genes, which may be associated with the development of asthma [13, 14]. Another epigenetic mark in inflammatory lung disease is the role of histone modification, which is another mechanism that can regulate gene expression. Histone modifications can also be associated with the growth and improvement of ILDs. Several functional RNAs including lncRNAs and miRNAs are capable of enhancing protein expression are other epigenetic marks that induce change in the chromosome by altering the DNA sequences [15, 16].

15.2.1 DNA Methylation

Methylation of DNA is a notable epigenetic mechanism involving the highly specific biochemical process of CpG island Methylation refers to the process of adding a methyl group to the 5′ end of a CpG island with the aid of enzyme DNMTs which mediates what causes genes to be silenced [5, 14, 17]. Methyl groups alter DNA's chemical structure, altering gene molecule–cell transcription machinery interactions [5, 9].

15.2.2 Histone Modifications

Another epigenetic mark is the post-translational modification of some highly conserved core chromatin-forming proteins, histones (H4, H3, H2B, and H2A). Methylation and Acetylation which are chemical and biological reactions that add either methyl or acetyl group to lysine residues are two main mechanisms by which histones and chromatin are modified. Acetylation of Histone which is controlled in a bidirectional fashion by a pair of enzymes called histone deacetylases and acetyltransferases leads to relatively uncondensed chromatin with increased permeability to transcription factors; while histone methylation which is catalyzed by histone methyltransferases or demethylases mediates relaxation of chromatin [18–20]. Some other post-translational modifications that directly alter the chromatin structure other than acetylation and methylation include ubiquitination and phosphorylation; the former which takes place on tyrosine, serine, and threonine. Phosphorylation of histones plays a major role in cell cycle, transcription regulation, DNA damage response, and programmed cell death [21]. Histone ubiquitination remains the least understood histone modification mechanism of epigenetics, which regulates transcription through proteasome-dependent mechanisms or destruction factors [5, 22].

15.2.3 Non-coding RNA Regulation

Current advances in technology have proven that RNAs do not only function as transmitters of genetic information to the ribosome, but several functional RNAs including lncRNAs and miRNAs can enhance protein appearance [15, 16]. mRNAs are Long noncoding RNA, lncRNAs that alters the expression of target gene either by degrading or silencing through methylation and alteration of factors of transcription. The miRNAs are reportedly capable of upregulating a disease's phenotype by influencing the activity of genes across the network, such as disruption of lung fibrosis, making them a case study for alternative therapy for complex diseases [5, 15, 23].

lncRNAs are non-coding transcript that has been functionally classified as: protein–protein interactions facilitated by scaffolds, tethering of proteins to a cis-transcription region, and transport of protein complexes to a trans-region of the genome [24]. Some other regulatory functions such as in gene expression, DNA metabolism, and chromatin structure in complex diseases have been attributed to lncRNAs [15, 16, 25–27].

15.3 Epigenetics in Inflammatory Lung Diseases

Basically, all inflammatory lung diseases have two key similar processes associated with their pathophysiology, lung inflammation, and lung remodeling. Though, inflammation in the lung in most cases are controllable with the use of steroidal compound, corticosteroid in the case of asthma, but the severe case may require other approaches like antibody-based anti-inflammatory therapy that targets the IgE, interleukin-5, or thymic stromal lymphopoietin [28]. Tissue remodeling in the other hand is a key characteristic of ILD of the airways including restrictive disorder such as IPF, COPD, and asthma ([1–7]; Gauveau et al. 2014).

15.3.1 Lung Inflammation

Inflammatory lung diseases as the term sounds generally have a distinctive feature that shows a potential effect in the pathogenesis, development, and prognosis of all lethal pulmonary illnesses called inflammation. Inflammation, a means by which the body uses different molecular mediators, vascular system, and immune system to fight external agents, typically involves removing the cause, limiting any harm to a minimum, purging any affected cells or tissues, and promoting recovery [29]. However, when inflammation is short lived and rapid or slow and prolonged, it stimulates the progression and development of acute ILDs like ARDS or chronic diseases like COPD, pulmonary fibrosis, and asthma [30, 31], associated to increasing mortality [4, 32, 33]. The inflammatory response in the lung often induced by toxins, pathogens, irritants, pollutants, and allergens, is a reaction that involves a complex defense mechanism toward external agents and restoration of damaged tissues. External agents induce the activation of inflammatory cells like neutrophils, lymphocytes, and macrophages and the release of pro- and anti-inflammatory factors such as cytokines and mediators like IL-6, IL-5, IL-4, and IL-1β, histamine, PGs, TNF-α, and leukotrienes, required to initiate inflammation in the lung [34, 35].

For instance, COPD is an obstructive inflammatory disease resulting from joint effect of the breakdown of the alveolar, emphysema, and chronic inflammation of the bronchial tubes, bronchitis. Such inflammation induces innate and adaptive responses with multiple inflammatory genes that are characterized by an increase in inflammatory cells such as T- and B-lymphocytes, macrophages, neutrophils in the lumen, and pro-inflammatory cytokines such as IL-1β, IL-6, and TNF-α [36, 37]. Asthma on the other hand only involves inflammation in the bronchial tube which results in narrow airways, contrary to COPD that involves both lung and airways.

15.3.2 Lung Remodeling

Tissue remodeling is the other key process in the progression of chronic ILDs, yet the most challenging. Increasing progress is being recorded for a therapeutic approach against significant inflammation in the lung, effort to reverse the defective structure and function of tissue secondary to inflammation is still to no avail. The structural and functional changes resulting from inflammation along with cellular hypertrophy, fibrosis, and cell proliferation in the airways are common in COPD and asthma. Furthermore, both are obstructive lung disorders that involve remodeling of the airways resulting from obstruction of airflow due to amplified inflammation. Although, these diseases manifest in different conditions, but both target the smooth muscle, fibroblasts, epithelial cells, and immune cells, indicating that common molecular mechanisms could be required to remodel and repair the affected tissues, which can be explored to develop anti-remodeling drug [2, 38].

15.4 Epigenetic Factor in Inflammatory Lung Diseases

15.4.1 Obstructive Lung Diseases

Among notable inflammatory lung diseases that resulted from obstructive disorders in the lung are asthma and COPD. These morbidities as a representative of obstructive disease paradigm, have similar pathogeneses that stem from environmental factors such as smoking and air pollution, which can be delineated by epigenetic regulations [8, 13, 39]. Several epigenetic mechanisms have been reportedly associated with these obstructive lung diseases. Asthma development and progression has been reported to alter the methylation of DNA status of a variety of subpopulations of cell such as T-, B-, and dendritic cells, and genes such as interleukins-2 and interferon-receptor genes [14]. Exposure to tobacco, allergens, and air pollution has also been strongly associated with alteration in DNA methylation in asthma and COPD, with the latter resulting in hypomethylation of CpG islands close to gene encoding a1-antithrypsin [13, 39]. The activity of enzymes histone acetylates and deacetylases is hampered in the presence of smoke-related oxidative stress, which triggers the expression of NF-dependent gene [40]. Generally, ncRNAs of the let-7 family have been reportedly downregulated in different chronic ILDs, leading to induction of massive immune responses [13, 41–43].

15.4.2 Restrictive Lung Diseases

A notable restrictive inflammatory lung disease is the IPF involving diffuse damage of the alveoli, the cause of which is yet clearly established. Several studies reported IPF to involve different etiologies such as intestinal pneumonia and a repeated bout of alveolitis [44–47]. As a result, the knowledge of epigenetic factors associated with the progression and development of IPF is scarce. However, the role of gene–environment interactions and post-transcriptional modification in lung fibrogenesis has been major tool in studying the epigenetic regulation of IPF. Multiple responses to environmental stimuli such as gastroesophageal reflux, viral pathogen, and smoke have been epigenetically linked with IPF [48–53]. Furthermore, the decrease of activity in co-stimulatory pathway genes of T-cell, immune compromise has also been linked with IPF [54].

15.5 Future Prospect of Epigenetics in Inflammatory Lung Diseases

Interestingly, after several studies on the genetic and molecular bases of inflammation, the role of epigenetic marks as associative mediators of inflammation is becoming clearer as a connection between genetic and environmental factors of multifaceted diseases [3, 4]. Then, scientists considered the possibility of targeting the epigenetic changes responsible for inflammations and remodeling responses as a novel and effective approach to treating complex lung disease. This is hypothesized on the premise that when the features of epigenetic response are antagonized, it is expected that regular function and structure of the lung tissues be reversed [2].

The appealing fact about the prospect of targeting epigenetic mechanisms of inflammatory lung diseases for potential therapy is that, epigenetic mechanisms are relatively reversible biochemical processes that are subject to manipulations using small molecular compounds such as oligonucleotides, compare to genomic mutations that are mostly uncontrollable [2]. Therefore, identifying epigenetic targets and developing a suitable therapeutic approach including dogged targets on tissue remodeling are the most promising prospective tools to combat such lethal diseases as pulmonary embolism, IPF, ARDS, PH, lung tuberculosis, COPD, and asthma.

15.6 Challenges of Epigenetics in Inflammatory Lung Diseases

The study and understanding of epigenetic mechanisms associated with inflammatory lung diseases have no doubt served as an eye opener to the progression and development of such lethal illnesses and delineated novel pathogenesis and effective therapeutic approaches. However, a number of challenges still bother the adequate exploitation of epigenetics in combating these morbidities, thereby requiring further research. Few of factors that serve as bottleneck to this promising approach include:

Heterogeneity of epigenetic factors in complex diseases such as IPF is a major challenge as it results in discrepancies in the approach adopted in the study of such dynamic factors [14]. The progression of each of these complex diseases may involve different phenotypes and specificity at different stages and molecular alterations, which will require that different epigenetic signatures be tracked at every stage of progression. Therefore, studies based on a single tissue from a diseased sample will not facilitate a conclusive and complete understanding of the epigenetic marks [3, 5]. However, Next generation sequencing at different time-point coupled with advanced imaging technologies such as magnetic resonance spectrophotometer can help a complete understanding of epigenomes and their signatures. Also, the lack of intersection between gene expression and epigenetic modification is a challenge to several previous methodical approaches in epigenomic profiling.

15.7 Conclusion

Epigenetic studies have provided a novel insight into understanding the development and progression of different ILDs and have become a promising tool in delineating new pathogenetic pathways with potential therapeutic advances. However, more research is required to completely understand the multifaceted interactions among epigenetics, genetics, and the environment in the development of these diseases.

References

1. Avci E, Sarvari P, Savai R, Seeger W, Pullamsetti SS. Epigenetic mechanisms in parenchymal lung diseases: bystanders or therapeutic targets? Int J Mol Sci. 2022;23:546. https://doi.org/10.3390/ijms23010546.
2. Comer BS, Mariam A, Cherie BA, Singer A, William A, Gerthoffer T. Epigenetic targets for novel therapies of lung diseases. Pharmacol Ther. 2014;2014 https://doi.org/10.1016/j.pharmthera.2014.11.006.
3. Raghuraman S, Donkin I, Versteyhe S, Barres R, Simar D. The emerging role of epigenetics in inflammation and immunometabolism. Trends Endocrinol Metab. 2016;27(11):782–95.
4. Ahmad S, Manzoor S, Siddiqui S, Mariappan N, Zafar I, Ahmad A, Ahmad A. Epigenetic underpinnings of inflammation: connecting the dots between pulmonary diseases, lung cancer and COVID-19. Semin Cancer Biol. 2021; https://doi.org/10.1016/j.semcancer.2021.01.003.
5. Tzouvelekis A, Kaminski N. Epigenetics in idiopathic pulmonary fibrosis. Biochem Cell Biol. 2015;93:1–12.
6. Rajendrasozhan S, Yao H, Rahman I. Current perspectives on role of chromatin modifications and deacetylases in lung inflammation in COPD. COPD: J Chron Obstruct Pulmon Dis. 2009;6(4):291–7.
7. Stylianou E. Epigenetics of chronic inflammatory diseases. J Inflamm Res. 2019;12:1–14.
8. Yang IV, Schwartz DA. Epigenetic control of gene expression in the lung. Am J Respir Crit Care Med. 2011;183(10):1295–301.
9. Goldberg AD, Allis CD, Bernstein E. Epigenetics: a landscape takes shape. Cell. 2007;128(4):635–8.
10. Schembri F, Sridhar S, Perdomo C, Gustafson AM, Zhang X, Ergun A, et al. MicroRNAs as modulators of smoking-induced gene expression changes in human airway epithelium. Proc Natl Acad Sci U S A. 2009;106(7):2319–24.
11. Perdomo C, Campbell JD, Gerrein J, Tellez CS, Garrison CB, Walser TC, et al. MicroRNA 4423 is a primate-specific regulator of airway epithelial cell differentiation and lung carcinogenesis. Proc Natl Acad Sci U S A. 2013;110(47):18946–51.
12. Suter M, Ma J, Harris A, Patterson L, Brown KA, Shope C, et al. Maternal tobacco use modestly alters correlated epigenome-wide placental DNA methylation and gene expression. Epigenetics. 2011;6(11):1284–94.
13. Liu XH, Liu ZL, Sun M, Liu J, Wang ZX, De W. The long non-coding RNA HOTAIR indicates a poor prognosis and promotes metastasis in non-small cell lung cancer. BMC Cancer. 2013;13:464.
14. Yang IV, Schwartz DA. Epigenetic mechanisms and the development of asthma. J Allergy Clin Immunol. 2012;130(6):1243–55.

15. Huang B, Zhang R. Regulatory non-coding RNAs: revolutionizing the RNA world. Mol Biol Rep. 2014;41:3915–23.
16. Karapetyan AR, Buiting C, Kuiper RA, Coolen MW. Regulatory roles for long ncRNA and mRNA. Cancers (Basel). 2013;5(2):462–90.
17. Magnusdottir E, Gillich A, Grabole N, Surani MA. Combinatorial control of cell fate and reprogramming in the mammalian germline. Curr Opin Genet Dev. 2012;22(5):466–74.
18. Helin K, Dhanak D. Chromatin proteins and modifications as drug targets. Nature. 2013;502 (7472):480–8.
19. Kimmins S, Sassone-Corsi P. Chromatin remodelling and epigenetic features of germ cells. Nature. 2005;434(7033):583–9.
20. Klose RJ, Zhang Y. Regulation of histone methylation by demethylimination and demethylation. Nat Rev Mol Cell Biol. 2007;8(4):307–18.
21. Rossetto D, Avvakumov N, Cote J. Histone phosphorylation: a chromatin modification involved in diverse nuclear events. Epigenetics. 2012;7(10):1098–108.
22. Zhang Y. Transcriptional regulation by histone ubiquitination and deubiquitination. Genes Dev. 2003;17(22):2733–40.
23. Nana-Sinkam SP, Hunter MG, Nuovo GJ, Schmittgen TD, Gelinas R, Galas D, Marsh CB. Integrating the MicroRNome into the study of lung disease. Am J Respir Crit Care Med. 2009;179(1):4–10.
24. Mercer TR, Dinger ME, Mattick JS. Long non-coding RNAs: insights into functions. Nat Rev Genet. 2009;10(3):155–9.
25. Maass PG, Luft FC, Bahring S. Long non-coding RNA in health and disease. J Mol Med (Berl). 2014;92:337–46.
26. Shi X, Sun M, Liu H, Yao Y, Song Y. Long non-coding RNAs: a new frontier in the study of human diseases. Cancer Lett. 2013;339(2):159–66.
27. Zhang H, Chen Z, Wang X, Huang Z, He Z, Chen Y. Long non-coding RNA: a new player in cancer. J Hematol Oncol. 2013;6:37.
28. Bel EH, Wenzel SE, Thompson PJ, Prazma CM, Keene ON, Yancey SW, et al. Oral glucocorticoid-sparing effect of mepolizumab in eosinophilic asthma. N Engl J Med. 2014;371:1189–97.
29. Rock KL, Kono H. The inflammatory response to cell death. Annu Rev Pathol. 2008;3:99–126.
30. Cukic V, Lovre V, Dragisic D, Ustamujic A. Asthma and chronic obstructive pulmonary disease (COPD) - differences and similarities, Mater. Sociomed. 2012;24(2):100–5.
31. Pahwa R, Goyal A, Bansal P, Jialal I. Chronic Inflammation. Treasure Island, FL: StatPearls; 2020.
32. Bateman ED, Hurd SS, Barnes PJ, Bousquet J, Drazen JM, FitzGerald JM, Gibson P, Ohta K, O'Byrne P, Pedersen SE, Pizzichini E, Sullivan SD, Wenzel SE, Zar EJ. Global strategy for asthma management and prevention: GINA executive summary. Eur Respir J. 2018;51(2): 143–78.
33. Ware LB, Matthay MB. The acute respiratory distress syndrome. N Engl J Med. 2000;342(18): 1334–49.
34. Aghasafari P, George U, Pidaparti R. A review of inflammatory mechanism in airway diseases, Inflamm. Res. 2019;68(1):59–74.
35. Robb CT, Regan KH, Dorward DA, Rossi AG. Key mechanisms governing resolution of lung inflammation. Semin Immunopathol. 2016;38(4):425–48.
36. Brusselle GG, Joos GF, Bracke KR. New insights into the immunology of chronic obstructive pulmonary disease. Lancet. 2011;378(9795):1015–26.
37. Givi ME, Redegeld FA, Folkerts G, Mortaz E. Dendritic cells in pathogenesis of COPD. Curr Pharm Des. 2012;18(16):2329–35.
38. Chipps BE, Zeiger RS, Borish L, Wenzel SE, Yegin A, Hayden ML, et al. Key findings and clinical implications from The Epidemiology and Natural History of Asthma: outcomes and Treatment Regimens (TENOR) study. J Allergy Clin Immunol. 2012;130:332–42.

39. Schwartz DA. Epigenetics and environmental lung disease. Proc Am Thorac Soc. 2010;7(2): 123–5.
40. Andresen E, Günther G, Bullwinkel J, Lange C, Heine H. Increased expression of beta-defensin 1 (DEFB1) in chronic obstructive pulmonary disease. PLoS One. 2011;6:e21898.
41. Leidinger P, Keller A, Borries A, Huwer H, Rohling M, Huebers J, et al. Specific peripheral miRNA profiles for distinguishing lung cancer from COPD. Lung Cancer. 2011;74:41–7.
42. Lu TX, Hartner J, Lim EJ, Fabry V, Mingler MK, Cole ET, et al. MicroRNA-21 limits in vivo immune response-mediated activation of the IL12/IFN-gamma pathway, Th1 polarization, and the severity of delayed-type hypersensitivity. J Immunol. 2011;187(6):3362–73.
43. Van Pottelberge GR, Mestdagh P, Bracke KR, Thas O, van Durme YM, Joos GF, et al. MicroRNA expression in induced sputum of smokers and patients with chronic obstructive pulmonary disease. Am J Respir Crit Care Med. 2011;183:898–906.
44. Barkauskas CE, Noble PW. Cellular mechanisms of tissue fibrosis. 7. New insights into the cellular mechanisms of pulmonary fibrosis. Am J Physiol Cell Physiol. 2014;306:C987–96.
45. Baroke E, Gauldie J, Kolb M. New treatment and markers of prognosis for idiopathic pulmonary fibrosis: lessons learned from translational research. Expert Rev Respir Med. 2013;7:465–78.
46. Blackwell TS, Tager AM, Borok Z, Moore BB, Schwartz DA, Anstrom KJ, et al. Future directions in idiopathic pulmonary fibrosis research. An NHLBI workshop report. Am J Respir Crit Care Med. 2014;189:214–22.
47. Lota HK, Wells AU. The evolving pharmacotherapy of pulmonary fibrosis. Expert Opin Pharmacother. 2013;14:79–89.
48. Antoniou KM, Hansell DM, Rubens MB, Marten K, Desai SR, Siafakas NM, et al. Idiopathic pulmonary fibrosis: outcome in relation to smoking status. Am J Respir Crit Care Med. 2008;177(2):190–4.
49. Konishi K, Gibson KF, Lindell KO, Richards TJ, Zhang Y, Dhir R, et al. Gene expression profiles of acute exacerbations of idiopathic pulmonary fibrosis. Am J Respir Crit Care Med. 2009;180(2):167–75.
50. Lee JS, Collard HR, Anstrom KJ, Martinez FJ, Noth I, Roberts RS, Yow E, IPFnet Investigators, et al. Anti-acid treatment and disease progression in idiopathic pulmonary fibrosis: an analysis of data from three randomised controlled trials. Lancet Respir Med. 2013;1(5):369–76.
51. Raghu G. Idiopathic pulmonary fibrosis: increased survival with "gastroesophageal reflux therapy": fact or fallacy? Am J Respir Crit Care Med. 2011;184(12):1330–2.
52. Raghu G, Meyer KC. Silent gastro-oesophageal reflux and microaspiration in IPF: mounting evidence for anti-reflux therapy? Eur Respir J. 2012;39(2):242–5.
53. Raghunath A, Hungin AP, Wooff D, Childs S. Prevalence of Helicobacter pylori in patients with gastro-oesophageal reflux disease: systematic review. BMJ. 2003;326(7392):737.
54. Kotsianidis I, Nakou E, Bouchliou I, Tzouvelekis A, Spanoudakis E, Steiropoulos P, et al. Global impairment of CD4+CD25+FOXP3+ regulatory T cells in idiopathic pulmonary fibrosis. Am J Respir Crit Care Med. 2009;179(12):1121–30.

MIX
Papier aus verantwortungsvollen Quellen
Paper from responsible sources
FSC® C105338

If you have any concerns about our products,
you can contact us on
ProductSafety@springernature.com

In case Publisher is established outside the EU,
the EU authorized representative is:
Springer Nature Customer Service Center GmbH
Europaplatz 3, 69115 Heidelberg, Germany

Printed by Libri Plureos GmbH
in Hamburg, Germany